Nonlinear Dispersive Waves and Fluids

CONTEMPORARY MATHEMATICS

725

Nonlinear Dispersive Waves and Fluids

AMS Special Sessions on Spectral Calculus and Quasilinear Partial Differential Equations, and PDE Analysis on Fluid Flows
January 5–7, 2017
Atlanta, Georgia

Shijun Zheng
Marius Beceanu
Jerry Bona
Geng Chen
Tuoc Van Phan
Avy Soffer
Editors

AMS AMERICAN MATHEMATICAL SOCIETY

2010 *Mathematics Subject Classification.* Primary 35-XX, 37-XX, 47-XX, 58-XX, 76-XX, 93-XX.

Library of Congress Cataloging-in-Publication Data

Names: Zheng, Shijun, editor. | Beceanu, Marius, editor. | Bona, J. L., editor. | Chen, Geng, 1980- editor. | Phan, Tuoc (Tuoc Van), editor. | Soffer, Avy (Avraham), editor. | AMS Special Session on Spectral Calculus and Quasilinear Partial Differential Equations (2017 : Atlanta, Georgia) | AMS Special Session on PDE Analysis on Fluid Flows (2017 : Atlanta, Georgia)

Title: Nonlinear dispersive waves and fluids : AMS Special Session on Spectral Calculus and Quasilinear Partial Differential Equations, Atlanta, Georgia, January 5-6, 2017 : AMS Special Session on PDE Analysis on Fluid Flows, Atlanta, Georgia, January 7, 2017 / edited by Shijun Zheng, Marius Beceanu, Jerry Bona, Geng Chen, Tuoc Van Phan, Avy Soffer.

Description: Providence, Rhode Island : American Mathematical Society, [2019] | Series: Contemporary mathematics ; Volume 725 | Includes bibliographical references.

Identifiers: LCCN 2018040209 | ISBN 9781470441098 (alk. paper)

Subjects: LCSH: Wave equation–Congresses. | Nonlinear waves–Congresses. | Nonlinear wave equations–Congresses.

Classification: LCC QC174.26.W28 N653 2019 | DDC 530.12/4–dc23

LC record available at https://lccn.loc.gov/2018040209

Contemporary Mathematics ISSN: 0271-4132 (print); ISSN: 1098-3627 (online)

DOI: https://doi.org/10.1090/conm/725

∞ The paper used in this book is acid-free and falls within the guidelines
established to ensure permanence and durability.
Visit the AMS home page at `https://www.ams.org/`

10 9 8 7 6 5 4 3 2 1 24 23 22 21 20 19

Contents

Preface vii

Combined list of speakers ix

Blowup rate for mass critical rotational nonlinear Schrödinger equations
NYLA BASHARAT, YI HU, and SHIJUN ZHENG 1

Gradient estimates for weak solutions of linear elliptic systems with singular-degenerate coefficients
DAT CAO, TADELE MENGESHA, and TUOC PHAN 13

Virial estimates for hard spheres
RYAN DENLINGER 35

The almost global existence to classical solution for a 3-D wave equation of nematic liquid-crystals
YI DU, GENG CHEN, and JIANLI LIU 53

Instability of solitons–revisited, I: The critical generalized KdV equation
LUIZ GUSTAVO FARAH, JUSTIN HOLMER, and SVETLANA ROUDENKO 65

Instability of solitons–revisited, II: The supercritical Zakharov-Kuznetsov equation
LUIZ GUSTAVO FARAH, JUSTIN HOLMER, and SVETLANA ROUDENKO 89

Stabilization of dispersion-generalized Benjamin-Ono
CYNTHIA FLORES, SEUNGLY OH, and DEREK SMITH 111

Uniqueness of standing-waves for a non-linear Schrödinger equation with three pure-power combinations in dimension one
DANIELE GARRISI and VLADIMIR GEORGIEV 137

Below-threshold solutions of a focusing energy-critical heat equation in $\mathbb{R}^4$
STEPHEN GUSTAFSON and DIMITRIOS ROXANAS 149

A regularity upgrade of pressure
DONG LI and XIAOYI ZHANG 163

On large future-global-in-time solutions to energy-supercritical nonlinear wave equation
SHUANG MIAO 187

The nonlinear Schrödinger equation with an inverse-square potential
JASON MURPHY 215

The harmonic map heat flow on conic manifolds
YUANZHEN SHAO and CHANGYOU WANG 227

On the global regularity issue of the two-dimensional magnetohydrodynamics system with magnetic diffusion weaker than a Laplacian
KAZUO YAMAZAKI 251

Orbital stability of standing waves for fractional Hartree equation with unbounded potentials
JIAN ZHANG, SHIJUN ZHENG, and SHIHUI ZHU 265

Preface

This volume stems from the two special sessions "Spectral Calculus and Quasilinear Partial Differential Equations" and "PDE Analysis on Fluid Flows" that took place at the AMS-MAA Joint Mathematics Meetings held in Atlanta, Georgia, January 4-7, 2017, where the participants include many well-known mathematicians as well as promising junior researchers in the field. The contributions therein review some of the recent developments in the mathematical theory of quantum waves and fluids. In particular, they reveal a small fraction of the facet of the current research topics that have stayed in the heart of the flow at its golden age.

The special proceedings features invited contributions of research papers and survey articles reflecting the frontier studies in evolutionary differential equations. The topics of the expositions include spectral and nonlinear analysis of PDEs arising in quantum mechanics, plasma, fluid mechanics, water waves (oceanology), crystallization, stellar dynamics (astrophysics) and other applied science.

The paper of Y. Du, G. Chen and J. Liu studies the global existence for wave equation of liquid crystals. S. Miao considers the existence of large data solution for NLW equation in the energy-supercritical regime. The paper of J. Murphy gives a survey on the scattering versus blowup results for NLS with an inverse square potential in the energy-subcritical case, while the paper of S. Gustafson and D. Roxanas surveys on global existence and blowup for nonlinear heat equation in the energy-critical case.

D. Garrisi and V. Georgiev prove the uniqueness of standing waves for NLS with combined power nonlinearities. J. Zhang, S. Zhu and S. Zheng show the orbital stability for standing waves for the Hartree equation. The paper of N. Basharat, Y. Hu and S. Zheng elaborates the blowup rate for solutions of rotational NLS near the ground state profile. For instability of solitons, L. Farah, J. Holmer and S. Roudenko give simplified approach in the proof of the case for gKdV and Zakharov-Kuznetsov equations.

The stabilization of generalized Benjamin-Ono equations for water wave model is studied by C. Flores, S. Oh and D. Smith. K. Yamazaki discusses the global regularity for MHD in two dimensions. D. Li and X. Zhang obtain some optimal regularity estimates for incompressible Euler equations. For the conormal derivative problem, D. Cao, T. Mengesha and T. Phan obtain Calderón-Zygmund type regularity estimates for a class of elliptic systems with singular, degenerated coefficients. The paper of Y. Shao and C. Wang proves certain local and global regularity results for harmonic map heat flow on manifolds with conic singularities. R. Denlinger presents a short 'determininistic' approach to the proof of a virial type estimate concerning the derivation of the Boltzmann's equation from a large system of interacting particles. All papers are peer-reviewed.

Recent two decades have seen rapidly growing interest in the study of the long time dynamics for dispersive evolutionary waves and fluid flows. These two broad areas might share some common mathematical challenges. For instance, how the large data theory and low regularity theory for equations of dispersive type are connected to the equations of fluids models, where some of the same open problems remain mysteries? It has become crucial to bring together researchers in the emerging areas to share, discuss and comprehend the fascinating and meanwhile hidden structures of the equations under investigation. They might address problems in connection with the central questions around regularity, long time behavior, scattering and resonance, and the spectral theory behind for the associated linear and nonlinear operators, existence of solitary waves, orbital and asymptotic stability, vorticity and wave turbulence, and singularity formation. We believe that the present proceedings with *Contemporary Mathematics* would fit in such a mission.

The volume might provide researchers a valuable reference on relevant problems in Analysis and PDEs. It might also serve as good and inspirational literature for graduate students and young mathematicians working in the area. The co-editors are content with the event that were well-attended, and especially delighted with the outcome for which it has come to fruition from the contributions and efforts.

We thank the American Mathematical Society for coordinating the special sessions at the JMM conference in Atlanta. We thank the AMS Editorial for sponsoring the publication of this special volume. We thank all the speakers, authors and participants for their enthusiasm and support, who have made the event possible and memorable.

Editors: Marius Beceanu, Jerry Bona, Geng Chen, Tuoc V. Phan, Avy Soffer

Principal Editor: Shijun Zheng

Combined List of Speakers

The underlying theme of the invited talks concentrated on latest development in the areas of analysis, evolutionary partial differential equations and mathematical physics that mainly arise in quantum mechanics, general relativity and fluid dynamics. The theory and the methods showcase recent trends and perspectives of the active program.

Here are some links to the two AMS special sessions at the JMM.

https://jointmathematicsmeetings.org/jmm2017/2180_program_ss3.html
https://jointmathematicsmeetings.org/jmm2017/2180_program_ss39.html

The following are name lists of the organizers and the speakers.

Two special sessions' organizers: Shijun Zheng, Marius Beceanu and Tuoc Van Phan; Xiang Xu, Geng Chen, and Ronghua Pan

Speakers at **Spectral Calculus and Quasilinear PDEs**:

Benjamin Harrop-Griffiths, New York University
Sung-Jin Oh, Korea Institute for Advanced Study
Jerry Bona, University of Illinois at Chicago
Jason Murphy, UC Berkeley
Zhiwu Lin, Georgia Institute of Technology
Xiaoyi Zhang, University of Iowa
Tadele Mengesha, University of Tennessee, Knoxville
Daniele Garrisi, Inha University
Hao Jia, Institute for Advanced Study
Gigliola Staffilani, Massachusetts Institute of Technology
Dong Li, University of British Columbia
Svetlana Roudenko, George Washington University
Casey Jao, UC Berkeley

Speakers at **PDE Analysis on Fluid Flows**:

Tao Huang, New York University, Shanghai
Changyou Wang, Purdue University
Wujun Zhang, Rutgers University
Tiziana Giorgi, New Mexico State University
Sookyung Joo, Old Dominion University
Andres Contreras, New Mexico State University
Geng Chen, University of Kansas

Michele Coti Zelati, University of Maryland
Mikhail Feldman, University of Wisconsin-Madison
Charis Tsikkou, West Virginia University
Dehua Wang, University of Pittsburgh
Qingtian Zhang, University of California, Davis
Kun Zhao, Tulane University
Xiaoming Wang, Florida State University
Oleksandr Misiats, Courant Institute, NYU
Qingshan Chen, Clemson University

Contemporary Mathematics
Volume **725**, 2019
https://doi.org/10.1090/conm/725/14556

Blowup rate for mass critical rotational nonlinear Schrödinger equations

Nyla Basharat, Yi Hu, and Shijun Zheng

Abstract. We consider the blowup rate for blowup solutions to L^2-critical, focusing NLS with a harmonic potential and a rotation term. Under a suitable spectral condition we prove that there holds the "log-log law" when the initial data is slightly above the ground state. We also construct minimal mass blowup solutions near the ground state level with distinct blowup rates.

1. Introduction

Consider the focusing nonlinear Schrödinger equation (NLS) with an angular momentum term on $\mathbb{R}^{1+n}$:

$$\begin{cases} iu_t = -\Delta u + Vu - |u|^{p-1}u + iA\cdot\nabla u \\ u(0,x) = u_0 \in \mathscr{H}^1, \end{cases} \tag{1}$$

where $u = u(t,x)$: $\mathbb{R}\times\mathbb{R}^n \to \mathbb{C}$ denotes the wave function, $V(x) = \gamma^2|x|^2$, $\gamma > 0$ is a trapping harmonic potential that confines the movement of particles, and $A(x) = Mx$, with $M = -M^T$ being an $n\times n$ real-valued skew-symmetric matrix. The linear hamiltonian $H_{A,V} := -\Delta + V + iA\cdot\nabla$ is essentially self-adjoint in L^2, whose eigenvalues are associated to the Landau levels as quantum numbers. The angular momentum operator $L_A u := iA\cdot\nabla u$ generates the rotation $e^{tA\cdot\nabla}u = u(e^{tM}x)$ in $\mathbb{R}^n$. The space $\mathscr{H}^1 = \mathscr{H}^{1,2}$ denotes the weighted Sobolev space given by: for $r\in(1,\infty)$,

$$\mathscr{H}^{1,r}(\mathbb{R}^n) := \{f\in L^r : \nabla f,\, xf \in L^r\},$$

which is endowed with the norm $|f|_{\mathscr{H}^{1,r}} = |\nabla f|_r + |xf|_r + |f|_r$, here $|\cdot|_r := |\cdot|_{L^r}$ denoting the L^r-norm.

When $n = 3$, equation (1) is also known as Gross-Pitaevskii equation which models rotating Bose-Einstein condensation (BEC) with attractive particle interactions in a dilute gaseous ultra-cold superfluid. The operator L_A are usually denoted by $-\Omega\cdot L$, where $\Omega = (\Omega_1,\Omega_2,\Omega_3)\in\mathbb{R}^3$ is a given angular velocity vector and $L := -ix\wedge\nabla$, in which case $M = \begin{pmatrix} 0 & -\Omega_3 & \Omega_2 \\ \Omega_3 & 0 & -\Omega_1 \\ -\Omega_2 & \Omega_1 & 0 \end{pmatrix}$.

2010 *Mathematics Subject Classification.* Primary 35Q55, 35B44; Secondary 35P30.
Key words and phrases. Blowup rate, harmonic potential, angular momentum.

Such a system describing rotating particles in a harmonic trap has acquired significance in connection with optics and atomic physics in theoretical and experimental physics [**1, 3, 4, 13, 18, 25**]. Meanwhile, it demands rigorous mathematical analysis on the evolution and dynamics of the quantized flow. For $p \in (1, 1+4/(n-2))$, the local in time existence and uniqueness of (1) has been established in [**2, 6, 14, 15**], see also [**8, 11, 28**] for the treatment in a general magnetic setting. The purpose of this article is to study how the rotation affects the wave collapse as well as energy concentration under a trapping potential. In particular, we will address the blowup rate for the blowup solution of the L^2-critical focusing equation in (1) where $p = 1 + 4/n$.

Let $H^1 = \{u \in L^2 : \nabla u \in L^2\}$ be the usual Sobolev space. Recall that the standard NLS reads

$$\begin{cases} i\varphi_t = -\Delta\varphi - |\varphi|^{p-1}\varphi, \\ \varphi(0,x) = \varphi_0 \in H^1, \end{cases} \tag{2}$$

and the well-posedness problem for $p \in (1, 1+4/(n-2)]$ has been studied for a few decades and is quite well understood in the euclidean space. Let $Q \in H^1(\mathbb{R}^n)$ be the unique positive radial function that satisfies ([**17, 26**])

$$-Q = -\Delta Q - Q^{1+\frac{4}{n}}. \tag{3}$$

When $p = 1+\frac{4}{n}$ and $|\varphi_0|_2 = |Q|_2$, Merle [**19**] was able to determine the profile for all blowup solutions with minimal mass at the ground state level, which are obtained from pseudo-conformal transform. Hence all blowup solutions have blowup rate $(T-t)^{-1}$. In the mass critical and supercritical case $p \in [1+4/n, \infty)$, the wave collapse dynamics appears very subtle issue for $|u_0|_2 = |Q|_2$. Within an arbitrarily small neighborhood of Q, there always exist φ_0 and ψ_0 such that the flow $\varphi_0 \mapsto \varphi$ blowups in finite time, and, $\psi_0 \mapsto \psi$ exists global in time and scatters as $t \to \infty$ in H^1.

When $p = 1+\frac{4}{n}$ and $|\varphi_0|_2 > |Q|_2$, Bourgain and Wang [**5**] constructed solutions of positive energy having blowup rate $(T-t)^{-1}$ in dimensions $n = 1, 2$, which was later shown unstable however. Perelman [**23**] gave the first rigorous demonstration of the existence and stability for the log-log speed for generic blowup solutions in 1d. More recently, under certain spectral condition in the Spectral Property (Section 3), Merle and Raphaël [**21**] proved the sharp blowup rate of the solutions for (2), i.e., there exists a universal constant α^* such that if $\varphi_0 \in B_{\alpha^*}$ with negative energy, then $\varphi(t,x)$ is a blowup solution to (2) with maximal interval of existence $[0,T)$ satisfying the log-log blowup rate as $t \to T$

$$|\nabla\varphi(t)|_2 \approx \sqrt{\frac{\log|\log(T-t)|}{T-t}},$$

where

$$B_\alpha := \left\{\phi \in H^1 : \int |Q|^2 dx < \int |\phi|^2 dx < \int |Q|^2 dx + \alpha\right\},$$

see Theorem 3.1. Such a blowup rate is also known to be stable in H^1.

In the presence of a rotational term, we will show how to prove such a "log-log law" for the NLS (1). Our proof is based on a virial identity for (1), the $\mathcal{R}$-transform (7) that maps solutions of (2) to solutions of (1), and an application of the above result of Merle and Raphaël's. This treatment is motivated by [**29**], where the analogous result is obtained for the case $A = 0$ and V being a harmonic potential. Our main result is stated as follows. Let α^* be the above-mentioned universal constant.

THEOREM 1.1. *Let $p = 1 + \frac{4}{n}$ in (1), $1 \leq n \leq 5$. Suppose $u_0 \in B_{\alpha^*}$ satifies*

$$\int |\nabla u_0|^2 - \frac{n}{n+2} \int |u_0|^{2+\frac{4}{n}} < 0. \tag{4}$$

Then $u \in C([0,T); \mathscr{H}^1)$ is a blowup solution of (1) in finite time $T < \infty$, with the log-log *blowup rate*

$$\lim_{t \to T} \frac{|\nabla u(t)|_2}{|\nabla Q|_2} \sqrt{\frac{T-t}{\log|\log(T-t)|}} = \frac{1}{\sqrt{2\pi}},$$

where Q is the unique solution of (3).

2. Local wellposedness for (1)

For $\varphi_0 \in \mathscr{H}^1$, the local well-posedness of (1) was proved for $1 \leq p < 1 + \frac{4}{n-2}$, see e.g. in [**11, 28**]. The proof for the local well-posedness relies on local in time dispersive estimates for $U(t) = e^{-itH_{A,V}}$, the fundamental solution on $[0, \delta)$ (for some small $\delta > 0$) constructed in [**27**]. The vectorial function A represents a magnetic potential that induces the Coriolis effect or centrifugal force for the spinor particles. Alternatively, this can also be done by means of the explicit formula in (12) for $U(t)$ defined in $(0, \pi/2\gamma)$. This formula is obtained from the so-called $\mathcal{R}$-transform in (7), a type of pseudo-conformal transform.

PROPOSITION 2.1. *Let $1 + \frac{4}{n} \leq p < 1 + \frac{4}{n-2}$. Suppose $u_0 \in \mathscr{H}^1$.*

(a) [Blowup Alternative]
  (i) *If $1 \leq p < 1 + \frac{4}{n-2}$, then there exists $T^* > 0$ such that (1) has a unique maximal solution $u \in C([0,T^*), \mathscr{H}^1) \cap L^q_{loc}([0,T^*), \mathscr{H}^{1,r})$, where $r = p + 1$ and $q = \frac{4(p+1)}{n(p-1)}$.*
  (ii) *If T^* is finite, then $|\nabla u|_2 \to \infty$ as $t \to T^*$ with a lower bound:*

$$|\nabla u(t)|_2 \geq \frac{C}{\sqrt{T^* - t}}.$$

(b) [Conservation Laws] *The following are conserved on the maximal lifespan $[0, T^*)$.*

$$\text{(mass) } M(u) = \int |u|^2$$

$$\text{(energy) } E(u) = \int \left(|\nabla u|^2 + V|u|^2 - \frac{2}{p+1}|u|^{p+1} + i\bar{u}A \cdot \nabla u \right)$$

$$\text{(angular momentum) } \ell_A(u) = \int i\bar{u}A \cdot \nabla u.$$

In the critical case $p = 1 + 4/n$, from [**6**] we know that $|Q|_2$ is the sharp threshold such that:

(a) If $|u_0| < |Q|_2$, then (1) has a unique global in time solution.
(b) For all $c \geq |Q|_2$, there exists u_0 with $|u_0|_2 = c$ so that u is a finite time blowup solution of (1).

As we mentioned in the introduction section, if $|u_0| = |Q|_2$, from Merle's characterization for the blowup profile of (2), all such blowup solutions have the blowup rate $|\nabla u(t)|_2 \approx \frac{C}{T^*-t}$ as $t \to T^*$, see Proposition 4.5.

3. A spectral property and the log-log law

To show the blowup rate for initial data above the ground state Q as stated in Theorem 1.1, we need the following Spectral Property. Let y denote the spatial variable in $\mathbb{R}^n$.

SPECTRAL PROPERTY. Consider the two Schrödinger operators

$$L_1 := -\Delta + \frac{2}{n}\left(\frac{4}{n}+1\right) Q^{\frac{4}{n}-1} y \cdot \nabla Q, \qquad L_2 := -\Delta + \frac{2}{n} Q^{\frac{4}{n}-1} y \cdot \nabla Q,$$

and the real-valued quadratic form for $\varepsilon = \varepsilon_1 + i\varepsilon_2 \in H^1$

$$H(\varepsilon,\varepsilon) := (L_1\varepsilon_1, \varepsilon_1) + (L_2\varepsilon_2, \varepsilon_2).$$

Let

$$Q_1 := \frac{n}{2} Q + y \cdot \nabla Q, \qquad Q_2 := \frac{n}{2} Q_1 + y \cdot \nabla Q_1.$$

Then there exists a universal constant $\delta_0 > 0$ such that for every $\varepsilon \in H^1$, if

$$(\varepsilon_1, Q) = (\varepsilon_1, Q_1) = (\varepsilon_1, y_j Q)_{1\le j\le n} = (\varepsilon_2, Q_1) = (\varepsilon_2, Q_2) = (\varepsilon_2, \partial_{y_j} Q)_{1\le j\le n} = 0,$$

then

$$H(\varepsilon,\varepsilon) \geq \delta_0 \left(\int |\nabla\varepsilon|^2 dy + \int |\varepsilon|^2 e^{-|y|} dy \right).$$

The proof of the Spectral Property in any dimension is not complete. It has been proved in [**20**] for dimension $n = 1$ by using the explicit solution $Q(x) = \left(\frac{3}{\cosh^2(2x)}\right)^{\frac{1}{4}}$ to (3). One can also find a computer assisted proof of the Spectral Property in dimensions $n = 2, 3, 4$ in [**12**]. For dimension 5 and higher see Remark 3.2.

The Spectral Property is equivalent to the coercivity for L_1 and L_2 on quadratic forms, the study of which involving the ground state solution Q naturally appears in a perturbation setting when dealing with stability problem. These two operators are related to the Lyapounov functionals $L_\pm$, where $L_+ = -\Delta + 1 - (1+\frac{4}{n})Q^{\frac{4}{n}}$ and $L_- = -\Delta + 1 - Q^{\frac{4}{n}}$, see [**12**].

Using the Spectral Property, Merle and Raphaël obtained the following blowup rate for (2) in the absence of potentials, see [**12, 21, 22**].

THEOREM 3.1. *Let $p = 1 + \frac{4}{n}$ in (2). Let $1 \le n \le 5$. There exists a universal constant $\alpha^* > 0$ such that the following is true. Suppose $\varphi_0 \in B_{\alpha^*}$ satisfies*

$$\int |\nabla\varphi_0|^2 - \frac{n}{n+2}\int |\varphi_0|^{2+\frac{4}{n}} < 0.$$

Then $\varphi \in C([0,T); H^1)$ is a blowup solution of (2) in finite time $T < \infty$, which admits the log-log *blowup rate*

$$\lim_{t\to T} \frac{|\nabla\varphi(t)|_2}{|\nabla Q|_2} \sqrt{\frac{T-t}{\log|\log(T-t)|}} = \frac{1}{\sqrt{2\pi}}. \tag{5}$$

REMARK 3.2. From [**12**] we know that the Spectral Property is true in dimensions $n = 1, 2, 3, 4$. If $n = 5$, Theorem 3.1 continues to hold true as soon as the Spectral Property verifies the orthogonality condition with $(\varepsilon_1, Q_1) = 0$ replaced by $(\varepsilon_1, |y|^2 Q) = 0$, which is numerically verified in [**12**]. For the above-mentioned reason, Theorem 3.1 remains open in dimensions $n \geq 6$.

REMARK 3.3. It is well-known that the log-log law is a generic behavior for those blowup solutions in the theorem, whose proof relies on algebraic cancellations related to the topological degeneracy of the linear operators L_1 and L_2 around Q. Such blowup rate is stable in the sense that the set $\mathfrak{U}_0$ in the log-log regime is open in H^1, where $\mathfrak{U}_0$ denotes the set of all initial data φ_0 in B_{α^*} so that the flow $\varphi_0 \mapsto \varphi(t)$ of (2) collapses in finite time $T^* < \infty$ with the log-log speed given in (5), see [**12, 24**].

4. The blowup rate for the rotational NLS

In this section we prove Theorem 1.1. We will always assume $p = 1 + \frac{4}{n}$ in both (1) and (2). We will need a virial identity for (2) and the $\mathfrak{R}$-transform introduced in Proposition 4.3. This transform gives a relation between the two solutions of (1) and (2), which is coined as a combination of the lens transform and the rotation $e^{tA\cdot\nabla}$. One can view it as certain pseudo-conformal symmetry in the rotational case, see [**7, 29**] in the presence of a quadratic potential only, i.e., $M = 0$ and $\gamma \neq 0$.

The following is a standard virial identity for (2) in the weighted Sobolev space $\mathscr{H}^1$, which can be proved by a direct calculation.

LEMMA 4.1. *Let φ be a solution to the problem (2) in $C([0,T), \mathscr{H}^1)$. Define $J(t) := \int |x|^2 |\varphi|^2 dx$. Then*

$$J'(t) = 4\Im \int x\overline{\varphi} \cdot \nabla \varphi dx, \qquad J''(t) = 8\mathcal{E}(\varphi_0),$$

where

$$\mathcal{E}(\varphi) = \int \left(|\nabla \varphi|^2 - \frac{n}{n+2} |\varphi|^{\frac{4}{n}+2} \right) dx.$$

LEMMA 4.2. (a) *Given any real n by n matrix M and any function f in $C_0^\infty \cap L^2(\mathbb{R}^n)$, we have*

$$e^{t(Mx)\cdot\nabla} f(x) = f(e^{tM} x). \tag{6}$$

(b) *If M is a real skew-symmetric matrix, then $e^{tM} \in SO(n)$ for all t, where $SO(n)$ is the group of n by n orthogonal matrices with determinant 1. Moreover, $(Mx)\cdot\nabla$ and Δ commute, i.e.,*

$$[(Mx)\cdot\nabla, \Delta] = 0.$$

Part (a) can be proven by showing that both sides of (6) obey the ODE

$$\partial_t F(t,x) = (Mx)\cdot\nabla F(t,x), \quad F(0,x) = f(x).$$

Part (b) follows from a straightforward calculation.

PROPOSITION 4.3. *Let $\varphi(t,x) \in C([0,T), H^1)$ be a solution to (2) where $T > 0$. Define the $\mathcal{R}$ transform $\varphi \mapsto \mathcal{R}(\varphi)$ to be*

$$\mathcal{R}\varphi(t,x) := \frac{1}{\cos^{\frac{n}{2}}(2\gamma t)} e^{-i\frac{\gamma}{2}|x|^2 \tan(2\gamma t)} \varphi\left(\frac{\tan(2\gamma t)}{2\gamma}, \frac{e^{tM}x}{\cos(2\gamma t)}\right). \tag{7}$$

Then $u = \mathcal{R}\varphi$ is a solution to (1) in $C([0, \frac{\arctan(2\gamma T)}{2\gamma}), \mathscr{H}^1)$.

Conversely, let $u(t,x) \in C([0,T^*), \mathscr{H}^1)$ be a solution to (1) where $T^* \in (0, \frac{\pi}{4\gamma}]$. Then $\varphi = \mathcal{R}^{-1}u$, given by

$$\varphi(t,x) := \frac{1}{(1+(2\gamma t)^2)^{\frac{n}{4}}} e^{i\frac{\gamma^2|x|^2 t}{1+(2\gamma t)^2}} u\left(\frac{\arctan(2\gamma t)}{2\gamma}, \frac{e^{-tM}x}{\sqrt{1+(2\gamma t)^2}}\right), \tag{8}$$

is a solution to (2) in $C([0, \frac{\tan(2\gamma T^*)}{2\gamma}), H^1)$, where $\mathcal{R}^{-1}$ is the inverse of $\mathcal{R}$.

PROOF. We will only briefly check (7) for $u = \mathcal{R}\phi$. The other one $u \mapsto \varphi$ is the inverse transform. By direct computation, we have

(9)

$$\begin{aligned}
u_t(t,x) = {} & n\gamma \frac{\sin(2\gamma t)}{\cos^{\frac{n}{2}+1}(2\gamma t)} e^{-i\frac{\gamma}{2}|x|^2 \tan(2\gamma t)} \varphi\left(\frac{\tan(2\gamma t)}{2\gamma}, \frac{e^{tM}x}{\cos(2\gamma t)}\right) \\
& - i\gamma^2|x|^2 \frac{1}{\cos^{\frac{n}{2}+2}(2\gamma t)} e^{-i\frac{\gamma}{2}|x|^2 \tan(2\gamma t)} \varphi\left(\frac{\tan(2\gamma t)}{2\gamma}, \frac{e^{tM}x}{\cos(2\gamma t)}\right) \\
& + \frac{1}{\cos^{\frac{n}{2}+2}(2\gamma t)} e^{-i\frac{\gamma}{2}|x|^2 \tan(2\gamma t)} \varphi_t\left(\frac{\tan(2\gamma t)}{2\gamma}, \frac{e^{tM}x}{\cos(2\gamma t)}\right) \\
& + \frac{1}{\cos^{\frac{n}{2}+1}(2\gamma t)} e^{-i\frac{\gamma}{2}|x|^2 \tan(2\gamma t)} (e^{tM}Mx) \cdot \nabla\varphi\left(\frac{\tan(2\gamma t)}{2\gamma}, \frac{e^{tM}x}{\cos(2\gamma t)}\right) \\
& + 2\gamma \frac{\sin(2\gamma t)}{\cos^{\frac{n}{2}+2}(2\gamma t)} e^{-i\frac{\gamma}{2}|x|^2 \tan(2\gamma t)} (e^{tM}x) \cdot \nabla\varphi\left(\frac{\tan(2\gamma t)}{2\gamma}, \frac{e^{tM}x}{\cos(2\gamma t)}\right).
\end{aligned}$$

To compute Δu and $i(Mx) \cdot \nabla u$, first note that

$$\begin{aligned}
\nabla\left(\varphi\left(\frac{\tan(2\gamma t)}{2\gamma}, \frac{e^{tM}x}{\cos(2\gamma t)}\right)\right) &= \frac{1}{\cos(2\gamma t)} (e^{tM})^T \nabla\varphi\left(\frac{\tan(2\gamma t)}{2\gamma}, \frac{e^{tM}x}{\cos(2\gamma t)}\right), \\
\Delta\left(\varphi\left(\frac{\tan(2\gamma t)}{2\gamma}, \frac{e^{tM}x}{\cos(2\gamma t)}\right)\right) &= \frac{1}{\cos^2(2\gamma t)} \Delta\varphi\left(\frac{\tan(2\gamma t)}{2\gamma}, \frac{e^{tM}x}{\cos(2\gamma t)}\right),
\end{aligned}$$

where we used

(a) If $B \in M_{n\times n}$ is a constant matrix, then $\nabla(\varphi(Bx)) = B^T(\nabla\varphi)(Bx)$;
(b) $\Delta = \operatorname{div}(\nabla)$, $\operatorname{div}\mathbf{F} = \operatorname{tr}(\frac{\partial(F_1\cdots,F_n)}{\partial(x_1,\cdots,x_n)})$ = trace of the Jacobian of $\mathbf{F}$;
(c) If C is a constant square matrix, $\mathbf{W} = [w_1, \cdots, w_n]^T$ is a vector-valued function of $x \in \mathbb{R}^n$, then

$$\operatorname{div}(C\mathbf{W}) = \operatorname{tr}([\nabla w_1, \cdots, \nabla w_n]C^T).$$

(d) $\operatorname{tr}(U^*\Lambda U) = \operatorname{tr}(\Lambda)$ if U is a unitary matrix and $\Lambda \in M_{n\times n}$.

Thus we obtain

$$
\begin{aligned}
\Delta u(t,x) = & -in\gamma \frac{\sin(2\gamma t)}{\cos^{\frac{n}{2}+1}(2\gamma t)} e^{-i\frac{\gamma}{2}|x|^2\tan(2\gamma t)} \varphi\left(\frac{\tan(2\gamma t)}{2\gamma}, \frac{e^{tM}x}{\cos(2\gamma t)}\right) \\
& - \gamma^2|x|^2 \frac{\sin^2(2\gamma t)}{\cos^{\frac{n}{2}+2}(2\gamma t)} e^{-i\frac{\gamma}{2}|x|^2\tan(2\gamma t)} \varphi\left(\frac{\tan(2\gamma t)}{2\gamma}, \frac{e^{tM}x}{\cos(2\gamma t)}\right) \\
& - i2\gamma \frac{\sin(2\gamma t)}{\cos^{\frac{n}{2}+2}(2\gamma t)} e^{-i\frac{\gamma}{2}|x|^2\tan(2\gamma t)} (e^{tM}x)\cdot\nabla\varphi\left(\frac{\tan(2\gamma t)}{2\gamma}, \frac{e^{tM}x}{\cos(2\gamma t)}\right) \\
& + \frac{1}{\cos^{\frac{n}{2}+2}(2\gamma t)} e^{-i\frac{\gamma}{2}|x|^2\tan(2\gamma t)} \Delta\varphi\left(\frac{\tan(2\gamma t)}{2\gamma}, \frac{e^{tM}x}{\cos(2\gamma t)}\right),
\end{aligned} \tag{10}
$$

and, noting that $(Mx)\cdot x = 0$,

$$
i(Mx)\cdot\nabla u = \frac{i}{\cos^{\frac{n}{2}+1}(2\gamma t)} e^{-i\frac{\gamma}{2}|x|^2\tan(2\gamma t)} (Me^{tM}x)\cdot\nabla\varphi\left(\frac{\tan(2\gamma t)}{2\gamma}, \frac{e^{tM}x}{\cos(2\gamma t)}\right). \tag{11}
$$

Bring (7), (9), (10), and (11) into Cauchy problem (1), and recall that φ satisfies (2) with $p = 1+4/n$, hence, we see that u is a solution to (1) with $u(0,x) = \varphi(0,x)$.

The above virtually shows that under the relation $u = \mathcal{R}\varphi \iff \varphi = \mathcal{R}^{-1}u$, u satisfies (1) if and only if φ satisfies (2). Therefore the second part of the proposition is also true. □

REMARK 4.4. The $\mathcal{R}$ transform also allows us to solve the linear equation for (1). The equation $i\partial_t\varphi = -\Delta\varphi$ has the fundamental solution

$$
e^{it\Delta}(x,y) = \frac{1}{(4\pi it)^{\frac{n}{2}}} e^{i\frac{|x-y|^2}{4t}}.
$$

Applying (7) we then obtain the fundamental solution to $i\partial_t u = H_{A,V}u$:

$$
e^{-itH_{A,V}}(x,y) = \left(\frac{\gamma}{2\pi i\sin(2\gamma t)}\right)^{\frac{n}{2}} e^{i\frac{\gamma}{2}(|x|^2+|y|^2)\cot(2\gamma t)} e^{-i\gamma\frac{(e^{tM}x)\cdot y}{\sin(2\gamma t)}}. \tag{12}
$$

This expression is significantly simpler than the one in [**10**] per Mehler's formula. To our best knowledge, (12) might be the first simply unified explicit formula compared with [**2, 14–16**].

Now we are ready to prove Theorem 1.1.

PROOF OF THEOREM 1.1. Let $u \in C([0,T), \mathscr{H}^1)$ be the blowup solution to the problem (1), where $[0,T)$ is the maximal interval of existence. Then by (8), there is a $\varphi(t,x) \in C([0, \frac{\tan(2\gamma T)}{2\gamma}), \mathscr{H}^1)$ that solves (2), where $[0, \frac{\tan(2\gamma T)}{2\gamma})$ is the maximal interval of existence. Note that $u_0 = \varphi_0$, and according to Theorem 3.1, the condition (4) suggests that φ is a blowup solution. Recall from (7), for $T \in (0, \frac{\pi}{4\gamma}]$

$$
u(t,x) = \frac{1}{\cos^{\frac{n}{2}}(2\gamma t)} e^{-i\frac{\gamma}{2}|x|^2\tan(2\gamma t)} \varphi\left(\frac{\tan(2\gamma t)}{2\gamma}, \frac{e^{tM}x}{\cos(2\gamma t)}\right).
$$

Then for all $t \in [0,T)$, we have

$$
\begin{aligned}
\nabla_x u(t,x) &= -i\gamma x \frac{\sin(2\gamma t)}{\cos^{\frac{n}{2}+1}(2\gamma t)} e^{-i\frac{\gamma}{2}|x|^2 \tan(2\gamma t)} \varphi\left(\frac{\tan(2\gamma t)}{2\gamma}, \frac{e^{tM}x}{\cos(2\gamma t)}\right) \\
&\quad + \frac{1}{\cos^{\frac{n}{2}+1}(2\gamma t)} e^{-i\frac{\gamma}{2}|x|^2 \tan(2\gamma t)} (e^{tM})^T \nabla\varphi\left(\frac{\tan(2\gamma t)}{2\gamma}, \frac{e^{tM}x}{\cos(2\gamma t)}\right) \\
&:= I_1 + I_2.
\end{aligned} \tag{13}
$$

For I_1, a change of variable gives

$$|I_1|_2 = \gamma \sin(2\gamma t) \left| x\varphi\left(\frac{\tan(2\gamma t)}{2\gamma}, x\right)\right|_2.$$

Let $J(t) := |x\varphi(t,x)|_2^2$. Then

$$J(t) = J(0) + J'(0)t + \int_0^t J''(\tau)\,(t-\tau)d\tau.$$

Note that

$$|J(0)| = |x\varphi_0|_2^2 \leq |\varphi_0|_{\mathscr{H}^1}^2,$$

and by Lemma 4.1, we have

$$
\begin{aligned}
|J'(0)| &= \left|4\Im \int x\overline{\varphi_0} \cdot \nabla\varphi_0 dx\right| \leq 4\|x\varphi_0\|_2 \|\nabla\varphi_0\|_2 \leq \|\varphi_0\|_{\mathscr{H}^1}^2, \\
J''(t) &= 8\mathcal{E}(\varphi_0).
\end{aligned}
$$

Thus

$$
\begin{aligned}
&\left| x\varphi\left(\frac{\tan(2\gamma t)}{2\gamma}, x\right)\right|_2^2 = J\left(\frac{\tan(2\gamma t)}{2\gamma}\right) \\
&\leq |J(0)| + |J'(0)| \frac{\tan(2\gamma t)}{2\gamma} + 4\mathcal{E}(\varphi_0)\left(\frac{\tan(2\gamma t)}{2\gamma}\right)^2 \\
&\leq |\varphi_0|_{\mathscr{H}^1}^2 + |\varphi_0|_{\mathscr{H}^1}^2 \frac{\tan(2\gamma T)}{2\gamma} + 4\mathcal{E}(\varphi_0)\left(\frac{\tan(2\gamma T)}{2\gamma}\right)^2,
\end{aligned}
$$

and so

$$|I_1|_2 \leq C(\varphi_0, T). \tag{14}$$

For I_2, in view of Lemma 4.2, $e^{tM^T} \in SO(n)$, a change of variable gives for $t \in [0,T)$, $(T \leq \pi/4\gamma)$

$$|I_2|_2 = \frac{1}{\cos(2\gamma t)} \left|\nabla\varphi\left(\frac{\tan(2\gamma t)}{2\gamma}, x\right)\right|_2.$$

As $t \to T$, $\frac{\tan(2\gamma t)}{2\gamma} \to \frac{\tan(2\gamma T)}{2\gamma}$, so by Theorem 3.1,

$$\lim_{t\to T} \frac{\left|\nabla\varphi\left(\frac{\tan(2\gamma t)}{2\gamma}, x\right)\right|_2}{|\nabla Q|_2} \sqrt{\frac{\frac{\tan(2\gamma T)}{2\gamma} - \frac{\tan(2\gamma t)}{2\gamma}}{\log\left|\log\left(\frac{\tan(2\gamma T)}{2\gamma} - \frac{\tan(2\gamma t)}{2\gamma}\right)\right|}} = \frac{1}{\sqrt{2\pi}}.$$

Note that as $t \to T$, there are

$$\frac{\sin(2\gamma(T-t))}{2\gamma(T-t)} \to 1 \qquad \text{and} \qquad \frac{\log\sin(2\gamma(T-t))}{\log(T-t)} \to 1,$$

so the above blowup rate can be simplified as

$$\lim_{t\to T}\frac{\left|\nabla\varphi\left(\frac{\tan(2\gamma t)}{2\gamma},x\right)\right|_2}{|\nabla Q|_2}\sqrt{\frac{T-t}{\log|\log(T-t)|}}=\frac{\cos(2\gamma T)}{\sqrt{2\pi}}$$

to yield

$$\lim_{t\to T}\frac{|I_2|_2}{|\nabla Q|_2}\sqrt{\frac{T-t}{\log|\log(T-t)|}}=\frac{1}{\sqrt{2\pi}}. \tag{15}$$

Therefore, combining (14) and (15) we obtain

$$\lim_{t\to T}\frac{|\nabla u|_2}{|\nabla Q|_2}\sqrt{\frac{T-t}{\log|\log(T-t)|}}=\frac{1}{\sqrt{2\pi}}.$$

□

4.1. Blowup rate at the ground state Q. We conclude with some discussions on the wave collapse rates for (1) when the initial data is near Q, which could be a subtle issue. Notice that this Q is not the ground state for (1), instead, it is the one for (2). If $|u_0|_2 = |Q|_2$, then the wave collapse for (1) is different than in the case $|u_0|_2 > |Q|_2$. Applying the transform (7) to the solitary wave $\varphi = e^{it}Q$ we can construct a blowup solution with blowup rate $(T-t)^{-1}$:

$$u(t,x)=\frac{1}{\cos^{\frac{n}{2}}(2\gamma t)}e^{-i\frac{\gamma}{2}|x|^2\tan(2\gamma t)}e^{i\frac{\tan(2\gamma t)}{2\gamma}}Q\left(\frac{e^{tM}x}{\cos(2\gamma t)}\right). \tag{16}$$

One easily checks that u blows up at $T=\frac{\pi}{4\gamma}$ satisfying

$$|\nabla u|_2\approx\frac{1}{(\frac{\pi}{2}-2\gamma t)}|\nabla Q|_2 \qquad \text{as } t\to T=\frac{\pi}{4\gamma}.$$

Note that solutions of the form (16) with such blowup time and singularity can also be obtained with other nonpositive, non-radial bound states Q_b as the profile in place of Q, where $|Q_b|_2 > |Q|_2$ and $n \geq 2$.

Suppose that u is a blowup solution to (1) with blowup time $T^* < \frac{\pi}{4\gamma}$ and $|u_0|_2 = |Q|_2$. Then by means of (8) we may define a blowup solution to (2) with the same initial data which blows up at $T_0 = \frac{\tan(2\gamma T)}{2\gamma}$. Merle [**19**] showed that up to the scaling and phase invariances of (2), the only minimal mass blowup solutions are of the form

$$\varphi(t,x)=\frac{1}{(T_0-t)^{\frac{n}{2}}}e^{-\frac{i|x|^2}{4(T_0-t)}}e^{\frac{i}{T_0-t}}Q\left(\frac{x}{T_0-t}-x_0\right)$$

for some $x_0\in\mathbb{R}^n$. By the $\mathfrak{R}$-transform (7) and the uniqueness of (1) (Proposition 2.1), we then establish a characterization for all minimal mass blowup solutions of (1).

PROPOSITION 4.5. *Let $|u_0|_2 = |Q|_2$. Let u be a blowup solution of (1) on $[0,T)$ with $T := T^* < \frac{\pi}{4\gamma}$. Then u must assume the following form (up to scaling and* *phase invariance): There exists* $x_1 \in \mathbb{R}^n$ *such that*

$$
\begin{aligned}
u(t,x) = &\left(\frac{2\gamma\cos(2\gamma T)}{\sin(2\gamma(T-t))}\right)^{\frac{n}{2}} e^{-i\frac{\gamma}{2}|x|^2\cot(2\gamma(T-t))} e^{i2\gamma\frac{\cos(2\gamma T)\cos(2\gamma t)}{\sin(2\gamma(T-t))}} \\
&\times Q\left(\frac{2\gamma\cos(2\gamma T)e^{tM}x}{\sin(2\gamma(T-t))} - x_1\right).
\end{aligned} \tag{17}
$$

Moreover, $|\nabla u|_2 \approx (T-t)^{-1}$ *as* $t \to T$.

Note that the blowup solution given in (16) is not covered by (17). For example, if $T^* = \pi/8\gamma$, $x_1 = 0$, then $u_0(x) = (2\gamma)^{n/2} e^{-i\frac{\gamma}{2}|x|^2} e^{i2\gamma} Q(2\gamma x)$. Rather, (16) can be viewed as a bordering case of the assertion in Proposition 4.5 corresponding to $T^* = \pi/4\gamma$.

REMARK 4.6. If the initial value is of the form $u_0 = (1+\varepsilon)Q$ with $0 < \varepsilon < \sqrt{1 + \frac{\alpha^*}{\|Q\|_2^2}} - 1$, then the corresponding solution to (1) will blowup at the rate stated in Theorem 1.1. Indeed, by the range for ε and the Pohozaev identity $\int |\nabla Q|^2 = \frac{n}{n+2}\int |Q|^{2+\frac{4}{n}}$, it is easy to verify $u_0 \in B_{\alpha^*}$ and condition (4).

REMARK 4.7. For large initial data one can also derive a general lower bound for the collapse rate. If the solution of (1) satisfies $\lim_{t\to T^*} |\nabla u|_2 = \infty$, then there exists $C = C_{p,n} > 0$ such that

$$|\nabla u(t)|_2 \geq C(T^* - t)^{-(\frac{1}{p-1} - \frac{n-2}{4})}.$$

This follows from quite standard argument as in [**9**] that is used to show the l.w.p and blowup alternative for the Cauchy problem (1) on $[0, T^*)$. In the L^2-critical case $p = 1 + \frac{4}{n}$, the lower bound becomes

$$|\nabla u(t)|_2 \geq C(T^* - t)^{-\frac{1}{2}}.$$

Acknowledgment. The authors would like to thank Remi Carles and Chenjie Fan for helpful comments and communications.

References

[1] A. Aftalion, *Vortices in Bose-Einstein condensates*, Progress in Nonlinear Differential Equations and their Applications, vol. 67, Birkhäuser Boston, Inc., Boston, MA, 2006. MR2228356

[2] P. Antonelli, D. Marahrens, and C. Sparber, *On the Cauchy problem for nonlinear Schrödinger equations with rotation*, Discrete Contin. Dyn. Syst. **32** (2012), no. 3, 703–715, DOI 10.3934/dcds.2012.32.703. MR2851876

[3] W. Bao and Y. Cai, *Ground states and dynamics of spin-orbit-coupled Bose-Einstein condensates*, SIAM J. Appl. Math. **75** (2015), no. 2, 492–517, DOI 10.1137/140979241. MR3323558

[4] W. Bao, H. Wang, and P. A. Markowich, *Ground, symmetric and central vortex states in rotating Bose-Einstein condensates*, Commun. Math. Sci. **3** (2005), no. 1, 57–88. MR2132826

[5] J. Bourgain and W. Wang, *Construction of blowup solutions for the nonlinear Schrödinger equation with critical nonlinearity*, Ann. Scuola Norm. Sup. Pisa Cl. Sci. (4) **25** (1997), no. 1-2, 197–215 (1998). Dedicated to Ennio De Giorgi. MR1655515

[6] N. Basharat, H. Hajaiej, Y. Hu, S. Zheng, Threshold for blowup and stability for nonlinear Schrödinger equation with rotation. Preprint.

[7] R. Carles, *Critical nonlinear Schrödinger equations with and without harmonic potential*, Math. Models Methods Appl. Sci. **12** (2002), no. 10, 1513–1523, DOI 10.1142/S0218202502002215. MR1933935

[8] T. Cazenave and M. J. Esteban, *On the stability of stationary states for nonlinear Schrödinger equations with an external magnetic field* (English, with Portuguese summary), Mat. Apl. Comput. **7** (1988), no. 3, 155–168. MR994761

[9] T. Cazenave and F. B. Weissler, *The Cauchy problem for the critical nonlinear Schrödinger equation in H^s*, Nonlinear Anal. **14** (1990), no. 10, 807–836, DOI 10.1016/0362-546X(90)90023-A. MR1055532

[10] A. Cheskidov, D. Marahrens, and C. Sparber, *Global attractor for a Ginzburg-Landau type model of rotating Bose-Einstein condensates*, Dyn. Partial Differ. Equ. **14** (2017), no. 1, 5–32, DOI 10.4310/DPDE.2017.v14.n1.a2. MR3607053

[11] A. De Bouard, *Nonlinear Schroedinger equations with magnetic fields*, Differential Integral Equations **4** (1991), no. 1, 73–88. MR1079611

[12] G. Fibich, F. Merle, and P. Raphaël, *Proof of a spectral property related to the singularity formation for the L^2 critical nonlinear Schrödinger equation*, Phys. D **220** (2006), no. 1, 1–13, DOI 10.1016/j.physd.2006.06.010. MR2252148

[13] E. P. Gross, *Structure of a quantized vortex in boson systems* (English, with Italian summary), Nuovo Cimento (10) **20** (1961), 454–477. MR0128907

[14] C. Hao, L. Hsiao, and H.-L. Li, *Global well posedness for the Gross-Pitaevskii equation with an angular momentum rotational term in three dimensions*, J. Math. Phys. **48** (2007), no. 10, 102105, 11, DOI 10.1063/1.2795218. MR2362770

[15] C. Hao, L. Hsiao, and H.-L. Li, *Global well posedness for the Gross-Pitaevskii equation with an angular momentum rotational term*, Math. Methods Appl. Sci. **31** (2008), no. 6, 655–664, DOI 10.1002/mma.931. MR2400070

[16] H. Kitada, *On a construction of the fundamental solution for Schrödinger equations*, J. Fac. Sci. Univ. Tokyo Sect. IA Math. **27** (1980), no. 1, 193–226. MR573337

[17] M. K. Kwong, *Uniqueness of positive solutions of $\Delta u - u + u^p = 0$ in $\mathbf{R}^n$*, Arch. Rational Mech. Anal. **105** (1989), no. 3, 243–266, DOI 10.1007/BF00251502. MR969899

[18] M. Matthews, B. Anderson, P. Haljan, D. Hall, C. Wiemann, E. A. Cornell, *Vortices in a Bose-Einstein condensates*, Phys. Rev. Lett. **83** (1999), 2498–2501.

[19] F. Merle, *Determination of blow-up solutions with minimal mass for nonlinear Schrödinger equations with critical power*, Duke Math. J. **69** (1993), no. 2, 427–454, DOI 10.1215/S0012-7094-93-06919-0. MR1203233

[20] F. Merle and P. Raphael, *The blow-up dynamic and upper bound on the blow-up rate for critical nonlinear Schrödinger equation*, Ann. of Math. (2) **161** (2005), no. 1, 157–222, DOI 10.4007/annals.2005.161.157. MR2150386

[21] F. Merle and P. Raphael, *Profiles and quantization of the blow up mass for critical nonlinear Schrödinger equation*, Comm. Math. Phys. **253** (2005), no. 3, 675–704, DOI 10.1007/s00220-004-1198-0. MR2116733

[22] F. Merle, P. Raphaël, and J. Szeftel, *Stable self-similar blow-up dynamics for slightly L^2 super-critical NLS equations*, Geom. Funct. Anal. **20** (2010), no. 4, 1028–1071, DOI 10.1007/s00039-010-0081-8. MR2729284

[23] G. Perelman, *On the formation of singularities in solutions of the critical nonlinear Schrödinger equation*, Ann. Henri Poincaré **2** (2001), no. 4, 605–673, DOI 10.1007/PL00001048. MR1852922

[24] P. Raphael, *Stability of the log-log bound for blow up solutions to the critical non linear Schrödinger equation*, Math. Ann. **331** (2005), no. 3, 577–609, DOI 10.1007/s00208-004-0596-0. MR2122541

[25] A. Recati, F. Zambelli, S. Stringari, *Overcritical rotation of a trapped Bose-Einstein condensate*, Phys. Rev. Lett. **86** (2001), 377–380.

[26] M. I. Weinstein, *Nonlinear Schrödinger equations and sharp interpolation estimates*, Comm. Math. Phys. **87** (1982/83), no. 4, 567–576. MR691044

[27] K. Yajima, *Schrödinger evolution equations with magnetic fields*, J. Analyse Math. **56** (1991), 29–76, DOI 10.1007/BF02820459. MR1243098

[28] S. Zheng, *Fractional regularity for nonlinear Schrödinger equations with magnetic fields*, Recent advances in harmonic analysis and partial differential equations, Contemp. Math., vol. 581, Amer. Math. Soc., Providence, RI, 2012, pp. 271–285, DOI 10.1090/conm/581/11533. MR3088496

[29] S.-h. Zhu and J. Zhang, *Profiles of blow-up solutions for the Gross-Pitaevskii equation*, Acta Math. Appl. Sin. Engl. Ser. **26** (2010), no. 4, 597–606, DOI 10.1007/s10255-010-0027-9. MR2720564

Department of Population Health Sciences, Augusta University, Augusta, Georgia 30912
Email address: nylabasharat12@gmail.com

Department of Mathematical Sciences, Georgia Southern University, Statesboro, Georgia 30460
Email address: yihu@georgiasouthern.edu

Department of Mathematical Sciences, Georgia Southern University, Statesboro, Georgia 30460
Email address: szheng@georgiasouthern.edu

Contemporary Mathematics
Volume **725**, 2019
https://doi.org/10.1090/conm/725/14553

Gradient estimates for weak solutions of linear elliptic systems with singular-degenerate coefficients

Dat Cao, Tadele Mengesha, and Tuoc Phan

ABSTRACT. This paper establishes Calderón-Zygmund type regularity estimates for solutions of the conormal derivative problem for a class of linear elliptic systems in divergence-form with singular, degenerate coefficients in bounded domains. In our class of equations, the principal terms are fourth order tensors of measurable functions that behave as some weight function in the Muckenhoupt class of A_2-weights. Regularity estimates for gradient of weak solutions in weighted Lebesgue spaces are established under some natural smallness conditions on the mean oscillation of coefficients. The results obtained recover known results when the coefficients are uniformly elliptic. These results can be considered as the Sobolev counterparts of the classical Hölder's regularity estimates established by B. Fabes, C. E. Kenig, and R. P. Serapioni.

1. Introduction

This work studies regularity estimates in weighted Sobolev spaces for weak solutions of a class of linear systems of elliptic equation with prescribed degenerate, singular coefficients in bounded domains with conormal boundary conditions. The system we are studying is given by

$$\begin{cases} \operatorname{div}[\mathcal{A}(x)\nabla u] & = \operatorname{div}[\mathbf{F}] \quad \text{in} \quad \Omega, \\ \langle \mathcal{A}(x)\nabla u - \mathbf{F}, \mathbf{n}\rangle & = 0 \quad\quad\quad\ \text{on} \quad \partial\Omega, \end{cases} \tag{1.1}$$

where $\Omega \subset \mathbb{R}^n$ is an open bounded domain with C^1 boundary, $\mathbf{F} : \Omega \to \mathbb{R}^{n\times N}$ is a given vector field, $u : \Omega \to \mathbb{R}^N$, and $n, N \in \mathbb{N}$, and $\mathbf{n}$ is the outward unit normal vector at points on the boundary of Ω. The coefficient $\mathcal{A} : \Omega \to \mathbb{R}^{nN\times nN}$ is a fourth order tensor $(A_{ij}^{\alpha\beta})_{i,j=1}^{N}{}_{\alpha,\beta=1}^{n}$ of measurable functions that could be degenerate or singular as some weight function in some Muckenhoupt class of weight. Precisely, we assume that there exists a constant $\Lambda > 0$ and a non-negative measurable function μ on $\mathbb{R}^n$

$$\Lambda\mu(x)|\xi|^2 \leq \langle \mathcal{A}(x)\xi, \xi\rangle = \sum_{\alpha,\beta=1}^{N}\sum_{i,j=1}^{n} A_{ij}^{\alpha\beta}\xi_\alpha^j\xi_\beta^i, \quad \forall\ \xi = (\xi_\alpha^i) \in \mathbb{R}^{nN}, \text{ a.e. } x \in \Omega, \text{ and} \tag{1.2}$$

$$\left|A_{ij}^{\alpha\beta}(x)\right| \leq \Lambda^{-1}\mu(x) \quad \text{a.e. } x \in \Omega, \text{ and all } \alpha, \beta = 1, \cdots, N;\ i, j = 1, \cdots n. \tag{1.3}$$

Our goal is to establish the regularity of weak solutions of the degenerate system (1.1) in weighted Sobolev spaces. Regularity estimates for weak solutions of uniformly elliptic equations and systems of equations ($\mu = 1$) for both Dirichlet and conormal derivative boundary value problems in Sobolev spaces is commonplace these days. One can find results in [**1**–**3**, **5**, **12**, **15**–**18**, **23**, **26**] in which uniform elliptic coefficients are studied. Essentially one expects that the matrix $\mathbf{F}$ and ∇u have the same integrability property. However by now it is well known that the mere assumption on the uniform ellipticity of the tensor of coefficients $\mathcal{A}$ is not sufficient for the gradient of the weak solution of (1.1) to have the same integrability as that of the data $\mathbf{F}$. This fact can be seen from the counterexample provided by N. G. Meyers in [**19**] . In the event that $\mathcal{A}$ is uniformly elliptic and continuous, the L^p-norm of ∇u can be controlled by the L^p-norm of the datum $\mathbf{F}$ and this is achieved via the Calderón-Zygmund theory of singular integrals and a perturbation technique, see [**5**, **12**, **15**–**18**] for this classical results for both elliptic and parabolic equations.

Our interest lies on equations with coefficients that may be degenerate or singular. The class of second order linear elliptic equations with degenerate coefficients with general case $\mu \in A_2$ was investigated by B. Fabes, C. E. Kenig, and R. P. Serapioni for the first time in the pioneering paper [**7**] in 1982. In this classical paper [**7**], among other important results, existence, and uniqueness of weak solutions in weighted Sobolev space $W_0^{1,2}(\Omega, \mu)$ were established; Harnack's inequality and Hölder's regularity of weak solutions were also obtained by adapting the Möser's iteration technique to the degenerate, non-uniformly elliptic equations (1.1). See also [**22**, **25**] for some other earlier results on elliptic equations with measurable degenerate coefficients.

Recently, in [**4**], we have obtained estimates in Lebesgue spaces for gradient of solutions to zero Dirichlet boundary value problems for linear degenerate equations of type (1.1) with general $\mu \in A_2$. Two weighted Calderón-Zygmund type regularity estimates for quasilinear elliptic equations with prescribed singular-degenerate coefficients and non-homogeneous Dirichlet boundary conditions are also established in [**24**]. We aim to extend the above mentioned results in [**4**, **24**] and establish the corresponding results for weak solutions of the conormal derivative problem for linear systems (1.1) by giving the right and optimal conditions on the coefficient tensor. As already indicated, the regularity estimate we obtain requires that coefficient matrix must not oscillate too much. To describe the condition precisely, we want to introduce a means of measuring oscillation. The following definition is a weighted version of functions of bounded mean oscillation that is compatible with degenerate and singular coefficients. This definition can be found in [**9**, **10**, **20**, **21**].

DEFINITION 1.1. Given $R_0 > 0$, we say $f : \Omega \to \mathbb{R}$ is function of bounded mean oscillation with weight μ in Ω if $[f]_{\mathrm{BMO}_{R_0}(\Omega,\mu)} < \infty$ where

$$[f]_{\mathrm{BMO}_{R_0}(\Omega,\mu)} = \sup_{\substack{x\in\Omega \\ 0<\rho<R_0}} \frac{1}{\mu(B_\rho(x)\cap\Omega)} \int_{B_\rho(x)\cap\Omega} |f(y) - \langle f\rangle_{B_\rho(x)\cap\Omega}| dy,$$

and

$$\langle f\rangle_{B_\rho(x)\cap\Omega} = \frac{1}{|B_\rho(x)\cap\Omega|}\int_{B_\rho(x)\cap\Omega} f(y)dy.$$

Observe that the classical John-Nirenberg BMO in Ω corresponds to $\mu = 1$ and $R_0 = \mathrm{diam}(\Omega)$. The class of Muckenhoupt A_p- weights will be recalled in Definition 2.2. Our first main result of this paper is the following theorem.

THEOREM 1.2. *Let $\Lambda > 0, M_0 \geq 1$, and $p \geq 2$ be given. There exists a sufficiently small constant $\delta = \delta(\Lambda, M_0, n, p) > 0$ such that the following statement holds. Suppose also that $\mu \in A_2$ with $[\mu]_{A_2} \leq M_0$, Ω is a C^1-domain,* (1.2)-(1.3) *hold on Ω, and $[\mathcal{A}]_{BMO_{R_0}(\Omega,\mu)} \leq \delta$ for some $R_0 > 0$. Then every weak solution $u \in W^{1,2}(\Omega, \mu, \mathbb{R}^N)$ of* (1.1) *corresponding to $|\mathbf{F}|/\mu \in L^p(\Omega, \mu, \mathbb{R}^N)$ satisfies the estimate*

$$\|\nabla u\|_{L^p(\Omega,\mu)} \leq C \left\| \frac{\mathbf{F}}{\mu} \right\|_{L^p(\Omega,\mu)},$$

where C is a constant depending only on n, Λ, p, M_0, R_0 and Ω.

Local regularity estimates for weak solutions of (1.1) are not only of great interest by themselves, but also important in many applications since they only require local information on the data. This paper also establishes interior regularity estimates, and local boundary estimates. To state the local results we use the notation B_r to denote for a ball of radius r centered at the origin, B_r^+ its upper part and T_r the flat part of the boundary of B_r^+. The next result presents the interior local gradient estimate.

THEOREM 1.3. *Let $\Lambda > 0, M_0 \geq 1$, and $p \geq 2$ be given. There exists a sufficiently small constant $\delta = \delta(\Lambda, M_0, n, p) > 0$ such that the following statement holds. Suppose that $\mu \in A_2$ with $[\mu]_{A_2} \leq M_0$, and* (1.2)-(1.3) *hold for a given tensor matrix $\mathcal{A}$ on B_2. Moreover, assume that $[\mathcal{A}]_{BMO_1(B_1,\mu)} \leq \delta$. Then, for every $\mathbf{F}$ such that $\mathbf{F}/\mu \in L^p(B_2, \mu, \mathbb{R}^N)$, if $u \in W^{1,2}(B_2, \mu, \mathbb{R}^N)$ is a weak solution to*

$$div[\mathcal{A}\nabla u] = div(\mathbf{F}) \quad in\ B_2, \tag{1.4}$$

then $\nabla u \in L^p(B_1, \mu)$ and

$$\|\nabla u\|_{L^p(B_1,\mu)} \leq C \left((\mu(B_1))^{\frac{1}{p}-\frac{1}{2}} \|\nabla u\|_{L^2(B_2,\mu)} + \|\mathbf{F}/\mu\|_{L^p(B_2,\mu)} \right).$$

where C is a constant depending only on Λ, p, n, M_0.

Our last result is a local boundary regularity one.

THEOREM 1.4. *Let $\Lambda > 0, M_0 \geq 1$, and $p \geq 2$ be given. There exists a sufficiently small constant $\delta = \delta(\Lambda, M_0, n, p) > 0$ such that the following statement holds. Suppose that $\mu \in A_2$ with $[\mu]_{A_2} \leq M_0$, and* (1.2)-(1.3) *hold for the given tensor matrix $\mathcal{A}$ on B_2^+. Moreover, assume that $[\mathcal{A}]_{BMO_1(B_1^+,\mu)} \leq \delta$. Then, for every $\mathbf{F}$ such that $\mathbf{F}/\mu \in L^p(B_2^+, \mu)$, if $u \in W^{1,2}(B_2^+, \mu, \mathbb{R}^N)$ is a weak solution to*

$$\begin{cases} div[\mathcal{A}\nabla u] = div(\mathbf{F}) & in\ B_2^+. \\ \langle \mathcal{A}\nabla u - \mathbf{F}, \mathbf{n} \rangle = 0 & on\ T_2, \end{cases} \tag{1.5}$$

then $\nabla u \in L^p(B_1^+, \mu)$ and

$$\|\nabla u\|_{L^p(B_1^+,\mu)} \leq C \left((\mu(B_1))^{\frac{1}{p}-\frac{1}{2}} \|\nabla u\|_{L^2(B_2^+,\mu)} + \|\mathbf{F}/\mu\|_{L^p(B_2^+,\mu)} \right),$$

where C is a constant depending only on Λ, p, n, M_0.

We now comment how we prove the main result. It is well-known that Theorem 1.2 can be obtained from Theorem 1.3 and Theorem 1.4 via standard arguments using partition of unity, and flattening the boundary. We therefore skip the proof of Theorem 1.2. Also, the proof of Theorem 1.3 can be done along the same line of arguments as in [**4**, Theorem 2.5] after making the necessary adjustment for systems. As such, we will not focus much on it but rather we state and use estimates that we may need in the proof of Theorem 1.4. To prove the latter, we will implement the approximation method of Caffarelli and Peral in [**3**]. The main idea in the approach is to locally consider the equation (1.5) as the perturbation of an equation for which the regularity of its solution is well understood. Key ingredients include Vitali's covering lemma, and the weak and strong (p, p) estimates of the weighted Hardy-Littlewood maximal operators. To be able to compare the solutions of the perturbed and un-perturbed equations, we prefer to use compactness argument as has been used in [**1**], but on weighted spaces. Such argument has been used in [**4**], similar regularity results as Theorem 1.4 are proved for zero-Dirichlet boundary value problem of equations. Essential properties of A_2 weights such as reverse Hölder's inequality and doubling property are properly utilized in dealing with technical issues arising from the degeneracy and singularity of the coefficients.

We remark that obtaining estimates as in Theorems 1.2-1.4 for large values of p is not always possible even for smooth but degenerate coefficient matrix $\mathcal{A}$ and $\mu \in A_2$. We refer [**4**] for a counterexample. In light of this and compared to [**6**], this paper provides the correct minimal conditions on the coefficients so that the linear map $\frac{\mathbf{F}}{\mu} \mapsto \nabla u$ remain continuous on the smaller space $L^p(\Omega, \mu)$, $p \geq 2$.

As in the Dirichlet boundary value problem studied in [**4**], to establish the weighted L^p-regularity estimates, we follow the approximation method of Cafarrelli and Peral in [**3**] where we view (1.1) locally as a perturbation of an elliptic homogeneous equation with constant coefficients. The key to the success of this approach to equations with degenerate coefficients is the novel way of measuring mean oscillation of coefficients which is found to be compatible with the degeneracy of the coefficients. The smallness of the measure, which is just a smallness of the mean oscillation with weights introduced in [**20, 21**] to study Hilbert transforms and characterize the dual of weighted Hardy space. This condition is optimal in the sense that it coincided with the well known result in [**2**] when $\mu = 1$. The counterexample given in [**4**] demonstrates the necessity of the smallness condition $[\mathcal{A}]_{\mathrm{BMO}}$ in Theorems 1.2-1.4.

The paper is organized as follows. In Section 2, we introduce notations, definitions, and provides some elementary estimates needed for the paper. Section 3 provides the approximation estimates. Level set estimates, and the proofs of the main theorems, Theorem 1.2-1.4 are given in Section 4, which is the last section of the paper.

2. Notations, definitions, and preliminary estimates

2.1. Definitions and existence of weak solutions. We start with some definitions, and notations.

DEFINITION 2.1. Let σ be a non-negative measure on $\mathbb{R}^n$ and a non-empty open set $U \subset \mathbb{R}^n$, we write

$$d\sigma = \sigma dx, \quad \sigma(U) = \int_U \sigma(x) dx.$$

For a locally integrable Lebesgue-measurable function f on $\mathbb{R}^n$, we always denote the average of f in U with respect of the measure $d\sigma$ as

$$\bar{f}_{\sigma,U} = \fint_U f(x)d\sigma = \frac{1}{\sigma(U)} \int_U f(x)\sigma dx.$$

For Lebesgue measure dx, we write

$$\bar{f}_U = \bar{f}_{dx,U}, \quad \text{and} \quad |U| = \int_U dx.$$

DEFINITION 2.2. Let $p \in (1, \infty)$ and $\mu \in L^1_{\text{loc}}(\mathbb{R}^n)$ be non-negative. The weight function μ is said to be of Muckenhoupt class A_p if

$$[\mu]_{A_p} := \sup_{B \subset \mathbb{R}^n} \left(\fint_B \mu(y)dy \right) \left(\fint_B \mu(y)^{-\frac{1}{p-1}} dy \right)^{p-1} < \infty,$$

where the supremum is taken over all balls $B \subset \mathbb{R}^n$.

It turns out that any $\mu \in A_p$ defines a measure $d\mu = \mu(x)dx$. Moreover, we can define the corresponding Lebesgue and Sobolev spaces with respect to the measure.

DEFINITION 2.3. Let $\mu \in A_p$ with $1 < p < \infty$, let $1 \leq q < \infty$ and $\Omega \subset \mathbb{R}^n$ be open, bounded. A locally integrable function f define on Ω is said to belong to the weighted Lebesgue space $L^q(\Omega, \mu)$ if

$$\|f\|_{L^q(\Omega,\mu)} = \left(\int_\Omega |f(x)|^q \mu(x)dx \right)^{1/q} < \infty.$$

Let $k \in \mathbb{N}$. A locally integrable function f defined on Ω is said to belong to the weighted Sobolev space $W^{k,q}(\Omega, \mu)$ if all of its distributional derivatives $D^\alpha f$ are in $L^q(\Omega, \mu)$ for $\alpha \in (\mathbb{N} \cup \{0\})^n$ with $|\alpha| \leq k$. The space $W^{k,q}(\Omega, \mu)$ is equipped the the norm

$$\|f\|_{W^{k,q}(\Omega,\mu)} = \left(\sum_{|\alpha| \leq k} \|D^\alpha f\|^q_{L^q(\Omega,\mu)} \right)^{1/q}.$$

We also denote $W^{1,q}_0(\Omega, \mu)$ be the closure of $C^\infty_0(\Omega)$ in $W^{1,q}(\Omega, \mu)$.

Observe that both $L^q(\Omega, \mu)$, and $W^{1,q}(\Omega, \mu)$ are Banach spaces with their natural norm. When $q = 2$, $W^{1,2}(\Omega, \mu)$ is a Hilbert space.

We now recall what we mean by weak solution of (1.1).

DEFINITION 2.4. Assume that (1.2),(1.3) hold and $|\mathbf{F}|/\mu \in L^2(\Omega, \mu)$. A function $u \in W^{1,2}(\Omega, \mu, \mathbb{R}^N)$ is a weak solution of (1.1) if

$$\int_\Omega \langle \mathcal{A}\nabla u, \nabla\varphi \rangle dx = \int_\Omega \langle \mathbf{F}, \nabla\varphi \rangle dx, \quad \forall\, \varphi \in W^{1,2}(\Omega, \mu, \mathbb{R}^N). \tag{2.1}$$

Existence of a weak solution can be shown following the standard Hilbert Space method via the application of the theorem of Lax-Milgram. For Dirichlet boundary value problems, this has been done in [**7**]. Similar argument can be applied for the conormal derivative problem as well.

LEMMA 2.5. *Suppose $\mathcal{A}$ satisfes (1.2) and (1.3). Then, for each $\mathbf{F}$ with $|\mathbf{F}|/\mu \in L^2(\Omega, \mu)$, there exists a weak solution $u \in W_0^{1,2}(\Omega, \mu, \mathbb{R}^N)$ to (1.1). Moreover,*

$$\int_\Omega |\nabla u|^2 d\mu \leq C(\Lambda) \int_\Omega \left|\frac{\mathbf{F}}{\mu}\right|^2 d\mu. \tag{2.2}$$

The solution u is unique up to a constant vector.

2.2. Weights and weighted norm inequalities. In this section we review and collect some results needed in the paper. The first lemma is a standard result in measure theory.

LEMMA 2.6. *Assume that $g \geq 0$ is a measurable function in a bounded subset $U \subset \mathbb{R}^n$. Let $\theta > 0$ and $K > 1$ be given constants. If μ is a weight in $\mathbb{R}^n$, then for any $1 \leq p < \infty$*

$$g \in L^p_\mu(U) \Leftrightarrow S := \sum_{j \geq 1} K^{qj} \mu(\{x \in U : g(x) > \theta K^j\}) < \infty.$$

Moreover, there exists a constant $C > 0$ such that

$$C^{-1} S \leq \|g\|^p_{L^p_\mu(U)} \leq C(\mu(U) + S),$$

where C depends only on θ, K and p.

For a given locally integrable function f we define the weighted maximal function as

$$\mathcal{M}^\mu f(x) = \sup_{\rho>0} ⨍_{B_\rho(x)} |f| d\mu = \sup_{\rho>0} \frac{1}{\mu(B_\rho(x))} \int_{B_\rho(x)} |f|\, \mu(x) dx$$

For functions f that are defined on a bounded domain, we define

$$\mathcal{M}^\mu_\Omega f(x) = \mathcal{M}^\mu(f\chi_\Omega)(x)$$

For $M_0 > 0$ given, assume that $\mu \in A_2$ such that $[\mu]_{A_2} \leq M_0$. The following two continuity results are well known for the maximal function.

- (Strong $p-p$) For any $1 < p < \infty$, there exists a constant $C = C(p, n, M_0)$ such that for any weight μ with a strong doubling property we have

$$\|\mathcal{M}^\mu\|_{L^p_\mu \to L^p_\mu} \leq C.$$

- (Weak $1-1$) When $p = 1$, there exists a constant C that depends on $C(p, n, M_0)$ such that for any weight μ with a strong doubling property and $\lambda > 0$

$$\mu(x \in \mathbb{R}^n : \mathcal{M}^\mu(f) > \lambda) \leq \frac{C}{\lambda} \int_{\mathbb{R}^n} |f| dx.$$

The proof of these estimates can be found in ([**14**, Proof of Lemma 7.1.10]). In this paper we will be using mostly A_2 weights. From the definition, it is immediate that $\mu \in A_2$, then so is μ^{-1} with

$$[\mu]_{A_2} = [\mu^{-1}]_{A_2}.$$

The following lemma is what is called reverse Hölder's inequality that holds for A_p weights.

LEMMA 2.7 (Theorem 9.2.2, [**14**], Remark 9.2.3 [**14**]). *For any $M_0 > 0$, there exist positive constants $C = (n, M_0)$ and $\gamma = \gamma(n, M_0)$ such that for all $\mu \in A_2$ satisfying $[\mu]_{A_2} \leq M_0$ the reverse Hölder condition holds:*

$$\left(\frac{1}{|B|}\int_B \mu^{(1+\gamma)}(x)dx\right)^{\frac{1}{1+\gamma}} \leq \frac{C}{|B|}\int_B \mu(x)dx, \quad \text{and}$$

for every ball $B \subset \mathbb{R}^n$. The inequality holds true as well if μ is replaced by μ^{-1}.

As a consequence of Lemma 2.7 we have the following two important inequalities which will be used frequently in this paper.

LEMMA 2.8. [**4**, Lemma 3.4] *For any $M_0 > 0$, let $\beta = \frac{\gamma}{2+\gamma} > 0$ where γ is a constant as given in Lemma 2.7. Then for any $\mu \in A_2$ satisfying $[\mu]_{A_2} \leq M_0$, for any ball $B \subset \mathbb{R}^n$, we have that*

- *if $u \in L^2(B, \mu)$, then $u \in L^{1+\beta}(B)$ and*

$$\left(\fint_B |u|^{1+\beta}dx\right)^{\frac{1}{1+\beta}} \leq C(n, M_0)\left(\fint_B |u|^2 d\mu\right)^{1/2},$$

- *if $u \in L^q(B)$ with $q \geq 1$, then*

$$\left(\fint_B |u|^\tau d\mu\right)^{1/\tau} \leq C(n, M_0)\left(\fint_B |u|^q dx\right)^{1/q}, \quad \text{with} \quad \tau = \frac{q\gamma}{1+\gamma}.$$

Next, we recall the weighted Sobolev-Poincaré inequality whose prove can be found in [**7**, Theorem 1.5, Theorem 1.6]

LEMMA 2.9. *Let $M_0 > 0$ and assume that $\mu \in A_2$ and $[\mu]_{A_2} \leq M_0$. Then, there exists a constant $C = C(n, M_0)$ and $\alpha = \alpha(n, M_0) > 0$ such that for every ball $B_R \subset \mathbb{R}^n$, and every $u \in W^{1,2}(B_R, \mu)$, $1 \leq \kappa \leq \frac{n}{n-1} + \alpha$, the following estimate holds*

$$\left(\frac{1}{\mu(B_R)}\int_{B_R} |u - A|^{2\kappa}\mu(x)dx\right)^{\frac{1}{2\kappa}} \leq CR\left(\frac{1}{\mu(B_R)}\int_{B_R} |\nabla u|^2 \mu(x)dx\right)^{1/2}, \tag{2.3}$$

where either

$$A = \frac{1}{\mu(B_R)}\int_{B_R} u(x)d\mu(x), \quad \text{or} \quad A = \frac{1}{|B_R|}\int_{B_R} u(x)dx.$$

The same result also holds if we replace the ball B_R with the half ball B_R^+.

2.3. Boundary Lipschitz regularity estimates. Consider the homogeneous system of equations with constant coefficients

$$\begin{cases} -\text{div}[\mathcal{A}_0 \nabla v] &= 0 \quad \text{in } B_4^+, \\ \langle \mathcal{A}_0 \nabla v, \mathbf{n}\rangle &= 0 \quad \text{on } T_4, \end{cases} \tag{2.4}$$

with an elliptic constant tensor $\mathcal{A}_0$ satisfying the inequality $\Lambda|\xi|^2 \leq \langle \mathcal{A}_0\xi, \xi\rangle \leq \Lambda^{-1}|\xi|^2$.

DEFINITION 2.10. $v \in W^{1,q}(B_4^+;\mathbb{R}^N)$ is a weak solution to (2.4) in B_4^+, for some $1 < q < \infty$, if

$$\int_{B_4^+} \langle \mathcal{A}_0 \nabla v, \nabla \varphi \rangle dx = 0, \quad \forall \varphi \in C_0^\infty(B_4;\mathbb{R}^N).$$

We remark the class of test functions used in Definition 2.10 can be enlarged to $W^{1,q'}(B_4^+,\mathbb{R}^N)$ whose trace vanish on the round part of ∂B_4^+ where $\frac{1}{q}+\frac{1}{q'}=1$. We recall the standard Lipschitz regularity estimate for weak solutions v of a system with uniformly elliptic constant coefficients. This result can be found in [**11**], see also [**8**, Theorem 4.1].

LEMMA 2.11. *Let $\mathcal{A}_0$ be a constant tensor satisfying conditions* (1.2) *and* (1.3) *with $\mu = 1$. Then there exists a constant $C = C(n,\Lambda)$ such that if $v \in W^{1,q}(B_4^+,\mathbb{R}^N)$ is a weak solution of* (2.4) *with $q > 1$, then*

$$\|\nabla v\|_{L^\infty(B_{\frac{7}{2}}^+)} \leq C \left(\fint_{B_4^+} |\nabla v|^q dx \right)^{\frac{1}{q}}.$$

2.4. Boundary weighted Caccioppoli's type estimates. We now study the main equation of interest

$$\left\{ \begin{array}{rcll} \operatorname{div}[\mathcal{A}(x)\nabla u] & = & \operatorname{div}[F] & \text{in} \quad B_4^+, \\ \langle \mathcal{A}(x)\nabla u - F, \mathbf{n}\rangle & = & 0 & \text{on} \quad T_4, \end{array} \right. \tag{2.5}$$

where
(2.6)

$$\Lambda \mu(x)|\xi|^2 \leq \langle \mathcal{A}(x)\xi,\xi\rangle \quad \text{for a.e. } x \in B_4^+, \quad \forall\, \xi \in \mathbb{R}^n \text{ and } |\mathcal{A}(x)| \leq \Lambda^{-1}\mu(x).$$

DEFINITION 2.12. $u \in W^{1,2}(B_4^+,\mu;\mathbb{R}^N)$ is a weak solution to (2.5) in B_4^+ if

$$\int_{B_4^+} \langle \mathcal{A}\nabla u, \nabla \varphi \rangle dx = \int_{B_4^+} \langle F, \nabla \varphi \rangle dx, \quad \forall \varphi \in C_0^\infty(B_4;\mathbb{R}^N).$$

We study the system of equations (2.5) as a local perturbation of (2.4) corresponding to some constant tensor $\mathcal{A}_0$ satisfying uniform ellipticity. In fact, if $v \in W^{1,1+\beta}(B_4^+,\mathbb{R}^N)$ is a weak solution of (2.4) we have the following weighted Caccioppoli estimate of for $u - v$ that is essential in the paper.

LEMMA 2.13. *Suppose that $M_0 > 0$ and c is a constant vector in $\mathbb{R}^N$. Let $\mathcal{A}_0$ and v be as in Lemma* 2.11, *and let $w = u - c - v$. Assume that* (2.6) *holds on B_4^+ and $[\mu]_{A_2} \leq M_0$. There exists a constant $C = C(\Lambda, M_0, n)$ such that for all non-negative function $\varphi \in C_0^\infty(B_4)$,*

$$\begin{aligned} \frac{1}{\mu(B_4)} \int_{B_4^+} |\nabla w|^2 \varphi^2 d\mu \leq C(\Lambda, M_0, n) \Bigg[& \frac{1 + \|\varphi \nabla v\|^2_{L^\infty(B_4^+)}}{\mu(B_4)} \int_{B_4^+} w^2 |\varphi|^2 d\mu \\ & + \frac{1}{\mu(B_4)} \int_{B_4^+} \left| \frac{F}{\mu} \right|^2 \varphi^2 d\mu \\ & + \frac{\|\varphi \nabla v\|^2_{L^\infty(B_4^+)}}{\mu(B_4)} \int_{B_4^+} |\mathcal{A} - \mathcal{A}_0|^2 \mu^{-1} dx \Bigg] \end{aligned}$$

PROOF. It is clear that w is a weak solution of the system of equations

$$\left\{\begin{array}{ll} \operatorname{div}[\mathcal{A}\nabla w] = \operatorname{div}[F - (\mathcal{A} - \mathcal{A}_0)\nabla v] & \text{in} \quad B_4^+, \\ \langle \mathcal{A}\nabla w - F + (\mathcal{A} - \mathcal{A}_0)\nabla v, \mathbf{n}\rangle = 0 & \text{on} \quad T_4. \end{array}\right.$$

For $\varphi \in C_0^\infty(B_4), 0 \leq \varphi \leq 1$ such that $\varphi = 1$ on B_2. Since $w\varphi^2 \in W^{1,2}(B_4^+, \mu, \mathbb{R}^N)$ and the trace of $w\varphi^2$ on the round part of B_4^+ is zero, by the remark stated above we can use $w\varphi^2$ as a test function to the above equation. Now, noting that $\nabla(w\varphi^2) = \varphi^2\nabla w + w \otimes \nabla(\varphi^2)$, we obtain

$$\begin{aligned} \int_{B_4^+} \langle \mathcal{A}\nabla w, \nabla w\rangle \varphi^2 dx = & - \int_{B_4^+} \langle \mathcal{A}\nabla w, w \otimes \nabla(\varphi^2)\rangle dx + \int_{B_4^+} \langle F, \nabla(w\varphi^2)\rangle dx \\ & - \int_{B_4^+} \langle (\mathcal{A} - \mathcal{A}_0)\nabla v, \nabla(w\varphi^2)\rangle dx. \end{aligned}$$

We then have the following estimate

$$\tag{2.7} \begin{aligned} \left| \int_{B_4^+} \langle \mathcal{A}\nabla w, \nabla w\rangle \varphi^2 dx \right| \leq & \left| \int_{B_4} \langle \mathcal{A}\nabla w, w \otimes \nabla(\varphi^2)\rangle dx \right| \\ & + \int_{B_4^+} |F| \left(|\nabla w||\varphi|^2 + 2|\nabla\varphi||\varphi||w|\right) dx \\ & + \int_{B_4^+} |(\mathcal{A} - \mathcal{A}_0)| \left(|\nabla v||\nabla w||\varphi|^2 + 2|\nabla v||\nabla\varphi||w||\varphi|\right) dx. \end{aligned}$$

Clearly, by (2.6),

$$\int_{B_4^+} \langle \mathcal{A}\nabla w, \nabla w\rangle \varphi^2 dx \geq \Lambda \int_{B_4^+} |\nabla w|^2 \varphi^2 d\mu.$$

For $\epsilon > 0$, using the boundedness of the coefficients assumed in (2.6) and Young's inequality, the first term on the right hand side can be estimated as

$$\begin{aligned} \left| \int_{B_4^+} \langle \mathcal{A}\nabla w, w \otimes \nabla(\varphi^2)\rangle dx \right| & \leq 2\Lambda^{-1} \int_{B_4^+} \mu|\nabla w|\varphi|\nabla\varphi||w| dx \\ & \leq \epsilon \int_{B_4^+} |\nabla w|^2 |\varphi|^2 d\mu + C(\Lambda, \epsilon) \int_{B_4^+} |\nabla\varphi|^2 |w|^2 d\mu. \end{aligned}$$

Similarly, we estimate the second term as

$$\begin{aligned} & \int_{B_4^+} |F| \left(|\nabla w||\varphi|^2 + 2|\nabla\varphi||\varphi||w|\right) dx \\ & \leq \int_{B_4^+} \frac{|F|}{\mu} \left(|\nabla w||\varphi|^2 + 2|\nabla\varphi||\varphi||w|\right) \mu dx \\ & \leq \epsilon \int_{B_4^+} |\nabla w|^2 \varphi^2 d\mu + C(\epsilon) \left[\int_{B_4^+} \left|\frac{F}{\mu}\right|^2 \varphi^2 d\mu + \int_{B_4^+} w^2 |\nabla\varphi|^2 d\mu \right]. \end{aligned}$$

To estimate the last term, we apply Hölder's inequality and Young's inequality to obtain

$$\int_{B_4^+} |\mathcal{A}-\mathcal{A}_0||\nabla v|\Big[\varphi^2|\nabla w| + 2|w|\varphi||\nabla \varphi|\Big] dx$$

$$\leq \|\varphi \nabla v\|_{L^\infty(B_4^+)} \int_{B_4^+} |\mathcal{A}-\mathcal{A}_0|\Big[\varphi|\nabla w| + 2|w||\nabla\varphi|\Big]\mu^{1/2}\frac{1}{\mu^{1/2}}dx$$

$$\leq \epsilon \int_{B_4^+} |\nabla w|^2\varphi^2(x)d\mu$$

$$+ C(\epsilon)\,\|\varphi\nabla v\|^2_{L^\infty(B_4^+)}\left[\int_{B_4^+}|\mathcal{A}-\mathcal{A}_0|^2\mu^{-1}dx + \int_{B_4^+} w^2|\nabla\varphi|^2 d\mu\right].$$

Therefore, combining the above estimates and choosing ϵ sufficiently small to absorb the term $\int_{B_4^+}|\nabla w|^2\varphi^2 d\mu$ on the right hand side yields the desired result. □

3. Approximation estimates

3.1. Interior approximation estimates. The following result is a modification of [**4**, Proposition 4.4] for systems, which will be needed in the paper.

PROPOSITION 3.1. *Let $\Lambda > 0, M_0 > 0$ be fixed and β be as in Lemma 2.8. For every $\epsilon > 0$ sufficiently small, there exists $\delta > 0$ depending on only $\epsilon, \Lambda, n, M_0$ such that the following statement holds true: If (1.2)-(1.3) hold on B_4 for $\mathcal{A}$, $[\mu]_{A_2} \leq M_0$, and*

$$\frac{1}{\mu(B_4)}\int_{B_4}|\mathcal{A}-\langle\mathcal{A}\rangle_{B_4}|^2\mu^{-1}dx + \fint_{B_4}\left|\frac{\mathbf{F}}{\mu}\right|^2 d\mu(x) \leq \delta^2,$$

for every weak solution $u \in W^{1,2}(B_4,\mu,\mathbb{R}^N)$ of

$$div[\mathcal{A}(x)\nabla u] = div[\mathbf{F}] \quad in \quad B_4$$

satisfying

$$\fint_{B_4}|\nabla u|^2 d\mu \leq 1,$$

then, there exist a tensor of constant coefficients $\mathcal{A}_0$ and a weak solution $v \in W^{1,1+\beta}(B_4,\mathbb{R}^N)$ of

$$div[\mathcal{A}_0\nabla v] = 0 \quad in\ B_4$$

such that

$$|\langle\mathcal{A}\rangle_{B_4} - \mathcal{A}_0| \leq \frac{\epsilon\mu(B_4)}{|B_4|}, \quad and \quad \fint_{B_2}|\nabla u - \nabla v|^2 d\mu \leq \epsilon.$$

Moreover, there is $C = C(\Lambda, n, M_0)$ such that

$$\fint_{B_3}|\nabla v|^2 dx \leq C. \tag{3.1}$$

PROOF. The proof is the same as that of Proposition 3.2 below. We therefore skip it. □

3.2. Boundary approximation estimates. Our main result of the subsection is the following proposition which is an up to the boundary approximation similar to Proposition 3.1.

PROPOSITION 3.2. *Let $\Lambda > 0, M_0 > 0$ be fixed and let β be as in Lemma 2.8. Suppose that $[\mu]_{A_2} \leq M_0$ and $\mathcal{A}$ satisfies (2.6). For every $\epsilon > 0$, there exists $\delta = \delta(\epsilon, \Lambda, n, M_0)$ such that: if*

$$\frac{1}{\mu(B_4)} \int_{B_4^+} |\mathcal{A} - \langle \mathcal{A} \rangle_{B_4}|^2 \mu^{-1} dx + \left(\frac{1}{\mu(B_4)} \int_{B_4^+} \left| \frac{F}{\mu} \right|^2 d\mu(x) \right)^{1/2} \leq \delta,$$

and $u \in W^{1,2}(B_4^+, \mu, \mathbb{R}^N)$ is a weak solution of (2.5) satisfying

$$\frac{1}{\mu(B_4)} \int_{B_4^+} |\nabla u|^2 d\mu \leq 1, \tag{3.2}$$

then there exist a constant matrix $\mathcal{A}_0$ and a weak solution $v \in W^{1,1+\beta}(B_4^+, \mathbb{R}^N)$ of (2.4) such that

$$\|\langle \mathcal{A} \rangle_{B_4} - \mathcal{A}_0\| \leq \frac{\epsilon \mu(B_4)}{|B_4|}, \tag{3.3}$$

and

$$\fint_{B_2^+} |\nabla u - \nabla v|^2 d\mu \leq \epsilon.$$

Moreover, there is $C = C(\Lambda, n, M_0)$ such that

$$\fint_{B_3^+} |\nabla v|^2 dx \leq C(\Lambda, n, M_0).$$

The proof of the lemma relies on the weighted Caccioppoli estimate, Lemma 2.13, and Lemma 3.1 below.

LEMMA 3.1. *Let $\Lambda > 0, M_0 > 0$ be fixed and let β be as in Lemma 2.8. Suppose that $[\mu]_{A_2} \leq M_0$ and $\mathcal{A}$ satisfies (2.6). For every $\epsilon > 0$ sufficiently small, there exists $\delta > 0$ depending on only $\epsilon, \Lambda, n,$ and M_0 such that: if*

$$\frac{1}{\mu(B_4)} \int_{B_4^+} |\mathcal{A} - \langle \mathcal{A} \rangle_{B_4}|^2 \mu^{-1} dx + \left(\frac{1}{\mu(B_4)} \int_{B_4^+} \left| \frac{\mathbf{F}}{\mu} \right|^2 d\mu(x) \right)^{1/2} \leq \delta,$$

and $u \in W^{1,2}(B_4^+, \mu, \mathbb{R}^N)$ is a weak solution of (2.5) satisfying

$$\frac{1}{\mu(B_4)} \int_{B_4^+} |\nabla u|^2 d\mu \leq 1, \tag{3.4}$$

then there exists a constant matrix $\mathcal{A}_0$ and a weak solution $v \in W^{1,1+\beta}(B_4^+, \mathbb{R}^N)$ of (2.4) such that

$$\|\langle \mathcal{A} \rangle_{B_4} - \mathcal{A}_0\| \leq \frac{\epsilon \mu(B_4)}{|B_4|},$$

and

$$\frac{1}{\mu(B_{7/2})} \int_{B_{7/2}^+} |\hat{u}_+ - v|^2 d\mu \leq \epsilon, \qquad \hat{u}_+ = u - \bar{u}_{\mu, B_4^+}.$$

Moreover, there is $C = C(\Lambda, n, M_0)$ such that

$$\fint_{B_3^+} |\nabla v|^2 dx \leq C(\Lambda, n, M_0). \tag{3.5}$$

PROOF. We first notice that for each $\lambda > 0$, if we use the scaling $\mathcal{A}_\lambda = \frac{1}{\lambda}\mathcal{A}$, $\mu_\lambda = \mu/\lambda$ and $\mathbf{F}_\lambda = \mathbf{F}/\lambda$, then for a weak solution u of (2.5), u is also a weak solution of the system

$$\begin{cases} \operatorname{div}[\mathcal{A}_\lambda \nabla u] = \operatorname{div}[\mathbf{F}_\lambda] & \text{in} \quad B_4^+, \\ \langle \mathcal{A}_\lambda \nabla u - \mathbf{F}, \mathbf{n} \rangle = 0 & \text{on } T_4. \end{cases}$$

Moreover, $[\mu_\lambda]_{A_2} = [\mu]_{A_2}$,

$$\Lambda \mu_\lambda(x)|\xi|^2 \leq \langle \mathcal{A}_\lambda(x)\xi, \xi\rangle, \quad \forall\, \xi \in \mathbb{R}^{nN}, \quad \text{for a.e. } x \in B_4, \text{ and } |\mathcal{A}_\lambda(x)| \leq \Lambda^{-1}\mu_\lambda(x),$$

and Lemma 3.1 is invariant with respect to this scaling. Therefore, without loss of generality, we may assume that

$$\bar{\mu}_{B_4} = \frac{1}{|B_4|}\int_{B_4} \mu(x)dx = 1. \tag{3.6}$$

In this case, it follows from (2.6) and (3.6) that

$$\Lambda|\xi|^2 \leq \langle \langle \mathcal{A}\rangle_{B_4}\xi, \xi\rangle, \forall \xi \in \mathbb{R}^{nN}, \qquad \text{and } |\langle \mathcal{A}\rangle_{B_4}| \leq \Lambda^{-1}. \tag{3.7}$$

We will use a contradiction argument to prove the lemma. Suppose that the conclusion is not true. Then there exists $\epsilon_0 > 0$ such that for each $k \in \mathbb{N}$, there are $\mu_k \in A_2$, $\mathcal{A}_k$ satisfying (2.6) with μ_k and $\mathcal{A}_k$ in place of μ and $\mathcal{A}$, and $\mathbf{F}_k$ and a weak solution $u_k \in W^{1,2}(B_4^+, \mu_k, \mathbb{R}^N)$ to

$$\begin{cases} \operatorname{div}[\mathcal{A}_k \nabla u_k] & = & \operatorname{div}[\mathbf{F}_k] & \text{in} & B_4^+, \\ \langle \mathcal{A}_k \nabla u_k - \mathbf{F}_k, \mathbf{n}\rangle & = & 0 & \text{on} & T_4, \end{cases} \tag{3.8}$$

$$\begin{cases} \dfrac{1}{\mu_k(B_4)}\displaystyle\int_{B_4^+} |\mathcal{A}_k - \langle \mathcal{A}_k\rangle_{B_4}|^2 \mu_k^{-1} dx + \left(\dfrac{1}{\mu_k(B_4)}\displaystyle\int_{B_4^+} \left|\dfrac{\mathbf{F}_k}{\mu_k}\right|^2 d\mu_k\right)^{1/2} \leq \dfrac{1}{k^2} \\ [\mu_k]_{A_2} \leq M_0, \quad \bar{\mu}_{k,B_4} = \dfrac{1}{|B_4|}\displaystyle\int_{B_4} \mu_k(x)dx = 1, \end{cases} \tag{3.9}$$

and

$$\frac{1}{\mu_k(B_4)}\int_{B_4^+} |\nabla u_k|^2 d\mu_k \leq 1, \tag{3.10}$$

but for all constant matrix $\mathcal{A}_0$ with $\|\langle \mathcal{A}_k\rangle_{B_4} - \mathcal{A}_0\| \leq \epsilon_0$, and for all weak solution $v \in W^{1,1+\beta}(B_4^+, \mathbb{R}^N)$ of (2.4) it holds that

$$\frac{1}{\mu_k(B_{7/2})}\int_{B_{7/2}^+} |\hat{u}_{k+} - v|^2 d\mu_k \geq \epsilon_0. \tag{3.11}$$

Since $\langle \mathcal{A}_k\rangle_{B_4}$ satisfies (3.7), $\langle \mathcal{A}_k\rangle_{B_4}$ is a bounded sequence in $\mathbb{R}^{nN\times nN}$. Thus, there exists a subsequence denoting again by $\langle \mathcal{A}_k\rangle_{B_4}$ and a constant tensor $\bar{\mathcal{A}} \in \mathbb{R}^{nN\times nN}$ such that

$$\lim_{k\to\infty} \langle \mathcal{A}_k\rangle_{B_4} = \bar{\mathcal{A}}. \tag{3.12}$$

From (3.10) and weighted Sobolev-Poincaré inequality Lemma 2.9 ,[**7**, Theorem 1.5] (which still applicable for half-balls), we see that

$$\frac{1}{\mu_k(B_4)}\int_{B_4^+}|\hat{u}_{k+}|^2 d\mu_k \leq \frac{C(n,M_0)}{\mu_k(B_4)}\int_{B_4^+}|\nabla u_k|^2 d\mu_k \leq C(n,M_0).$$

Since $\mu_k(B_4) = |B_4|$, it implies that $\|\hat{u}_{k+}\|_{W^{1,2}(B_4^+,\mu_k,\mathbb{R}^N)} \leq C(n,M_0)$, for all $k \in \mathbb{N}$. This together with Lemma 2.8 implies that

$$\|\hat{u}_{k+}\|_{W^{1,1+\beta}(B_4^+,\mathbb{R}^N)} \leq C(n,M_0)\,\|\hat{u}_{k+}\|_{W^{1,2}(B_4^+,\mu_k,\mathbb{R}^N)} \leq C(n,M_0).$$

Therefore, using the compact imbedding $W^{1,1+\beta}(B_4^+) \hookrightarrow L^{1+\beta}(B_4^+)$ and the diagonal argument, we find a subsequence of u_k denoted again by u_k, and $u \in W^{1,1+\beta}(B_4^+,\mathbb{R}^N)$ such that

(3.13)

$$\begin{cases}\hat{u}_{k+} \to u \text{ strongly in } L^{1+\beta}(B_4^+,\mathbb{R}^N), \quad \nabla u_k \rightharpoonup \nabla u \text{ weakly in } L^{1+\beta}(B_4^+,\mathbb{R}^N), \quad \text{and}\\ \hat{u}_{k+} \to u \text{ a.e. in } \; B_4^+.\end{cases}$$

As a consequence,

$$\|u\|_{W^{1,1+\beta}(B_4^+)} \leq C(n,M_0). \tag{3.14}$$

We claim that u is a weak solution of

$$\begin{cases}\operatorname{div}[\bar{\mathcal{A}}\nabla u] &= 0 \quad \text{in } \; B_4^+,\\ \langle \bar{\mathcal{A}}\nabla u, \mathbf{n}\rangle &= 0 \quad \text{on } \; T_4.\end{cases} \tag{3.15}$$

To do this, using $\varphi \in C_0^\infty(B_4,\mathbb{R}^N)$ as a test function for equation (3.8), we have

$$\int_{B_4^+}\langle \mathcal{A}_k\nabla \hat{u}_{k+}, \nabla\varphi\rangle dx = \int_{B_4^+}\langle \mathbf{F}_k, \nabla\varphi\rangle dx. \tag{3.16}$$

By Hölder's inequality, we estimate the right hand side of (3.16) as

$$\begin{aligned}\left|\fint_{B_4^+}\langle \mathbf{F}_k,\nabla\varphi\rangle dx\right| &\leq \left\{\fint_{B_4^+}\left|\frac{\mathbf{F}_k}{\mu_k}\right|^2\mu_k dx\right\}^{1/2}\left\{\fint_{B_4^+}|\nabla\varphi|^2\mu_k dx\right\}^{1/2}\\ &\leq \|\nabla\varphi\|_{L^\infty(B_4^+)}\left\{\frac{1}{\mu_k(B_4^+)}\int_{B_4^+}\left|\frac{\mathbf{F}_k}{\mu_k}\right|^2 d\mu_k(x)\right\}^{1/2}\frac{\mu_k(B_4^+)}{|B_4^+|}\\ &\leq C\frac{\|\nabla\varphi\|_{L^\infty(B_4)}}{k}.\end{aligned}$$

Therefore, letting $k \to \infty$ yields

$$\lim_{k\to\infty}\int_{B_4^+}\langle \mathbf{F}_k,\nabla\varphi\rangle dx = 0. \tag{3.17}$$

On the other hand, it follows from (3.9), (3.10), and Hölder's inequality that

$$
\begin{aligned}
&\left| \fint_{B_4^+} \langle (\mathcal{A}_k - \langle \mathcal{A}_k \rangle_{B_4}) \nabla u_k, \nabla \varphi \rangle dx \right| \\
&\quad \leq \fint_{B_4^+} |\mathcal{A}_k - \langle \mathcal{A}_k \rangle_{B_4}| |\nabla u_k| \mu_k^{1/2} |\nabla \varphi| \mu_k^{-1/2} dx \\
&\quad \leq \|\nabla \varphi\|_{L^\infty(B_4^+)} \left\{ \fint_{B_4^+} |\mathcal{A}_k - \langle \mathcal{A}_k \rangle_{B_4}|^2 \mu_k^{-1} dx \right\}^{1/2} \left\{ \frac{1}{|B_4^+|} \int_{B_4^+} |\nabla u_k|^2 d\mu_k \right\}^{1/2} \\
&\quad \leq \frac{\|\nabla \varphi\|_{L^\infty(B_4^+)}}{\sqrt{k}} \left\{ \fint_{B_4^+} |\nabla u_k|^2 d\mu_k \right\}^{1/2} \\
&\quad \leq C \frac{\|\nabla \varphi\|_{L^\infty(B_4^+)}}{\sqrt{k}} \to 0, \quad \text{as} \quad k \to \infty.
\end{aligned}
$$

Thus,

$$
\begin{aligned}
0 &= \lim_{k\to\infty} \int_{B_4^+} \langle (\mathcal{A}_k - \langle \mathcal{A}_k \rangle_{B_4}) \nabla u_k, \nabla \varphi \rangle dx \\
&= \lim_{k\to\infty} \left[\int_{B_4^+} \langle \mathcal{A}_k \nabla u_k, \nabla \varphi \rangle dx - \int_{B_4^+} \langle \langle \mathcal{A}_k \rangle_{B_4} \nabla u_k, \nabla \varphi \rangle dx \right].
\end{aligned}
$$

We also notice that ∇u_k converges weakly to ∇u in $L^{1+\beta}(B_4, \mathbb{R}^N)$ from (3.13), and $\langle \mathcal{A}_k \rangle_{B_4}$ converges strongly to constant tensor $\bar{\mathcal{A}}$, hence,

$$
\lim_{k\to\infty} \int_{B_4^+} \langle \langle \mathcal{A}_k \rangle_{B_4} \nabla u_k, \nabla \varphi \rangle dx = \int_{B_4^+} \langle \bar{\mathcal{A}} \nabla u, \nabla \varphi \rangle dx.
$$

Consequently,

$$
\lim_{k\to\infty} \int_{B_4^+} \langle \mathcal{A}_k \nabla u_k, \nabla \varphi \rangle dx = \int_{B_4^+} \langle \bar{\mathcal{A}} \nabla u, \nabla \varphi \rangle dx. \tag{3.18}
$$

Combining (3.17) and (3.18) yields

$$
\int_{B_4^+} \langle \bar{\mathcal{A}} \nabla u, \nabla \varphi \rangle dx = 0, \quad \forall\, \varphi \in C_0^\infty(B_4, \mathbb{R}^N).
$$

Now, since $\bar{\mathcal{A}} = \lim_{k\to\infty} \langle \mathcal{A}_k \rangle_{B_4}$, and by (3.7), we see that

$$
\Lambda |\xi|^2 \leq \langle \bar{\mathcal{A}} \xi, \xi \rangle, \quad \forall\, \xi \in \mathbb{R}^{nN},
$$

i.e., the constant tensor $\bar{\mathcal{A}}$ is uniformly elliptic. Hence, $u \in C^\infty(\overline{B}_{15/4}^+)$ by Lemma 2.11. Moreover, it follows from Lemma 2.11 and (3.14) that

$$
\begin{aligned}
\fint_{B_{7/2}^+} |\nabla u|^2 d\mu_k \leq \|\nabla u\|_{L^\infty(B_{7/2})}^2 &\leq C(n, \Lambda) \left(\fint_{B_4^+} |\nabla u|^{1+\beta} dx \right)^{\frac{2}{1+\beta}} \\
&\leq C(n, M_0, \Lambda), \quad \forall\, k \in \mathbb{N}.
\end{aligned} \tag{3.19}
$$

Using the above and following the argument used in [**4**, Lemma 4.3] we obtain that

$$
\lim_{k\to\infty} \fint_{B_{7/2}^+} |\hat{u}_k - u - c_k|^2 d\mu_k = 0, \quad \text{where} \quad c_k = \fint_{B_{7/2}^+} [\hat{u}_k - u] dx \tag{3.20}
$$

However, note that since $\langle \mathcal{A}_k \rangle_{B_4} \to \bar{\mathcal{A}}$ and $\epsilon_0 > 0$,

$$\left\| \langle \mathcal{A}_k \rangle_{B_4} - \bar{\mathcal{A}} \right\| \leq \epsilon_0$$

for sufficiently large k. But, this contradicts to (3.11) if we take $\mathcal{A}_0 = \bar{\mathcal{A}}$, $v = u - c_k$ and k large enough.

We now turn to prove (3.5). Suppose that we have β, $\mathcal{A}_0$ and $v \in W^{1,1+\beta}(B_4^+)$ satisfying the first part of the lemma. Then, by taking ϵ sufficiently small, we see that

$$\frac{\Lambda}{2}|\xi|^2 \leq \langle \mathcal{A}_0 \xi, \xi \rangle \leq 2\Lambda^{-1}|\xi|^2, \quad \forall\, \xi \in \mathbb{R}^n.$$

Hence, by the standard regularity theory for elliptic systems, Lemma 2.11, v is in $C^\infty(B_{15/4}^+)$. Moreover, from Lemma 2.8, we also have

$$\fint_{B_{16/5}^+} |v|^2 dx \leq C(n, \Lambda) \left\{ \fint_{B_{7/2}^+} |v|^{1+\beta} dx \right\}^{\frac{2}{1+\beta}}.$$

and as a consequence

$$\fint_{B_{16/5}^+} |v|^2 dx \leq C(n, \Lambda, M_0) \fint_{B_{7/2}^+} |v|^2 d\mu.$$

The preceding estimate together with the energy estimate for v implies that

$$\fint_{B_3^+} |\nabla v|^2 dx \leq C(n, \Lambda, M_0) \fint_{B_{\frac{16}{5}}^+} |v|^2 dx \leq C(n, \Lambda, M_0) \fint_{B_{7/2}^+} |v|^2 d\mu.$$

Therefore,

$$\begin{aligned}
\fint_{B_3^+} |\nabla v|^2 dx &\leq C(n, \Lambda, M_0) \left[\fint_{B_{7/2}^+} |\hat{u} - v|^2 d\mu + \fint_{B_{7/2}^+} |u - \bar{u}_{\mu, B_4}|^2 d\mu \right] \\
&\leq C(n, \Lambda, M_0) \left[\epsilon + \frac{\mu(B_4^+)}{\mu(B_{7/2}^+)} \fint_{B_4^+} |u - \bar{u}_{\mu, B_4}|^2 d\mu \right] \\
&\leq C(n, \Lambda, M_0) \left[\epsilon + \frac{\mu(B_4)}{\mu(B_{7/2})} \fint_{B_4^+} |\nabla u|^2 d\mu \right],
\end{aligned}$$

where we have used the Poincaré's inequality, Lemma 2.9. Thus, by choosing $\epsilon < 1$ and using (3.4), and the doubling property of $[\mu]$, we deduce that

$$\fint_{B_3^+} |\nabla v|^2 dx \leq C(n, \Lambda, M_0)\, [1 + C] = C(n, \Lambda, M_0).$$

Therefore, it completes the proof of Lemma 3.1. □

4. Level set estimates, and proofs of main theorems

4.1. Level set estimates.

We begin with the following result.

LEMMA 4.1. *Suppose that $M_0 > 0$ and $\mu \in A_2$ such that $[\mu]_{A_2} \leq M_0$. Then there exists a constant $\varpi > 1$ so that for every $\epsilon > 0$, there is $\delta = \delta(\Lambda, M_0, n, \epsilon) > 0$ sufficiently small such that if $\mathcal{A}$ satisfies* (2.6) *on B_2^+,*

$$[\mathcal{A}]_{BMO_1(B_1^+, \mu)} \leq \delta,$$

$u \in W^{1,2}(B_2^+, \mu)$ is a weak solution to

$$
(4.1) \qquad \begin{cases} div[\mathcal{A}(x)\nabla u] &= div[\mathbf{F}] \quad in \quad B_2^+, \\ \langle \mathcal{A}(x)\nabla u - \mathbf{F}, \mathbf{n}\rangle &= 0 \quad on \quad T_2, \end{cases}
$$

and if
(4.2)

$$
B_\rho^+(\bar{y}) \cap \{x \in B_2^+ : \mathcal{M}^\mu(\chi_{B_2^+}|\nabla u|^2) \leq 1\} \cap \{x \in B_2^+ : \mathcal{M}^\mu\left(\left|\frac{\mathbf{F}}{\mu}\right|^2 \chi_{B_2^+}\right) \leq \delta^2\} \neq \emptyset,
$$

for some $\bar{y} = (y', 0) \in T_1$ and some $\rho \in (0, 1/2)$, then

$$
\mu\Big(\Big\{x \in B_1^+ : \mathcal{M}^\mu(\chi_{B_2^+}|\nabla u|^2) > \varpi^2\Big\} \cap B_\rho^+(\bar{y})\Big) < \epsilon\mu(B_1).
$$

PROOF. The proof of this lemma is standard. We give it here for the sake of completeness. For a given $\epsilon > 0$, let $\eta = \epsilon/C^*$ where C^* is a positive number to be determined depending only on M_0, n. Then, let $\delta = \delta(\eta, \Lambda, M_0, n)$ be defined as in Proposition 3.2. We now prove the lemma with this choice of δ. Observe that from the hypothesis (4.2), there exists $x_0 \in B_\rho^+(\bar{y})$ such that for all $l > 0$,

$$
(4.3) \qquad \fint_{B_l(x_0)} \chi_{B_2^+}|\nabla u|^2 d\mu \leq 1, \quad \text{and} \quad \fint_{B_l(x_0)} \chi_{B_2^+}\left|\frac{\mathbf{F}}{\mu}\right|^2 d\mu \leq \delta^2.
$$

Since $B_{4\rho}^+(\bar{y}) \subset B_{5\rho}(x_0) \cap B_2^+$, it follows that

$$
\frac{1}{\mu(B_{4\rho}(\bar{y}))}\int_{B_{4\rho}(\bar{y})} \chi_{B_2^+}|\nabla u|^2 d\mu \leq \frac{\mu(B_{5\rho}(x_0))}{\mu(B_{4\rho}(\bar{y}))}\fint_{B_{5\rho}(x_0)} \chi_{B_2^+}|\nabla u|^2 d\mu \leq \frac{5^{2n}M_0}{4^{2n}}.
$$

Similarly, we have that

$$
\frac{1}{\mu(B_{4\rho})(\bar{y})}\int_{B_{4\rho}^+(\bar{y}} \chi_{B_2^+}\left|\frac{F}{\mu}\right|^2 d\mu \leq \delta^2 M_0\left(\frac{5}{4}\right)^{2n}.
$$

From these estimates, $[\mathbb{A}]_{\mathrm{BMO}_1(B_1^+,\mu)} \leq \delta$, and after some appropriate scaling, dilation, and translation, we can apply Proposition 3.2 to see that there exists a constant matrix $\mathcal{A}_0$ and a weak solution v to

$$
\begin{cases} \mathrm{div}[\mathcal{A}_0\nabla v] &= 0 \quad \text{in} \quad B_{4\rho}^+(\bar{y}), \\ \langle \mathcal{A}_0\nabla v, \mathbf{n}\rangle &= 0 \quad \text{on} \quad T_{4\rho}(\bar{y}), \end{cases}
$$

satisfying

$$
(4.4) \quad \frac{1}{\mu(B_{2\rho}(\bar{y})}\int_{B_{2\rho}^+(\bar{y})} |\nabla u - \nabla v|^2 d\mu < \eta M_0(5/4)^{2n}, \quad \text{and} \quad \|\nabla v\|_{L^\infty(B_{3\rho}^+(\bar{y}))} \leq C_0,
$$

for some positive constant C_0 that depends only n, Λ and M_0.

Now let $\varpi > 0$ such that $\varpi^2 = \max\{M_0 3^{2n}, 4C_0^2\}$, where C_0 is from (4.4). We will show that

$$
\{x : \mathcal{M}^\mu(\chi_{B_2^+}|\nabla u|)^2 > \varpi^2\} \cap B_\rho^+(\bar{y}) \subset \{x : \mathcal{M}^\mu(\chi_{B_{2\rho}^+(\bar{y})}|\nabla u - \nabla v|^2) > C_0^2\} \cap B_\rho^+(\bar{y}).
$$

To prove the claim, we consider $x \in B_\rho^+(\bar{y})$ such that

$$
(4.5) \qquad \mathcal{M}^\mu(\chi_{B_{2\rho}^+(\bar{y})}|\nabla u - \nabla v|^2)(x) \leq C_0^2,
$$

and we only need to show that for any $r > 0$

$$
\fint_{B_r(x)} \chi_{B_2^+}|\nabla u|^2 d\mu \leq \varpi^2.
$$

Indeed, if $r < \rho$, then $B_r(x) \cap B_2^+ \subset B_{2\rho}^+(\bar{y})$, using (4.4) and (4.5), we see that

$$\begin{aligned}\fint_{B_r(x)} \chi_{B_2^+} |\nabla u|^2 d\mu &\leq 2 \fint_{B_r(x)} \chi_{B_{2\rho}^+(\bar{y})} |\nabla u - \nabla v|^2 d\mu + 2 \fint_{B_r^+(x)} |\nabla v|^2 d\mu \\ &\leq 2\mathcal{M}^\mu(\chi_{B_{2\rho}^+(\bar{y})} |\nabla u - \nabla v|^2)(x) + 2C_0^2 \leq 4C_0^2 \leq \varpi^2.\end{aligned}$$

Also, if $r \geq \rho$, then note that $B_r(x) \subset B_{3r}(x_0)$ and by (4.3), we obtain that

$$\fint_{B_r(x)} \chi_{B_2^+} |\nabla u|^2 d\mu(x) \leq \frac{\mu(B_{3r}(x_0))}{\mu(B_r(x))} \fint_{B_{3r}(x_0)} \chi_{B_2^+} |\nabla u|^2 d\mu(x) \leq M_0 3^{2n} \leq \varpi^2,$$

and this proves the claim. From this claim, we can deduce that

$$\begin{aligned}&\mu(B_\rho^+(\bar{y}) \cap \{x \in B_2^+ : \mathcal{M}^\mu(\chi_{B_2^+} |\nabla u|^2) > \varpi^2\}) \\ &\leq \mu(\{x \in B_\rho^+(\bar{y}) : \mathcal{M}^\mu(\chi_{B_{2\rho}^+(\bar{y})} |\nabla u - \nabla v|^2) > C_0^2\}) \\ &\leq \frac{C(n, M_0)}{C_0^2} \mu(B_{2\rho}(\bar{y})) \frac{1}{\mu(B_{2\rho}(\bar{y}))} \int_{B_{2\rho}^+(\bar{y})} |\nabla u - \nabla v|^2 d\mu \\ &\leq C^* \eta \, \mu(B_1),\end{aligned}$$

where we have used the weak $(1,1)$ estimates of $\mathcal{M}^\mu$, (4.4) and the doubling property of μ. The proof is then complete once we choose $\eta > 0$ such that $C^*\eta = \epsilon$. □

LEMMA 4.2. *Suppose that $M_0 > 0$ and $\mu \in A_2$ such that $[\mu]_{A_2} \leq M_0$. Then there exists a constant $\varpi > 1$ so that for every $\epsilon > 0$, there is $\delta = \delta(\Lambda, M_0, n, \epsilon) > 0$ sufficiently small such that if $\mathcal{A}$ satisfies (2.6) on B_2^+,*

$$[\mathcal{A}]_{BMO_1(B_1^+, \mu)} \leq \delta,$$

$u \in W^{1,2}(B_2^+, \mu)$ is a weak solution to (4.1) and

$$\mu(\{x \in B_1^+ : \mathcal{M}^\mu(\chi_{B_2^+} |\nabla u|^2) > \varpi^2\} \cap B_\rho(y)) \geq \epsilon \mu(B_\rho(y)), \tag{4.6}$$

whenever $y \in B_1^+$, and $\rho < \frac{1}{8}$, then

$$B_\rho(y) \cap B_2^+ \subset \{x \in B_1^+ : \mathcal{M}^\mu(\chi_{B_2^+} |\nabla u|^2) > 1\} \cup \{x \in B_1^+ : \mathcal{M}^\mu\left(\left|\frac{\mathbf{F}}{\mu}\right|^2 \chi_{B_2^+}\right) > \delta^2\}. \tag{4.7}$$

PROOF. We begin by noting that in the case that $y \in B_1^+$, and $B_{4\rho}(y) \cap \{x_n = 0\} = \emptyset$, then we have $B_{2\rho}(y) \subset B_2^+$, and the situation is exactly as in [**4**, Proposition 4.7]. So we skip the proof of this case. In the event $B_{2\rho}(y) \cap \{x_n = 0\} \neq \emptyset$, we prove the lemma using a contradiction argument. Suppose that (4.6) holds but (4.7) fails. Then there exists $x_0 \in B_\rho(y) \cap B_2^+$ such that

$$\mathcal{M}^\mu(\chi_{B_2^+} |\nabla u|^2)(x_0) \leq 1 \quad \text{and} \quad \mathcal{M}^\mu\left(\left|\frac{\mathbf{F}}{\mu}\right|^2 \chi_{B_2^+}\right)(x_0) \leq \delta^2.$$

Set $\bar{y} = (y', 0) \in B_{2\rho}(y) \cap T_1$. Then we have that $B_{2\rho}(y) \cap B_2^+ \subset B_{4\rho}(\bar{y}) \cap B_2^+$. As a consequence we have that

$$x_0 \in B_\rho(y) \cap B_2^+ \subset B_{4\rho}(\bar{y}) \cap B_2^+,$$

where we have used $4\rho \in (0,1)$. Now all the hypotheses of Lemma 4.1 are satisfied with y replaced by $\bar{y}$ and ρ replaced by 4ρ. Applying using $\frac{\epsilon}{M_0(6)^{2n}}$

$$\begin{aligned}
&\mu(\{x \in B_1^+ : \mathcal{M}^\mu(\chi_{B_2^+}|\nabla u|^2) > \varpi^2\} \cap B_\rho^+(y)) \\
&\leq \mu(\{x \in B_1^+ : \mathcal{M}^\mu(\chi_{B_2^+}|\nabla u|^2) > \varpi^2\} \cap B_{4\rho}^+(\bar{y})) \\
&< \frac{\epsilon}{M_0(6)^{2n}}\mu(B_{4\rho}(\bar{y})) \\
&< \frac{\epsilon}{M_0(6)^{2n}}\mu(B_{6\rho}(y)) \leq \frac{\epsilon}{M_0(6)^{2n}} M_0(6)^{2n}\mu(B_\rho(y)) \\
&= \epsilon\mu(B_\rho(y)),
\end{aligned}$$

where we have used the inclusion $B_{4\rho}(\bar{y}) \subset B_{6\rho}(y)$. The last inequality obviously contradicts (4.6). □

4.2. Proof of main theorems. As we already discussed, Theorem 1.2 follows from Theorems 1.3-1.4, the partition of unity, the procedure of flattening the boundary, and the energy estimates. This is standard, and therefore we do not provide the proof of Theorem 1.2. Proof of Theorem 1.3 is similar, and much simpler than that of Theorem 1.4. Therefore, this proof is also skipped. It remains to provide the proof of Theorem 1.3.

PROOF OF THEOREM 1.3. Let us choose M large such that for $u_M = u/M$

$$\mu(\{x \in B_1^+ : \mathcal{M}^\mu(\chi_{B_2^+}|\nabla u_M|^2) > \varpi^2\}) \leq \epsilon\mu(B_1(y)), \quad \forall y \in B_1^+. \tag{4.8}$$

We can always choose such M because from the weak $(1,1)$ estimate for the maximal function $\mathcal{M}^\mu$

$$\mu(\{x \in B_2^+ : \mathcal{M}^\mu(\chi_{B_2^+}|\nabla u_M|^2) > \varpi^2\}) \leq \frac{C(n,M_0)}{M^2\varpi^2}\int_{B_2^+}|\nabla u|^2 d\mu.$$

Now take M according to the formula

$$\frac{C}{M^2\varpi^2}\int_{B_2^+}|\nabla u|^2 d\mu = \epsilon\frac{\mu(B_2)}{M_0\left(\frac{|B_2|}{|B_1|}\right)^2},$$

while keeping in mind that since $B_1(y) \subset B_2$, we have that $B_1(y) \cap B_2 = B_1(y)$. It follows from the above calculations that

$$M^2\mu(B_1) \leq M^2\mu(B_2) = M_0\,(2)^{2n}\frac{C}{\epsilon\varpi^2}\|\nabla u\|^2_{L^2(B_2^+,\mu)}. \tag{4.9}$$

Now, let $\varpi > 0$ sufficiently large defined as in Lemma 4.2. Let $\epsilon > 0$ sufficiently small such that $\epsilon_1\varpi^p \leq 1/2$, where $\epsilon_1 = M_0 10^{2n}\epsilon$. Then, with this epsilon, let $\delta > 0$ defined as in Lemma 4.2. From Lemma 4.2, and the modified Vitali covering lemma in [**15, 26**], it follows that

$$\begin{aligned}
&\mu\Big(B_1^+ : \mathcal{M}^\mu(\chi_{B_2^+}|\nabla u|^2) > \varpi^2\Big\}\Big) \\
&\leq \epsilon_1\left[\mu\Big(\Big\{B_1^+ : \mathcal{M}^\mu(\chi_{B_2^+}|\nabla u|^2) > 1\Big\}\Big) + \mu\Big(\Big\{B_1^+ : \mathcal{M}^\mu(\chi_{B_2^+}\left|\frac{\mathbf{F}}{\mu}\right|^2) > \delta^2\Big\}\Big)\right].
\end{aligned}$$

See, for examples [**4**, Lemma 5.10] or [**24**, Proposition 4.9] for the details on the proof of this claim. Using this claim, and by induction, we then infer that for any $k \in \mathbb{N}$,

$$
\begin{aligned}
&\mu(\{x \in B_1^+ : \mathcal{M}^\mu(\chi_{B_2^+}|\nabla u|^2) > \varpi^{2k}\}) \\
&\leq \sum_{i=1}^{k} \epsilon_1^i \mu\left(\{x \in B_1^+ : \mathcal{M}^\mu\left(\left|\frac{\mathbf{F}}{\mu}\right|^2 \chi_{B_2^+}\right) > \delta^2 \varpi^{2(k-i)}\}\right) \\
&\qquad + \epsilon_1^k \mu(\{x \in B_1^+ : \mathcal{M}^\mu(\chi_{B_2^+}|\nabla u|^2) > 1\}).
\end{aligned} \tag{4.10}
$$

Now consider the sum

$$
S = \sum_{k=1}^{\infty} \varpi^{pk} \mu(\{B_1^+ : \mathcal{M}^\mu(\chi_{B_2^+}|\nabla u_M|^2)(x) > \varpi^{2k}\}).
$$

Observe that, when finite, $S^{2/p}$ is comparable to the $L^{p/2}$- norm of $\mathcal{M}^\mu(\chi_{B_2^+}|\nabla u_M|)$ in B_1^+.

Applying (4.10) to the summand we have that

$$
\begin{aligned}
S \leq &\sum_{k=1}^{\infty} \varpi^{pk} \left[\sum_{i=1}^{k} \epsilon_1^i \mu\left(\{x \in B_1^+ : \mathcal{M}^\mu\left(\left|\frac{\mathbf{F}_M}{\mu}\right|^2 \chi_{B_2^+}\right) > \delta^2 \varpi^{2(k-i)}\}\right)\right] \\
&+ \sum_{k=1}^{\infty} \varpi^{pk} \epsilon_1^k \mu(\{B_1^+ : \mathcal{M}^\mu(\chi_{B_2^+}|\nabla u_M|^2)(x) > 1\}).
\end{aligned}
$$

Applying summation by parts we have that

$$
\begin{aligned}
S &\leq \sum_{j=1}^{\infty} (\varpi^p \epsilon_1)^j \left[\sum_{k=j}^{\infty} \varpi^{p(k-j)} \mu\left(\{x \in B_1^+ : \mathcal{M}^\mu\left(\left|\frac{\mathbf{F}_M}{\mu}\right|^2 \chi_{B_2^+}\right) > \delta^2 \varpi^{2(k-j)}\}\right)\right] \\
&\quad + \sum_{k=1}^{\infty} (\varpi^p \epsilon_1)^k \mu(\{B_1^+ : \mathcal{M}^\mu(\chi_{B_2^+}|\nabla u_M|^2)(x) > 1\}) \\
&\leq C \left(\left\|\mathcal{M}^\mu\left(\chi_{B_2^+}\left|\frac{\mathbf{F}_M}{\mu}\right|^2\right)\right\|_{L^{p/2}(B_1^+,\mu)}^{p/2} + \|\nabla u_M\|_{L^2(B_2^+,\mu)}^2\right) \sum_{k=1}^{\infty} (\varpi^p \epsilon_1)^k,
\end{aligned}
$$

where we have applied the weak $(1,1)$ estimate of the maximal function. Now chose ϵ small so that $\varpi^p \epsilon_1 < 1$ to obtain that

$$
\begin{aligned}
S &\leq C \left(\left\|\mathcal{M}^\mu\left(\chi_{B_2^+}\left|\frac{\mathbf{F}_M}{\mu}\right|^2\right)\right\|_{L^{p/2}(B_1^+,\mu)}^{p/2} + \|\nabla u_M\|_{L^2(B_2^+,\mu)}^2\right) \\
&\leq C \left(\left\|\frac{\mathbf{F}_M}{\mu}\right\|_{L^p(B_2^+,\mu)}^{p} + \|\nabla u_M\|_{L^2(B_2^+,\mu)}^2\right),
\end{aligned}
$$

where we have applied the strong (p,p) estimate for the maximal function operator $\mathcal{M}^\mu$.

Now applying Lemma 2.6 we have that

$$
\|\nabla u_M\|_{L^p(B_1^+,\mu)}^p \leq C \|\mathcal{M}^\mu(\chi_{B_2^+}|\nabla u_M|^2)\|_{L^{p/2}(B_1^+,\mu)}^{p/2} \leq C(S + \mu(B_1)),
$$

and therefore multiplying by M^p and applying (4.9) we obtain that

$$\|\nabla u\|^p_{L^p(B_1^+,\mu)} \leq C\left(\left\|\frac{\mathbf{F}}{\mu}\right\|^p_{L^p(B_2^+,\mu)} + (\mu(B_1))^{1-\frac{p}{2}}\|\nabla u\|^2_{L^2(B_2^+,\mu)}\right).$$

This completes the proof of the theorem. □

Acknowledgments. D. Cao would like to thank the Department of Mathematics, University of Tennessee at Knoxville, TN for the support and hospitality from which part of this work was done when he visited. T. Mengesha's research is supported by NSF grants DMS-1506512 and DMS-1615726. T. Phan's research is supported by the Simons Foundation, grant # 354889.

References

[1] S.-S. Byun and L. Wang, *The conormal derivative problem for elliptic equations with BMO coefficients on Reifenberg flat domains*, Proc. London Math. Soc. (3) **90** (2005), no. 1, 245–272, DOI 10.1112/S0024611504014960. MR2107043

[2] S.-S. Byun and L. Wang, *Elliptic equations with BMO coefficients in Reifenberg domains*, Comm. Pure Appl. Math. **57** (2004), no. 10, 1283–1310, DOI 10.1002/cpa.20037. MR2069724

[3] L. A. Caffarelli and I. Peral, *On $W^{1,p}$ estimates for elliptic equations in divergence form*, Comm. Pure Appl. Math. **51** (1998), no. 1, 1–21, DOI 10.1002/(SICI)1097-0312(199801)51:1⟨1::AID-CPA1⟩3.3.CO;2-N. MR1486629

[4] D. Cao, T. Mengesha and T. Phan, *Weighted $W^{1,p}$ estimates for weak solutions of degenerate and singular equations,* to appear in the Indiana University Mathematics Journal.

[5] L. C. Evans, *Partial differential equations*, 2nd ed., Graduate Studies in Mathematics, vol. 19, American Mathematical Society, Providence, RI, 2010. MR2597943

[6] G. Di Fazio, M. S. Fanciullo, and P. Zamboni, *L^p estimates for degenerate elliptic systems with VMO coefficients*, Algebra i Analiz **25** (2013), no. 6, 24–36, DOI 10.1090/s1061-0022-2014-01322-2; English transl., St. Petersburg Math. J. **25** (2014), no. 6, 909–917. MR3234838

[7] E. B. Fabes, C. E. Kenig, and R. P. Serapioni, *The local regularity of solutions of degenerate elliptic equations*, Comm. Partial Differential Equations **7** (1982), no. 1, 77–116, DOI 10.1080/03605308208820218. MR643158

[8] Q. Han and F. Lin, *Elliptic partial differential equations*, 2nd ed., Courant Lecture Notes in Mathematics, vol. 1, Courant Institute of Mathematical Sciences, New York; American Mathematical Society, Providence, RI, 2011. MR2777537

[9] J. García-Cuerva, *Weighted H('P) spaces*, ProQuest LLC, Ann Arbor, MI, 1975. Thesis (Ph.D.)–Washington University in St. Louis. MR2625255

[10] J. García-Cuerva, *Weighted H^p spaces*, Dissertationes Math. (Rozprawy Mat.) **162** (1979), 63. MR549091

[11] M. Giaquinta and L. Martinazzi, *An introduction to the regularity theory for elliptic systems, harmonic maps and minimal graphs*, 2nd ed., Appunti. Scuola Normale Superiore di Pisa (Nuova Serie) [Lecture Notes. Scuola Normale Superiore di Pisa (New Series)], vol. 11, Edizioni della Normale, Pisa, 2012. MR3099262

[12] D. Gilbarg and N. S. Trudinger, *Elliptic partial differential equations of second order*, Classics in Mathematics, Springer-Verlag, Berlin, 2001. Reprint of the 1998 edition. MR1814364

[13] E. Giusti, *Direct methods in the calculus of variations*, World Scientific Publishing Co., Inc., River Edge, NJ, 2003. MR1962933

[14] L. Grafakos, *Modern Fourier analysis*, 2nd ed., Graduate Texts in Mathematics, vol. 250, Springer, New York, 2009. MR2463316

[15] N. V. Krylov, *Lectures on elliptic and parabolic equations in Sobolev spaces*, Graduate Studies in Mathematics, vol. 96, American Mathematical Society, Providence, RI, 2008. MR2435520

[16] O. A. Ladyženskaja, V. A. Solonnikov, and N. N. Ural′ceva, *Linear and quasilinear equations of parabolic type* (Russian), Translated from the Russian by S. Smith. Translations of Mathematical Monographs, Vol. 23, American Mathematical Society, Providence, R.I., 1968. MR0241822

[17] G. M. Lieberman, *Second order parabolic differential equations*, World Scientific Publishing Co., Inc., River Edge, NJ, 1996. MR1465184

[18] A. Maugeri, D. K. Palagachev, and L. G. Softova, *Elliptic and parabolic equations with discontinuous coefficients*, Mathematical Research, vol. 109, Wiley-VCH Verlag Berlin GmbH, Berlin, 2000. MR2260015

[19] N. G. Meyers, *An L^p e-estimate for the gradient of solutions of second order elliptic divergence equations*, Ann. Scuola Norm. Sup. Pisa (3) **17** (1963), 189–206. MR0159110

[20] B. Muckenhoupt and R. L. Wheeden, *Weighted bounded mean oscillation and the Hilbert transform*, Studia Math. **54** (1975/76), no. 3, 221–237, DOI 10.4064/sm-54-3-221-237. MR0399741

[21] B. Muckenhoupt and R. L. Wheeden, *On the dual of weighted H^1 of the half-space*, Studia Math. **63** (1978), no. 1, 57–79, DOI 10.4064/sm-63-1-57-79. MR508882

[22] M. K. V. Murthy and G. Stampacchia, *Boundary value problems for some degenerate-elliptic operators* (English, with Italian summary), Ann. Mat. Pura Appl. (4) **80** (1968), 1–122, DOI 10.1007/BF02413623. MR0249828

[23] I. Peral and F. Soria, *A note on $W^{1,p}$ estimates for quasilinear parabolic equations*, Proceedings of the 2001 Luminy Conference on Quasilinear Elliptic and Parabolic Equations and System, Electron. J. Differ. Equ. Conf., vol. 8, Southwest Texas State Univ., San Marcos, TX, 2002, pp. 121–131. MR1990299

[24] T. Phan, *Weighted Calderon-Zygmund estimates for weak solutions of quasi-linear degenerate elliptic equations*, submitted, arXiv:1702.08622.

[25] E. W. Stredulinsky, *Weighted inequalities and applications to degenerate elliptic partial differential equations*, ProQuest LLC, Ann Arbor, MI, 1981. Thesis (Ph.D.)–Indiana University. MR2631634

[26] L. H. Wang, *A geometric approach to the Calderón-Zygmund estimates*, Acta Math. Sin. (Engl. Ser.) **19** (2003), no. 2, 381–396, DOI 10.1007/s10114-003-0264-4. MR1987802

Department of Mathematics and Statistics, Texas Tech University, Box 41042, Lubbock, Texas 79409-1042
Email address: dat.cao@ttu.edu

Department of Mathematics, University of Tennessee, Knoxville, 227 Ayres Hall, 1403 Circle Drive, Knoxville, Tennessee 37996
Email address: mengesha@math.utk.edu

Department of Mathematics, University of Tennessee, Knoxville, 227 Ayres Hall, 1403 Circle Drive, Knoxville, Tennessee 37996
Email address: phan@math.utk.edu

Contemporary Mathematics
Volume **725**, 2019
https://doi.org/10.1090/conm/725/14554

Virial estimates for hard spheres

Ryan Denlinger

ABSTRACT. We review a virial-type estimate which bounds the strength of interaction for a gas of N hard spheres (billiard balls) dispersing into Euclidean space $\mathbb{R}^d$. This type of estimate has been known for decades in the context of (semi-)dispersing billiards, and is essentially trivial in that context. Our goal, however, is to write virial estimates in a way which may lend insight into the problem of rigorously deriving Boltzmann's equation (cf. Lanford's theorem). Using virial estimates, we provide a short proof of lower bounds (sharp up to powers of logarithms) on the convergence rate of the first marginal in Lanford's theorem. Such lower bounds will often, but not always, follow trivially from energy conservation; the proof we present holds assuming only that the limiting dynamics is regular enough and does not reduce to free transport.

1. Introduction

The problem of interest to us is that of deriving various nonlinear partial differential equations (PDEs) starting from the Newtonian gas of N hard spheres. Depending on the chosen scalings, the relevant PDE could be the Navier-Stokes-Fourier equations, Boltzmann's equation, etc. (though fully nonlinear Navier-Stokes-Fourier is far out of reach by current methods). The existence and uniqueness of solutions to nonlinear PDEs is generally an open problem, except in the presence of very special conservation principles or perturbative assumptions. Even when solutions are known, the analysis tends to be quite complicated, depending on the strength of available *a priori* estimates for a hypothetical solution. For this reason, we are naturally led to the problem of deriving analogous *a priori* bounds on the particle model.

1.1. Hydrodynamic limits of interacting particle systems.

There is not one unique way to approach the derivation of hydrodynamic equations starting from mechanical laws. One possible strategy to attack this problem would be to set up a hydrodynamic scaling at the particle level and let $N \to \infty$ (with the hope that local Gibbs states will possess some ergodicity). This is sometimes a useful approach in the presence of *stochasticity* (e.g. see [**20**]) but has not been particularly fruitful in the deterministic case due to the limited understanding of dynamical systems in many dimensions. (Note however that some one-dimensional models are tractable, e.g. identical hard rods. [**5**]) A second possible strategy for hydrodynamic limits is to look for an intermediate (kinetic) description, retaining some of the microscopic information but not all of it. Kinetic descriptions operate on much smaller timescales than hydrodynamic descriptions because time averaging always washes some microscopic information away. Therefore, in order to pass from a kinetic description to a hydrodynamic description *at the particle level*, we need *quantitative bounds* on *long time intervals* (compactness is not enough!).

Despite striking advances in the passage from Boltzmann's equation to hydrodynamic equations in various low-density regimes (see [**23**] for an overview), the derivation of Boltzmann's equation from Newton's laws is still in its infancy. A classical theorem due to O. E. Lanford establishes the validity of Boltzmann's equation for a hard sphere gas, but only up to a fraction of the mean free time for a particle of gas. [**13, 17, 21**] Obviously Lanford's theorem is completely unsatisfactory because we need *many* collisions even to progress past $t = 0$ in a hydrodynamic description. R. Illner and M. Pulvirenti were able to obtain convergence globally in time, but only when the gas is so diffuse that particles mostly do not collide at all. [**15, 16**] H. van Beijeren, O. E. Lanford, J. L. Lebowitz and H. Spohn were able to derive the (non-conservative) *linear* Boltzmann equation for a tagged particle in an equilibrium background, as well as the linearized Boltzmann equation which is formally associated to the response of the background itself. [**18, 26**]

Much more recently, T. Bodineau, I. Gallagher, and L. Saint-Raymond were able to quantify the convergence from [**26**] on time scales T_N diverging like a power of $\log \log N$, thereby deriving Brownian motion in a suitable hydrodynamic scaling. [**3**] (Note that the $\log \log N$ timescale is still troublesome from a physical point of view but it is hard to avoid mathematically using the series-based methods of [**3, 17, 26**].) In a follow-up work, the same authors considered a symmetrized perturbation of size $1/N$ (this simulates the response of an equilibrium background to the influence of a tagged particle). [**4**] This leads to a rigorous derivation of the *linearized* Boltzmann equation on long timescales, and subsequently a derivation of linear hydrodynamic models (in two dimensions only).

1.2. Monotonicity, convexity, Morawetz, Bony. One of the classical problems for billiard systems (such as the hard sphere gas) is to estimate the number of collisions in a finite time interval. It is known that this number is finite for N hard spheres in $\mathbb{R}^d$ (see [**14, 27**] for two different proofs) but *a priori* it might depend on the initial condition. Actually it turns out that the number of collisions is bounded *uniformly* with respect to initial conditions, but might grow like N^N or worse. [**6**] Note that *even if we ignore high-order correlations* and simply consider clusters of $\mathcal{O}(\log N)$ particles, the function $(\log N)^{\log N}$ *still* grows faster than any power of N. On the other hand, with respect to the Lanford theorem, one does not *really* care about the total number of collisions. We care about *estimates* in *good function spaces*; we do not need to count all collisions the same way.

Workers in billiards theory have known for decades that *some* collisions can be estimated very efficiently by constructing monotonic or convex functions of phase space coordinates. (Monotonicity or convexity is here measured along a fixed trajectory.) This idea was stated explicitly in [**6**] and was used implicitly in both [**27**] and [**14**]. It turns out that these (monotonic or convex) functions are of the same type as the functions appearing in proofs of virial and Morawetz type estimates for dispersive PDEs and Vlasov-type equations. (In fact T. Tao points out the connection explicitly in his book [**25**]; his §1.5 Example 1.34 may be viewed as a caricature of our Corollary 3.4, whereas our Corollary 3.4 is all but written already by R. Illner in [**14**].)

There is no known analogue of virial or Morawetz identities for the Boltzmann equation in general (without assuming extra estimates above the energy level). The closest known results are set in one space dimension; technically, the physical setting is $\mathbb{R}^3$ with spatial variation along just one axis. In that case, for certain collision kernels, it is possible to write down an integral which effectively tracks the accumulation of collisions as the solution $f(t)$ interacts with itself. One can prove monotonicity in time, as well as uniform boundedness in large time, using conservation laws. [**2**, **7**–**9**] This technique is known as *Bony's functional* or *Glimm's functional*, by analogy with similar techniques for hyperbolic conservation laws in one space dimension. The point of this technique is that, in one dimension, two disturbances will perhaps pass through each other a few times and interact, but each time some part of the *potential for interaction* is expended and cannot be used again. This potential for interaction can only be measured directly due to the one-dimensional geometry. In higher dimensions, one would have to consider potentials along many possible trajectories of the system and this is just too difficult to quantify (compare the difficulty of tracking shocks in higher dimensions).

1.3. Main results and organization of this paper. The main focus of this work is Proposition 4.1, which is a virial-type spacetime estimate for hard spheres. As noted above, virial-type estimates are essentially classical in the billiards literature, and they play a prominent role in the derivation of Boltzmann's equation. (See [**15**, **16**], particularly the first lemma of [**16**], which is Lemma 3.2 in the present manuscript.) The main difference with Proposition 4.1 is that the classical virial bound is re-formulated to control a quantity closely associated with the hard sphere BBGKY hierarchy, for a wide class of initial data. Unfortunately, while these estimates are quite general, they do not lead directly to coercive estimates at the limit for any nontrivial scaling of which we are aware. We will show, however, that virial estimates can be used to place *lower* bounds on the convergence rate in Lanford's theorem. Such lower bounds may, but do not always, follow trivially from energy conservation; our result holds under essentially minimal assumptions on the initial data. The types of data which are newly covered by our result have a product structure at the initial time, $f_0(x,v) = \rho(x)m(v)$, or are convex combinations $f_0(x,v) = \int d\alpha \rho_\alpha(x) m_\alpha(v)$ where each m_α has the same conserved moments. (See Example 6.1.)

Section 2 introduces the basic notation of this work, which mostly follows the presentation of [**13**]. Section 3 gives an elementary derivation of an identity due to Illner [**14**]. In Section 4, we apply this identity in a heuristic manner to derive the virial-type spacetime estimate; a rigorous proof may be found in [**12**]. Section 5 gives a very concise overview of Lanford's theorem, [**13**, **17**], and some recent developments. Finally in Section 6 we use virial identities to prove a bound from below on the convergence rate in Lanford's theorem (including cases where such lower bounds would not follow directly from energy conservation).

2. Notation

Consider N non-overlapping hard spheres centered at positions $x_i \in \mathbb{R}^d$ with velocities $v_i \in \mathbb{R}^d$ for $i = 1, 2, \ldots, N$. The spheres are considered to have identical mass and radius, and are in all other ways physically indistinguishable. For convenience, we will assume without loss that all particles have unit diameter.

The collection of all positions is a tuple X_N,

$$X_N = (x_1, x_2, \dots, x_N) \in \mathbb{R}^{dN}$$

$$V_N = (v_1, v_2, \dots, v_N) \in \mathbb{R}^{dN}$$

The classical phase-space coordinates of ith particle are given by $z_i = (x_i, v_i)$, and the phase-space coordinates of the whole gas are denoted

$$Z_N = (z_1, z_2, \dots, z_N) \in \mathbb{R}^{2dN}$$

We may also write $Z_N = (X_N, V_N)$. The following function will play a central role in our analysis: for $t \in \mathbb{R}$ and $Z_N \in \mathbb{R}^{2dN}$, we define

$$r_N(t, Z_N) = \sum_{i=1}^{N} \left(x_i \cdot v_i - |v_i|^2 t\right) \tag{1}$$

Following [**13**], we may introduce the N-particle phase space $\mathcal{D}_N$, which is defined by

$$\mathcal{D}_N = \left\{ Z_N \in \mathbb{R}^{2dN} \middle| \ \forall 1 \le i < j \le N,\ |x_i - x_j| > 1 \right\} \tag{2}$$

The choice of $\mathcal{D}_N$ is motivated by requirement that the spheres be mutually disjoint at all times. The closure of $\mathcal{D}_N$ in $\mathbb{R}^{2dN}$ in the standard topology is denoted $\overline{\mathcal{D}}_N$, and we will also write $\partial\mathcal{D}_N = \overline{\mathcal{D}}_N \backslash \mathcal{D}_N$. We will use the notation a.e. $Z_N \in \mathcal{D}_N$ to refer to a typical point for the Lebesgue measure on $\mathcal{D}_N$. The notation a.e. $Z_N \in \partial\mathcal{D}_N$ will refer to a typical point for the induced surface measure arising from the natural embedding $\partial\mathcal{D}_N \subset \mathbb{R}^{2dN}$.

Formally speaking, we wish to solve Newton's laws with a hard core interaction. This means if $Z_N(t_0) = (X_N(t_0), V_N(t_0)) \in \mathcal{D}_N$ then

$$\frac{d}{dt} X_N(t)\bigg|_{t=t_0} = V_N(t_0)$$

$$\frac{d}{dt} V_N(t)\bigg|_{t=t_0} = 0$$

Hence the particles move freely between collisions. At each collision (that is, $Z_N(t_0) \in \partial\mathcal{D}_N$), the particles are required to interact elastically, thereby conserving momentum, energy, and angular momentum. The set of possible interactions for two-body elastic collisions is easy to parametrize explicitly. Suppose that there exists $i < j$ such that $x_j(t_0) = x_i(t_0) + \omega$ for some $\omega \in \mathbb{S}^{d-1}$; and, further suppose that $|x_{j'}(t_0) - x_{i'}(t_0)| > 1$ for any $i' < j'$ such that $(i', j') \neq (i, j)$. Let us denote

$$\lim_{t \to t_0^-} V_N(t) = (v_1, \dots, v_i, \dots, v_j, \dots, v_N)$$

$$\lim_{t \to t_0^+} V_N(t) = \left(v_1, \dots, v_i^*, \dots, v_j^*, \dots, v_N\right)$$

Then we have

$$v_i^* = v_i + \omega\omega \cdot (v_j - v_i)$$

$$v_j^* = v_j - \omega\omega \cdot (v_j - v_i)$$

Similarly for a.e. $Z_N \in \partial\mathcal{D}_N$ we will use the notation Z_N^* to refer to the image of the point Z_N through the collision transformation. The map $Z_N \mapsto Z_N^*$ is a measurable involution.

In the above "definition," we have neglected to specify uniquely what happens when more than two particles collide at the same time. Multiple particle interactions occur with zero probability, though this statement requires justification which we will not discuss. (See [**1**] or [**13**].) The hard sphere flow at time t defines a measurable map

$$\psi_N^t : \mathcal{D}_N \to \mathcal{D}_N$$

For each $t \in \mathbb{R}$, the map ψ_N^t preserves the Lebesgue measure on $\mathcal{D}_N \subset \mathbb{R}^{2dN}$. Complete proofs of the existence of the hard sphere flow ψ_N^t may be found in the literature. [**1**, **13**]

Following Boltzmann's great insight, we realize that it is not very interesting to discuss any *particular* trajectory $\{\psi_N^t Z_N\}_{t\geq 0}$, because it is physically infeasible (or impossible) to measure the positions and velocities of all the particles at a given instant. Therefore, the initial value problem for Newton's laws is *not* the correct problem for us to solve. The correct approach is to place a probability density $f_N(0, Z_N)$ on the set of possible initial states $Z_N \in \mathcal{D}_N$. The function $f_N(0, Z_N)$ represents our uncertainty about the actual state of the system. Since we have no physical means to distinguish between two particles in our model, the function $f_N(0, Z_N)$ must be *symmetric* with respect to interchange of particle indices.

We will denote by $\mathcal{S}_N$ the symmetric group on N letters. Any permutation $\sigma \in \mathcal{S}_N$ acts on the phase-space coordinates $Z_N = (z_1, z_2, \ldots, z_N) \in \mathcal{D}_N$ as follows:

$$\sigma Z_N = \left(z_{\sigma(1)}, z_{\sigma(2)}, \ldots, z_{\sigma(N)}\right) \in \mathcal{D}_N$$

Similarly, if $f_N(Z_N)$ is any function on $\mathcal{D}_N$, then σ acts on f_N by composition: $\sigma f_N = f_N \circ \sigma$. Let $\mathcal{P}(\mathcal{D}_N)$ denote the set of probability measures on $\mathcal{D}_N$, and furthermore let $\mathcal{P}_{\text{a.c.}}(\mathcal{D}_N)$ denote the set of probability measures which are absolutely continuous with respect to the Lebesgue measure on $\mathcal{D}_N$. Any element of $\mathcal{P}_{\text{a.c.}}(\mathcal{D}_N)$ may be represented uniquely (a.e. $Z_N \in \mathcal{D}_N$) by a non-negative function $f_N(Z_N)$ such that $\int_{\mathcal{D}_N} f_N(Z_N) dZ_N = 1$. Finally let $\mathcal{P}_{\text{a.c.}}^{\text{sym}}(\mathcal{D}_N)$ be the set of absolutely continuous measures on $\mathcal{D}_N$ such that the associated function f_N is invariant under the action of $\mathcal{S}_N$. Henceforth, when we write f_N, we will always mean an element of $\mathcal{P}_{\text{a.c.}}^{\text{sym}}(\mathcal{D}_N)$.

Let $f_N(0)$ be any element of $\mathcal{P}_{\text{a.c.}}^{\text{sym}}(\mathcal{D}_N)$, which we regard as the initial state of the N particle gas. For any $t \in \mathbb{R}$ we will let $f_N(t)$ be the pushforward of $f_N(0)$ under the hard sphere flow ψ_N^t; then, $f_N(t)$ is likewise an element of $\mathcal{P}_{\text{a.c.}}^{\text{sym}}(\mathcal{D}_N)$. Since ψ_N^t preserves the Lebesgue measure on $\mathcal{D}_N$, we may write the following expression for $f_N(t)$:

$$f_N(t, Z_N) = f_N\left(0, \psi_N^{-t} Z_N\right) \tag{3}$$

The functions $f_N(0)$ and $f_N(t)$ may be extended by zero so as to be defined on $\mathbb{R}^{2dN}$.

For any $1 \leq s \leq N$, we define the marginal $f_N^{(s)}(t)$ by partial integration:

$$f_N^{(s)}(t, Z_s) = \int_{\mathbb{R}^{2d(N-s)}} f_N(t, Z_N) dz_{s+1} \ldots dz_N \tag{4}$$

The evolution of the marginals $f_N^{(s)}(t)$ may be described explicitly via the so-called BBGKY hierarchy (Bogoliubov-Born-Green-Kirkwood-Yvon) [**13**], though we will *not* be making any use of the BBGKY hierarchy except in Section 6. The marginals $f_N^{(s)}(t)$ are non-negative symmetric functions on $\mathbb{R}^{2ds}$ with unit mass.

The main result we will show, Proposition 4.1, will control the trace of the marginals $f_N^{(s)}(t, Z_s)$ along a certain hypersurface in $\mathbb{R} \times \mathbb{R}^{2ds}$, with polynomial dependence on N for large values of N. This is slightly problematic because the trace of an L^1 function is simply not defined; moreover, even if the data $f_N(0)$ is *smooth*, the function $f_N(t)$ typically develops singularities. Nevertheless, due to technical arguments which we will not discuss, it is possible to show that if $f_N(0)$ is smooth and compactly supported in $\mathcal{D}_N$ then the required traces of $f_N^{(s)}(t)$ do, in fact, exist (at least for almost every $t \in \mathbb{R}$). See [**10, 13, 22, 24**] for more information on regularity issues for hard spheres. Our estimates do not depend on the choice of regularization, except insofar as the regularized marginals must be a sequence of symmetric non-negative functions which are indeed marginals in the sense of (4). Therefore, similar to the proof of the classical trace theorem in partial differential equations ($W^{1,p}(U) \subseteq L^p(\partial U)$ for sufficiently smooth bounded regions $U \subset \mathbb{R}^k$), the traces are actually meaningful *for solutions of Liouville's equation* even if the initial data is only L^1. We will not discuss further the issues of regularity.

3. A monotonicity formula

Let us fix an initial point $Z_N \in \mathcal{D}_N$ in the microscopic phase-space, and consider the trajectory $\{\psi_N^t Z_N\}_{t \in \mathbb{R}}$. Our analysis begins with a simple observation: with $r_N(t, Z_N)$ as in (1), if we define

$$r_{Z_N}(t) = r_N\left(t, \psi_N^t Z_N\right) \tag{5}$$

then for any t_0 such that $\psi_N^{t_0} Z_N \in \mathcal{D}_N$ we have

$$\left.\frac{d}{dt} r_{Z_N}(t)\right|_{t=t_0} = 0 \tag{6}$$

Indeed, we see that if $\dot{x}_i = v_i$ and $\dot{v}_i = 0$ then

$$\frac{d}{dt}\left(x_i \cdot v_i - |v_i|^2 t\right) = 0$$

Therefore, the difference $r_{Z_N}(t) - r_{Z_N}(0)$ is simply equal to a sum along *collisions* of incremental jumps in $r_{Z_N}(t)$. It will turn out that all of these jumps have the *same sign*, and we can compute the jumps explicitly in terms of collision parameters.

Let us compute the jump in $r_{Z_N}(t)$ across a collision taking place at time $t_0 \in \mathbb{R}$. We may assume that the interacting particles are simply those labelled $i = 1, 2$, since collisions are binary and particles are indistinguishable. The position coordinates are continuous in time, so we write them x_1, x_2, with $x_2 = x_1 + \omega$ for some $\omega \in \mathbb{S}^{d-1}$. The pre-collisional velocities will be denoted $v_1 \equiv v_1(t_0^-), v_2 \equiv v_2(t_0^-)$ and the post-collisional velocities will be denoted $v_1^* \equiv v_1(t_0^+), v_2^* \equiv v_2(t_0^+)$. We have

$$\begin{aligned} r_{Z_N}(t_0^+) - r_{Z_N}(t_0^-) = &\left\{\left(x_1 \cdot v_1^* - |v_1^*|^2 t_0\right) + \left(x_2 \cdot v_2^* - |v_2^*|^2 t_0\right)\right\} + \\ &+ \left\{-\left(x_1 \cdot v_1 - |v_1|^2 t_0\right) - \left(x_2 \cdot v_2 - |v_2|^2 t_0\right)\right\} \end{aligned}$$

Due to energy conservation,

$$|v_1^*|^2 + |v_2^*|^2 = |v_1|^2 + |v_2|^2$$

so we may eliminate the explicit dependence on t_0.

$$r_{Z_N}(t_0^+) - r_{Z_N}(t_0^-) = x_1 \cdot v_1^* + x_2 \cdot v_2^* - x_1 \cdot v_1 - x_2 \cdot v_2$$

Since $x_2 = x_1 + \omega$, this gives us

$$r_{Z_N}(t_0^+) - r_{Z_N}(t_0^-) = x_1 \cdot (v_1^* + v_2^* - v_1 - v_2) + \omega \cdot (v_2^* - v_2)$$

Due to momentum conservation,

$$v_1^* + v_2^* = v_1 + v_2$$

so we may eliminate the explicit dependence on the position coordinates. Hence

$$r_{Z_N}(t_0^+) - r_{Z_N}(t_0^-) = \omega \cdot (v_2^* - v_2)$$

This is the same as

$$r_{Z_N}(t_0^+) - r_{Z_N}(t_0^-) = -\omega \cdot (v_2 - v_1)$$

by the collisional change of variables from Section 2. But v_1, v_2 are the velocities of the two particles *coming into* a collision, so we must have

$$(x_2 - x_1) \cdot (v_2 - v_1) \leq 0$$

and therefore $\omega \cdot (v_2 - v_1) \leq 0$. Hence,

$$r_{Z_N}(t_0^+) - r_{Z_N}(t_0^-) = |\omega \cdot (v_2 - v_1)| \geq 0$$

Adding up all collisions along the trajectory we obtain the following identity, which was observed by Illner: [**14**]

PROPOSITION 3.1. *(Illner) For* a.e. $Z_N \in \mathcal{D}_N$ *and* a.e. $t \geq 0$ *there holds*

$$r_{Z_N}(t) - r_{Z_N}(0) = \sum_k \left| \omega_k \cdot \left(v_{j_k}(t_k^-) - v_{i_k}(t_k^-) \right) \right| \tag{7}$$

where the sum $\sum_k$ *is over all collisions along the trajectory* $\{\psi_N^\tau Z_N\}_\tau$ *for* $0 \leq \tau \leq t$.

We will require an auxiliary lemma due to Illner and Pulvirenti which follows easily from Proposition 3.1. [**15, 16**] In order to state the lemma, we introduce a new function on $\mathcal{D}_N$,

$$I_N(Z_N) = \sum_{i=1}^N |x_i|^2 \tag{8}$$

The proof is a computation, which we include for completeness.

LEMMA 3.2. *For* a.e. $Z_N = (X_N, V_N) \in \mathcal{D}_N$ *and all* $t \in \mathbb{R}$, *we have*

$$I_N\left(\psi_N^t Z_N\right) \geq I_N\left((X_N + V_N t, V_N)\right) \tag{9}$$

PROOF. By time-reversibility we may assume $t \geq 0$. The function $I_N(\psi_N^t Z_N)$ is globally continuous in t for a.e. $Z_N \in \mathcal{D}_N$. With this in mind, it suffices to point out that the desired inequality is true for $t = 0$, and between collisions (using energy conservation) we have

$$\frac{d}{dt}\left\{I_N\left(\psi_N^t Z_N\right) - I_N\left((X_N + V_N t, V_N)\right)\right\} = 2\left\{r_{Z_N}(t) - r_{Z_N}(0)\right\}$$

We conclude by Proposition 3.1. □

REMARK. Lemma 3.2 was the key estimate which Illner and Pulvirenti [**15, 16**] relied upon to rigorously derive Boltzmann's equation, globally in time, for a rarefied gas in $\mathbb{R}^d$. The theorem of Illner and Pulvirenti is analogous to "small data" results for nonlinear PDE, and therefore does not resolve the problem of deriving Boltzmann's equation near global Maxwellians. Note however that Proposition 3.1

is *more* general than Lemma 3.2 and generally contains more detailed information about the dynamics, including possible cancellations. Another important point is that Illner and Pulvirenti actually applied Lemma 3.2 to isolated *clusters* of particles, controlling separately the interactions between clusters. In the same way Proposition 3.1 is applicable to isolated clusters of particles, just as it is applicable to the gas as a whole.

The next lemma is technical, and again follows from Proposition 3.1.

LEMMA 3.3. *For* a.e. $Z_N = (X_N, V_N) \in \mathcal{D}_N$, *all* $t \geq 0$, *and all* $\lambda > 0$, *there holds*

$$|r_{Z_N}(t)| \leq \frac{1}{2}\lambda^{-1} \sum_{i=1}^{N} \left(\lambda^2 |x_i|^2 + |v_i|^2\right) \tag{10}$$

PROOF. Recall that $r_{Z_N}(t) \equiv r_N\left(t, \psi_N^t Z_N\right)$. On the other hand, by (1),

$$\begin{aligned}
|r_N(t, Z_N)| &\leq \sum_{i=1}^{N} \left|x_i \cdot v_i - |v_i|^2 t\right| \\
&= \sum_{i=1}^{N} |(x_i - v_i t) \cdot v_i| \\
&\leq \frac{1}{2} \sum_{i=1}^{N} \left(\lambda |x_i - v_i t|^2 + \lambda^{-1} |v_i|^2\right) \\
&= \frac{1}{2} \lambda I_N \left((X_N - V_N t, V_N)\right) + \frac{1}{2}\lambda^{-1} \sum_{i=1}^{N} |v_i|^2
\end{aligned}$$

We can bound the first term on the last line using Lemma 3.2. Hence,

$$|r_N(t, Z_N)| \leq \frac{1}{2} \lambda I_N \left(\psi_N^{-t} Z_N\right) + \frac{1}{2} \lambda^{-1} \sum_{i=1}^{N} |v_i|^2$$

Replace Z_N by $\psi_N^t Z_N$ on both sides and use the conservation of energy to conclude. □

Combining Proposition 3.1 and Lemma 3.3, we obtain:

COROLLARY 3.4. *For* a.e. $Z_N = (X_N, V_N) \in \mathcal{D}_N$ *and all* $\lambda > 0$, *we have*

$$\sum_k \left|\omega_k \cdot \left(v_{j_k}(t_k^-) - v_{i_k}(t_k^-)\right)\right| \leq 2\lambda^{-1} \sum_{i=1}^{N} \left(\lambda^2 |x_i|^2 + |v_i|^2\right) \tag{11}$$

where the sum $\sum_k$ *is over all collisions along the trajectory* $\{\psi_N^t Z_N\}_{t \in \mathbb{R}}$.

REMARK. It is interesting to compare Corollary 3.4 against the case where we count *all* collisions equally, without the weighting factor $|\omega \cdot (v_j - v_i)|$. In that case the best one can do with current technology is bound the number of collisions by super-exponential functions of the number of particles, e.g. growing faster than N^N; we refer to [**6**] for estimates of this type. Indeed, the authors of [**6**] rightly note that much better estimates can be proven if not all collisions are counted. Note that unweighted collision estimates are of direct interest for dynamical systems theory, whereas kinetic theory is more interested in finding good *function spaces* whose associated norms may well contain weights.

REMARK. Clearly, Corollary 3.4 represents the "worst case" behavior for a system of hard spheres. If the initial conditions Z_N are chosen "randomly" (with suitable scalings of x, v; cf. low-density limit, [**11**]) then the left-hand side should typically be *much smaller* than the right-hand side, at least when collisions are counted on finite time intervals. This statement can be formalized and proved (in an average sense for suitable f_0) on a small time interval in the Boltzmann-Grad scaling, using the bounds from the proof of Lanford's theorem. An interesting open question, which should be addressed, is whether improvements can be obtained (on average) for Corollary 3.4, locally in time and away from local equilibria, while assuming less than what is required to prove Lanford's theorem. Such a result by itself cannot be expected to allow improvement of the time of convergence in Lanford's theorem, but may provide relevant insights in that direction.

4. An averaging trick

The previous section was primarily concerned with weighted sums over collisions which occur along a single trajectory $\{\psi_N^t Z_N\}_t$. However, as has been explained in Section 2, we are really interested in ensemble averages over many trajectories. This is due to the physical fact that we cannot say with any precision what the initial state Z_N "really" is. We will "prove" a spacetime estimate by averaging both sides of (11) with respect to the *same* measure $f_N(0, Z_N)dZ_N$ and applying a change of variables on the left-hand side. The change of variables as presented here is not entirely rigorous, though we are confident that this approach can be converted into a rigorous proof. An alternative, completely rigorous, proof of the virial-type estimate (Proposition 4.1) has already been given. [**12**]

REMARK. We emphasize that the results of this section are not new, nor are they especially novel except perhaps in the style of presentation; indeed, estimates of the type shown here go back many decades. In particular, in this section we will prove a virial-type estimate for the second marginal which holds under finiteness of second moments, but this estimate is *not* uniform in the Boltzmann-Grad scaling. The novelty of our contribution is precisely the fact that virial estimates *can* sometimes provide nontrivial information in the Boltzmann-Grad scaling (not easily accessible by other means), but that discussion is deferred to Section 6.

We will find it helpful to define an auxiliary function,

$$W_N^{(i,j)}(Z_N) = |(x_j - x_i) \cdot (v_j - v_i)| \tag{12}$$

Observe that $W_N^{(i,j)}(Z_N) = |\omega \cdot (v_j - v_i)|$ if $Z_N \in \partial\mathcal{D}_N$ represents a collision between particles i and j with $x_j = x_i + \omega$. For any $Z_N \in \mathcal{D}_N$ let $\tilde{\imath}_N(Z_N), \tilde{\jmath}_N(Z_N)$ be chosen such that $0 < \left|x_{\tilde{\imath}_N} - x_{\tilde{\jmath}_N}\right| \leq |x_i - x_j|$ for all $i \neq j$. This uniquely defines $\tilde{\imath}_N,\tilde{\jmath}_N$ (up to switching the two indices) for a.e. $Z_N \in \mathcal{D}_N$ and also for a.e. $Z_N \in \partial\mathcal{D}_N$. Let us finally define

$$W_N(Z_N) = W_N^{(\tilde{\imath}_N(Z_N),\tilde{\jmath}_N(Z_N))}(Z_N) \tag{13}$$

so that $W_N(Z_N)$ is always equal to the correct collision parameter $|\omega \cdot (v_j - v_i)|$ (when binary collisions are well-defined) globally along $\partial\mathcal{D}_N$.

The "proof" of the spacetime estimate is based on the following observation: the collision sum $\sum_k$ on the left-hand side of (11) may be re-cast as an integral in time:

$$\sum_k W_N\left(\psi_N^{t_k} Z_N\right) \equiv \int_{\mathbb{R}} \delta_{\psi_N^t Z_N \in \partial\mathcal{D}_N} W_N\left(\psi_N^t Z_N\right) dt$$

Average both sides with respect to $f_N(0, Z_N)dZ_N$.

$$\int_{\mathcal{D}_N} \left\{ \sum_k W_N\left(\psi_N^{t_k} Z_N\right) \right\} f_N(0, Z_N)dZ_N = \\ = \int_{\mathcal{D}_N} \int_{\mathbb{R}} \delta_{\psi_N^t Z_N \in \partial\mathcal{D}_N} W_N\left(\psi_N^t Z_N\right) f_N(0, Z_N)dtdZ_N$$

The double integral on the right-hand side reduces (by Fubini) to an integral of "something" over $\partial\mathcal{D}_N$, due to the delta-function and the identity $f_N(t, Z_N) = f_N(0, \psi_N^{-t} Z_N)$. Unfortunately, making the change of variables precise requires a technical application of the divergence theorem and careful manipulation of delta functions. (The proof of [**12**] avoids any mention of delta functions.) Here we record the result of correct manipulations:

$$\int_{\mathcal{D}_N} \left\{ \sum_k W_N\left(\psi_N^{t_k} Z_N\right) \right\} f_N(0, Z_N)dZ_N = \\ = \int_{\mathbb{R}} \int_{\partial\mathcal{D}_N} \left[W_N\left(Z_N\right)\right]^2 f_N(t, Z_N)d\sigma_N dV_N dt \tag{14}$$

where $d\sigma_N dV_N$ represents the surface measure on $\partial\mathcal{D}_N$.

To conclude, we bound the left hand side of (14) using Corollary 3.4, then reduce both sides using the symmetry of $f_N(t)$ and the definition of the marginals of $f_N(t)$. We have also simplified the estimate by optimal choice of the parameter $\lambda > 0$.

PROPOSITION 4.1. *For each $N \in \mathbb{N}$, let $f_N(0)$ be an initial probability density on $\mathcal{D}_N$, which we assume to be symmetric under particle interchange, and let $f_N(t, Z_N) = f_N(0, \psi_N^{-t} Z_N)$. Let $f_N^{(s)}(t)$, $1 \leq s \leq N$, denote the s-marginal of $f_N(t)$. Further assume that $f_N(0)$ is smooth and compactly supported in the interior of $\mathcal{D}_N$. Then for all $2 \leq s \leq N$ there holds*

$$\sum_{1 \leq i < j \leq s} \int_{-\infty}^{\infty} \int_{\mathbb{R}^{ds} \times \mathbb{R}^{d(s-1)} \times \mathbb{S}^{d-1}} \left|\omega \cdot (v_j - v_i)\right|^2 \times \\ \times f_N^{(s)}(t, \dots, x_i, v_i, \dots, x_i + \omega, v_j, \dots) d\omega dX_s^{(j)} dV_s dt \leq \\ \leq C_d \frac{s(s-1)}{N} \left(\int |x|^2 f_N^{(1)}(0, x, v) dx dv \right)^{\frac{1}{2}} \left(\int |v|^2 f_N^{(1)}(0, x, v) dx dv \right)^{\frac{1}{2}} \tag{15}$$

where $dX_s^{(j)} = dx_1 \dots dx_{j-1} dx_{j+1} \dots dx_s$ and C_d is a constant depending only on the dimension d.

REMARK. It is possible to compare Proposition 4.1 with Lemma 1 of [**16**]; several other related formulas may also be found in [**14**].

REMARK. Note that, up to constants, Proposition 4.1 is formally equivalent to Corollary 3.4 because we can always fix a point $Z_N^0 \in \mathcal{D}_N$ (such that the trajectory

$\{\psi_N^t Z_N^0\}_t$ is globally defined) and let

$$f_N(0, Z_N) = \frac{1}{N!} \sum_{\sigma \in \mathcal{S}_N} \delta_{Z_N = \sigma Z_N^0} \tag{16}$$

Therefore, just as Corollary 3.4 is suboptimal for "many" choices of initial condition Z_N (under suitable scalings), Proposition 4.1 is suboptimal for most physically relevant densities $f_N(0)$. Again this is not a theorem but an empirical observation, rooted in the apparent success of kinetic theory in describing physical reality.

5. Factorized data

In this section, and for the remainder of the work, we will assume that spheres are re-scaled to have diameter $\varepsilon > 0$ instead of diameter 1. Thus the proper condition defining $\mathcal{D}_N$ is $|x_i - x_j| > \varepsilon$ instead of $|x_i - x_j| > 1$.

$$\mathcal{D}_N = \left\{ Z_N = (X_N, V_N) \in \mathbb{R}^{dN} \times \mathbb{R}^{dN} \,|\forall 1 \leq i < j \leq N,\ |x_i - x_j| > \varepsilon \right\} \tag{17}$$

Moreover, we assume the Boltzmann-Grad scaling $N\varepsilon^{d-1} = 1$; physically, this means that the mean free path for a typical particle is of order one. Note that the total volume occupied by all N particles is of order ε; hence, the Boltzmann-Grad limit describes a state of low density.

Suppose $f_0(x, v)$ is a measurable function on $\mathbb{R}^d \times \mathbb{R}^d$ with

$$0 \leq f_0(x, v) \in \left(L^1_{x,v} \cap L^\infty_x L^1_v \right) \left(\mathbb{R}^d \times \mathbb{R}^d \right) \tag{18}$$

Furthermore let us suppose, for convenience, that f_0 is a normalized probability distribution:

$$\int_{\mathbb{R}^{2d}} f_0(x, v) dx dv = 1 \tag{19}$$

Then it is natural to define "factorized" states on $\mathcal{D}_N$ in the following way:

$$f_N(0, Z_N) = \mathcal{Z}_N^{-1} f_0^{\otimes N}(Z_N) \mathbf{1}_{Z_N \in \mathcal{D}_N} \tag{20}$$

Here $\mathcal{Z}_N$ is the partition function,

$$\mathcal{Z}_N = \int_{\mathbb{R}^{2dN}} f_0^{\otimes N}(Z_N) \mathbf{1}_{Z_N \in \mathcal{D}_N} dZ_N \tag{21}$$

We will use the imprecise shorthand $f_N(0) \sim f_0^{\otimes N}$ for such "factorized" initial data. Then it is possible to prove the following *pointwise* estimate for the first marginal at $t = 0$, valid for almost every $(x, v) \in \mathbb{R}^{2d}$ under the Boltzmann-Grad scaling $N\varepsilon^{d-1} = 1$, for all small enough $\varepsilon > 0$ depending only on f_0 and d:

$$\left| f_N^{(1)}(0, x, v) - f_0(x, v) \right| \leq C_d \left\| f_0 \right\|_{L^\infty_x L^1_v} f_0(x, v) \varepsilon \tag{22}$$

Similar pointwise convergence estimates (also of order ε) are available for higher order marginals $f_N^{(s)}(0)$ as well, as long as s is fixed as $N \to \infty$. We refer to [**13**] or [**12**] for detailed proofs.

The $\mathcal{O}(\varepsilon)$ convergence rate for the first marginal at $t = 0$ arises from careful partition function estimates, and is intuitively due to the fact that the total volume occupied by all particles is roughly $N\varepsilon^d$, that is, $\mathcal{O}(\varepsilon)$ in the Boltzmann-Grad scaling $N\varepsilon^{d-1} = 1$. It is not hard to derive *worse* convergence rates under weaker regularity assumptions, e.g. $f_0 \in L^p_x L^1_v$ with $d < p < \infty$. However, as far as we are aware, one does not obtain an error which is *smaller* than $\mathcal{O}(\varepsilon)$ even if f_0 is in the Schwartz class, or jointly Gaussian in x and v.

Let us now assume that f_0 is in the Schwartz class and let $f(t)$ be the solution of Boltzmann's equation (on a small time interval) with hard sphere interaction and $f(0) = f_0$:

$$(\partial_t + v \cdot \nabla_x) f(t,x,v) = Q^+(f,f)(t,x,v) - Q^-(f,f)(t,x,v) \tag{23}$$

$$Q^+(f,f) = \int_{\mathbb{R}^d \times \mathbb{S}^{d-1}} d\omega dv_2 \, |\omega \cdot (v - v_2)| \, f(t,x,v^*) f(t,x,v_2^*) \tag{24}$$

$$Q^-(f,f) = \int_{\mathbb{R}^d \times \mathbb{S}^{d-1}} d\omega dv_2 \, |\omega \cdot (v - v_2)| \, f(t,x,v) f(t,x,v_2) \tag{25}$$

Here we define the collisional change of variables, for a unit vector $\omega \in \mathbb{S}^{d-1}$ and any $v, v_2 \in \mathbb{R}^d$,

$$\begin{aligned} v^* &= v + \omega\omega \cdot (v_2 - v) \\ v_2^* &= v_2 - \omega\omega \cdot (v_2 - v) \end{aligned} \tag{26}$$

In the mid 1970s, Oscar Lanford showed that if $f_0(x,v)$ is nice enough (say smooth with compact support on $\mathbb{R}^{2d}$) then the function $f_N(t, Z_N) = f_N(0, \psi_N^{-t} Z_N)$ with initial data $f_N(0) \sim f_0^{\otimes N}$ has first marginal converging to the solution of Boltzmann's equation on a small time interval in the Boltzmann-Grad scaling $N\varepsilon^{d-1} = 1$. [**17**] Namely, for some small $T > 0$ depending on f_0, any $t \in [0,T]$, and almost every $x, v \in \mathbb{R}^d$,

$$\lim_{N \to \infty} f_N^{(1)}(t,x,v) = f(t,x,v) \tag{27}$$

Moreover it follows from Lanford's proof that the higher order marginals, say $f_N^{(s)}(t)$, converge to the tensor products $f(t)^{\otimes s}$. Lanford's theorem was the first rigorous justification of kinetic theory from deterministic Newtonian mechanics, and his proof remains the basis of most of the more recent developments in first principles derivations of collisional kinetic equations.

Remark. We must point out that a completely different point of view, pioneered by Kac and McKean, is to start with a *stochastic* model in which the microscopic position coordinates are essentially hidden variables. In these models, the impact parameter is automatically random even in the N particle system, and much more detailed results are available. Most notably, unlike the Lanford theorem, the convergence to Boltzmann is often proven globally in time. We refer to [**19**] and references therein for more details.

Within the past few years, a number of authors have worked to make Lanford's theorem into a more quantitative result. Arguably the most notable contribution along these lines is [**13**], in which convergence rates of order $\mathcal{O}\left(\varepsilon^{\frac{d-1}{d+1}-}\right)$ were obtained. (The convergence can be proven in $L^\infty_{x,v}$ for the first marginal, but not for higher marginals, due to issues of irreversibility which we do not discuss here.) It is not established in [**13**] whether the convergence rate obtained therein is optimal (nor does there seem to be a particularly compelling reason that it should be optimal). Intuitively it should not be possible to obtain an error that is much smaller than $\mathcal{O}(\varepsilon)$, since such small errors cannot even be proven at $t = 0$. We are not aware of any proof in the literature that the convergence rate in Lanford's theorem can be $\mathcal{O}\left(\varepsilon^{1-}\right)$; however, what we will prove here is that the $L^\infty_{x,v}$ error (of the first marginal) certainly cannot be much *smaller* than $\mathcal{O}(\varepsilon)$. The proof is based on a virial-type inequality similar to (15).

6. On convergence rates in Lanford's theorem

In this section we will establish a lower bound on the convergence rate in Lanford's theorem; throughout our discussion, we will actually assume Maxwellian tails jointly in x and v, as in [**15**,**16**]. Before stating our main result, a few comments are in order. The first is that conservation laws may automatically imply bounds from below: after all, we certainly have

$$\int_{\mathbb{R}^{2d}} |v|^2 f_N^{(1)}(t,x,v)dxdv = \int_{\mathbb{R}^{2d}} |v|^2 f_N^{(1)}(0,x,v)dxdv \tag{28}$$

Therefore, in the event that $f_N(0)$ is chosen to satisfy the following inequality:

$$\liminf_{N\to\infty} \varepsilon^{-1} \left| \int_{\mathbb{R}^{2d}} |v|^2 f_N^{(1)}(0,x,v)dxdv - \int_{\mathbb{R}^{2d}} |v|^2 f_0(x,v)dxdv \right| > 0 \tag{29}$$

then $f_N^{(1)}(t)$ cannot be within $o(\varepsilon)$ of $f(t)$, since the kinetic energy witnesses a slower convergence rate. On the other hand, by the example below, it is not hard to construct factorized densities $f_N(0)$ which satisfy both

$$\int_{\mathbb{R}^{2d}} |v|^2 f_N^{(1)}(0,x,v)dxdv = \int_{\mathbb{R}^{2d}} |v|^2 f_0(x,v)dxdv \tag{30}$$

and

$$\int_{\mathbb{R}^{2d}} v f_N^{(1)}(0,x,v)dxdv = \int_{\mathbb{R}^{2d}} v f_0(x,v)dxdv \tag{31}$$

Therefore the conserved moments do not *always* place a lower bound on the convergence rate in Lanford's theorem.

EXAMPLE 6.1. Fix a large number K. Let $\mathcal{J}$ be a (possibly uncountable) index set and let μ be a probability measure on on $\mathcal{J}$. For μ-a.e. $\alpha \in \mathcal{J}$, measurably with respect to α, we pick a non-negative measurable function $\rho_\alpha(x)$ with $\int \rho_\alpha(x)dx = 1$ and $\|\rho_\alpha\|_\infty \leq K$, and a non-negative measurable function $m_\alpha(v)$ with $\int m_\alpha(v)dv = 1$, $\int v m_\alpha(v)dv = 0$ and $\int |v|^2 m_\alpha(v)dv = 1$. Then if we write

$$f_0(x,v) = \int_{\mathcal{J}} d\mu(\alpha)\rho_\alpha(x)m_\alpha(v) \tag{32}$$

and let $f_N(0) \sim f_0^{\otimes N}$ as in Section 5, then we have both (30) and (31).

The second observation is that (under natural decay assumptions) we can place lower bounds on the L^∞ error, up to powers of $\log \frac{1}{\varepsilon}$, by placing lower bounds on a weighted L^1 error. To see why, observe that if both $f_N^{(1)}(t)$ and $f(t)$ have Maxwellian tails jointly in x and v (as in the work of [**15**,**16**]) then there exists a number $k = k(t) > 0$ (depending on f) such that the following estimate holds for all $C > 0$:

$$\begin{aligned} &\int_{\mathbb{R}^{2d}} \left(|x|^2 + |v|^2\right) \left| f_N^{(1)}(t) - f(t) \right| dx dv \lesssim \\ (33) \qquad &\lesssim \left(C \log \frac{1}{\varepsilon} \right)^{d+1} \left\| f_N^{(1)}(t,x,v) - f(t,x,v) \right\|_{L^\infty_{x,v}} + \varepsilon^{kC} \end{aligned}$$

Hence we will not actually mention the $L^\infty_{x,v}$ norm in stating our main result.

Finally, we remark that for any solution $f(t)$ of Boltzmann's equation having enough regularity and decay in x and v, there holds

$$\frac{d}{dt} \int_{\mathbb{R}^{2d}} \left(x \cdot v - |v|^2 t \right) f(t,x,v) dx dv = 0 \tag{34}$$

We will not actually use any regularity properties of Boltzmann's equation aside from (34). Therefore, instead of trying to find optimal conditions which guarantee (34), we will simply include (34) as a hypothesis in the theorem.

THEOREM 6.1. *Let $f_N(t, Z_N)$ be a non-negative solution of the Liouville equation for N identical hard spheres of diameter $\varepsilon > 0$, for each N in the Boltzmann-Grad scaling $N\varepsilon^{d-1} = 1$. We suppose that f_N is symmetric under particle interchange, that f_N is a probability density, and that the following moment estimate holds:*

$$\sup_N \frac{1}{N} \int_{\mathcal{D}_N} \sum_{i=1}^N \left(|x_i|^2 + |v_i|^2 \right) f_N(0, Z_N) dZ_N < \infty \tag{35}$$

Additionally, assume that $f(t)$ is a solution of the Boltzmann equation which satisfies

$$\forall t \in [0,T], \quad \int_{\mathbb{R}^{2d}} \left(x \cdot v - |v|^2 t \right) f(t,x,v) dx dv = \int_{\mathbb{R}^{2d}} x \cdot v f(0,x,v) dx dv \tag{36}$$

Assume, moreover, that $f(t)$ does not satisfy the free transport equation on any open subinterval of $[0,T]$. Finally, assume that

$$\lim_{N\to\infty} f_N^{(1)}(t) = f(t) \tag{37}$$

holds in the sense of distributions for each $t \in [0,T]$. Then for any T_1, T_2 with $0 \leq T_1 < T_2 \leq T$ we have

$$\liminf_{N\to\infty} \varepsilon^{-1} \sup_{t\in[T_1,T_2]} \int_{\mathbb{R}^{2d}} \left(|x|^2 + |v|^2 \right) \left| f_N^{(1)}(t) - f(t) \right| dx dv > 0 \tag{38}$$

REMARK. The requirement that f is not a solution of free transport is a technical condition; it excludes local Maxwellian functions, e.g. $f(t) = ce^{-|x-vt|^2} e^{-|v|^2}$. It is expected that optimal $\mathcal{O}(\varepsilon)$ convergence rates hold for such solutions, but this is not a part of the theorem.

REMARK. The conditions of Theorem 6.1 are satisfied if $f_0 : \mathbb{R}^d \times \mathbb{R}^d \to \mathbb{R}$ is a smooth, compactly supported, non-negative probability density function, with associated solution $f(t)$ of Boltzmann's equation; and, $f_N(0) \sim f_0^{\otimes N}$ as in Section 5, and T is the small time appearing in the original theorem of Lanford. [**17**]

REMARK. The supremum over $[T_1, T_2]$ in (38) can actually be replaced by a supremum over the two-point set $\{T_1, T_2\}$ (the proof is the same).

PROOF. We will assume that (38) fails for some T_1, T_2 in order to reach a contradiction. We have the following virial identity (see the proof of Proposition 1.3.5 in [**12**], or average out (7) in the manner of Section 4):

$$\begin{aligned}\int_{\mathbb{R}^{2d}} \left(x\cdot v - |v|^2 T_2\right) f_N^{(1)}(T_2) dx dv - \int_{\mathbb{R}^{2d}} \left(x\cdot v - |v|^2 T_1\right) f_N^{(1)}(T_1) dx dv \\ = C_d \frac{N-1}{2}\varepsilon^d \int_{T_1}^{T_2}\int_{\mathbb{R}^d\times\mathbb{R}^d\times\mathbb{R}^d\times\mathbb{S}^{d-1}} |\omega\cdot(v_2-v_1)|^2 \times \\ \times f_N^{(2)}(t,x_1,v_1,x_1+\varepsilon\omega,v_2) d\omega dx_1 dv_1 dv_2 dt\end{aligned} \tag{39}$$

Therefore, using the Boltzmann-Grad scaling $N\varepsilon^{d-1}=1$ and (36), we obtain:

$$\begin{aligned}\int_{T_1}^{T_2}\int_{\mathbb{R}^d\times\mathbb{R}^d\times\mathbb{R}^d\times\mathbb{S}^{d-1}} |\omega\cdot(v_2-v_1)|^2 f_N^{(2)}(t,x_1,v_1,x_1+\varepsilon\omega,v_2) d\omega dx_1 dv_1 dv_2 dt \\ \leq C_{d,T}\varepsilon^{-1} \sup_{t\in[T_1,T_2]} \int_{\mathbb{R}^{2d}} \left(|x|^2+|v|^2\right)\left|f_N^{(1)}(t,x,v) - f(t,x,v)\right| dx dv\end{aligned} \tag{40}$$

Here we have used the triangle inequality and the fact that (36) holds with $t=T_1$ and $t=T_2$.

Hence if (38) is not true then there exists a subsequence N' (depending on T_1, T_2) such that

$$\lim_{N'} \int_{T_1}^{T_2}\int |\omega\cdot(v_2-v_1)|^2 f_{N'}^{(2)}(t,x_1,v_1,x_1+\varepsilon\omega,v_2) d\omega dx_1 dv_1 dv_2 dt = 0 \tag{41}$$

The remainder of the proof consists in showing that (41) combined with (37) implies that $f(t)$ satisfies the free transport equation on $[T_1, T_2]$.

Define the transport semigroup, which acts on a function $g(x,v)$, by the formula

$$\left[\mathcal{T}(\tau)g\right](x,v) = g(x-\tau v, v) \tag{42}$$

Also define the (lowest order) BBGKY collision operator

$$C_2 f_N^{(2)}(t,x_1,v_1) = \int_{\mathbb{R}^d\times\mathbb{S}^{d-1}} \omega\cdot(v_2-v_1) f_N^{(2)}(t,x_1,v_1,x_1+\varepsilon\omega,v_2) d\omega dv_2 \tag{43}$$

Then we have (see [**13**], note that there is a factor of $(N-1)\varepsilon^{d-1}$ which we ignore in view of the Boltzmann-Grad scaling):

$$\left(\partial_t + v\cdot\nabla_x\right) f_N^{(1)}(t) = C_2 f_N^{(2)}(t) \tag{44}$$

Therefore for $t\in[T_1,T_2]$ we have the Duhamel representation

$$f_N^{(1)}(t) = \mathcal{T}(t-T_1) f_N^{(1)}(T_1) + \int_{T_1}^t \mathcal{T}(t-\tau) C_2 f_N^{(2)}(\tau) d\tau \tag{45}$$

Let us employ the usual L^2 inner product

$$\langle\varphi, f\rangle = \int_{\mathbb{R}^{2d}} \varphi(x,v) f(x,v) dx dv \tag{46}$$

Then for any smooth compactly supported function φ and any $t\in[T_1,T_2]$ we have

$$\left\langle\varphi, f_N^{(1)}(t) - \mathcal{T}(t-T_1) f_N^{(1)}(T_1)\right\rangle = \int_{T_1}^t \left\langle\varphi, \mathcal{T}(t-\tau) C_2 f_N^{(2)}(\tau)\right\rangle d\tau \tag{47}$$

Now by (37) and the assumption that $f(t)$ does not solve the free transport equation on $[T_1, T_2]$, in order to reach a contradiction it suffices to show:

$$\lim_{N'} \int_{T_1}^t \left\langle \varphi, \mathcal{T}(t-\tau) C_2 f_N^{(2)}(\tau) \right\rangle d\tau = 0 \tag{48}$$

Using duality we have

$$\begin{aligned} \int_{T_1}^t \left\langle \varphi, \mathcal{T}(t-\tau) C_2 f_N^{(2)}(\tau) \right\rangle d\tau &= \int_{T_1}^t \left\langle \mathcal{T}(-(t-\tau))\varphi, C_2 f_N^{(2)}(\tau) \right\rangle d\tau \\ &= \int_{T_1}^t \left\langle \Phi(\tau), C_2 f_N^{(2)}(\tau) \right\rangle d\tau \end{aligned} \tag{49}$$

Here we have defined (considering t fixed)

$$\Phi(\tau) = \mathcal{T}(-(t-\tau))\varphi \tag{50}$$

Using the definition of the BBGKY collision operator, we have

$$\begin{aligned} &\int_{T_1}^t \left\langle \Phi(\tau), C_2 f_N^{(2)}(\tau) \right\rangle d\tau = \\ &\int_{T_1}^t \int d\tau d\omega dx_1 dv_1 dv_2 \, \omega \cdot (v_2 - v_1) \Phi(\tau, x_1, v_1) f_N^{(2)}(\tau, x_1, v_1, x_1 + \varepsilon\omega, v_2) \end{aligned} \tag{51}$$

We recall the collisional change of variables

$$\begin{aligned} v_1^* &= v_1 + \omega\omega \cdot (v_2 - v_1) \\ v_2^* &= v_2 - \omega\omega \cdot (v_2 - v_1) \end{aligned} \tag{52}$$

as well as the boundary condition

$$f_N^{(2)}(\tau, x_1, v_1^*, x_1 + \varepsilon\omega, v_2^*) = f_N^{(2)}(\tau, x_1, v_1, x_1 + \varepsilon\omega, v_2) \tag{53}$$

hence what we need to control is the *difference*

$$\Phi(\tau, x_1, v_1) - \Phi(\tau, x_1, v_1^*) \tag{54}$$

Of course for $T_1 \leq \tau \leq t$ we have

$$\begin{aligned} |\Phi(\tau, x_1, v_1) - \Phi(\tau, x_1, v_1^*)| &\leq \|\nabla_v \Phi(\tau)\|_{L^\infty_{x,v}} |v_1 - v_1^*| \\ &= \|\nabla_v \Phi(\tau)\|_{L^\infty_{x,v}} |\omega \cdot (v_2 - v_1)| \\ &\leq C_T \|\nabla_{x,v}\varphi\|_\infty |\omega \cdot (v_2 - v_1)| \end{aligned} \tag{55}$$

(This trick is inspired by related work due to Cercignani using the Bony functional. [**7**–**9**])

Hence the right hand side of (51) is controlled by the following quantity:

$$C_T \|\nabla_{x,v}\varphi\|_\infty \int_{T_1}^t \int d\tau d\omega dx_1 dv_1 dv_2 \, |\omega \cdot (v_2 - v_1)|^2 \, f_N^{(2)}(\tau, x_1, v_1, x_1 + \varepsilon\omega, v_2) \tag{56}$$

which is going to zero (along the subsequence N') due to (41). □

7. Acknowledgments

This paper is largely based on work completed for the author's dissertation at New York University. The partial manuscript was completed under a postdoctoral fellowship at the University of Texas at Austin, for which I am most appreciative. I would like to thank my PhD advisor, Nader Masmoudi, as well as Pierre Germain, for their advice and comments. I would also like to thank Nataša Pavlović for reading an early version of this manuscript and providing insightful feedback. Additionally, I wish to indicate my appreciation to the anonymous referee(s) for helpful comments following careful reading of the present version. Finally I would like to thank the organizers of this special session of JMM 2017 for the invitation, without which this manuscript would most likely never have reached its current state of completion.

References

[1] R. K. Alexander, *The infinite hard-sphere system*, ProQuest LLC, Ann Arbor, MI, 1975. Thesis (Ph.D.)–University of California, Berkeley. MR2625918

[2] A. Biryuk, W. Craig, and V. Panferov, *Strong solutions of the Boltzmann equation in one spatial dimension* (English, with English and French summaries), C. R. Math. Acad. Sci. Paris **342** (2006), no. 11, 843–848, DOI 10.1016/j.crma.2006.04.005. MR2224633

[3] T. Bodineau, I. Gallagher, and L. Saint-Raymond, *The Brownian motion as the limit of a deterministic system of hard-spheres*, Invent. Math. **203** (2016), no. 2, 493–553, DOI 10.1007/s00222-015-0593-9. MR3455156

[4] T. Bodineau, I. Gallagher, and L. Saint-Raymond, *From hard sphere dynamics to the Stokes-Fourier equations: an L^2 analysis of the Boltzmann-Grad limit*, Ann. PDE **3** (2017), no. 1, Art. 2, 118, DOI 10.1007/s40818-016-0018-0. MR3625187

[5] C. Boldrighini, R. L. Dobrushin, and Yu. M. Sukhov, *One-dimensional hard rod caricature of hydrodynamics*, J. Statist. Phys. **31** (1983), no. 3, 577–616, DOI 10.1007/BF01019499. MR711490

[6] D. Burago, S. Ferleger, and A. Kononenko, *Uniform estimates on the number of collisions in semi-dispersing billiards*, Ann. of Math. (2) **147** (1998), no. 3, 695–708, DOI 10.2307/120962. MR1637663

[7] C. Cercignani, *A remarkable estimate for the solutions of the Boltzmann equation*, Appl. Math. Lett. **5** (1992), no. 5, 59–62, DOI 10.1016/0893-9659(92)90065-H. MR1345903

[8] C. Cercignani, *Weak solutions of the Boltzmann equation and energy conservation*, Appl. Math. Lett. **8** (1995), no. 2, 53–59, DOI 10.1016/0893-9659(95)00011-E. MR1357252

[9] C. Cercignani, *Global weak solutions of the Boltzmann equation*, J. Stat. Phys. **118** (2005), no. 1-2, 333–342, DOI 10.1007/s10955-004-8786-4. MR2122558

[10] C. Cercignani, V. I. Gerasimenko, and D. Ya. Petrina, *Many-particle dynamics and kinetic equations*, Mathematics and its Applications, vol. 420, Kluwer Academic Publishers Group, Dordrecht, 1997. Translated from the Russian manuscript by K. Petrina and V. Gredzhuk. MR1472233

[11] C. Cercignani, R. Illner, and M. Pulvirenti, *The mathematical theory of dilute gases*, Applied Mathematical Sciences, vol. 106, Springer-Verlag, New York, 1994. MR1307620

[12] R. Denlinger, *The propagation of chaos for a rarefied gas of hard spheres in vacuum*, ProQuest LLC, Ann Arbor, MI, 2016. Thesis (Ph.D.)–New York University. MR3563990

[13] I. Gallagher, L. Saint-Raymond, and B. Texier, *From Newton to Boltzmann: hard spheres and short-range potentials*, Zurich Lectures in Advanced Mathematics, European Mathematical Society (EMS), Zürich, 2013. MR3157048

[14] R. Illner, *On the number of collisions in a hard sphere particle system in all space*, Transport Theory Statist. Phys. **18** (1989), no. 1, 71–86, DOI 10.1080/00411458908214499. MR1006667

[15] R. Illner and M. Pulvirenti, *Global validity of the Boltzmann equation for a two-dimensional rare gas in vacuum*, Comm. Math. Phys. **105** (1986), no. 2, 189–203. MR849204

[16] R. Illner and M. Pulvirenti, *Global validity of the Boltzmann equation for two- and three-dimensional rare gas in vacuum. Erratum and improved result: "Global validity of the Boltzmann equation for a two-dimensional rare gas in vacuum" [Comm. Math. Phys.* **105** *(1986), no. 2, 189–203; MR0849204 (88d:82061)] and "Global validity of the Boltzmann equation for a three-dimensional rare gas in vacuum" [ibid.* **113** *(1987), no. 1, 79–85; MR0918406 (89b:82052)] by Pulvirenti*, Comm. Math. Phys. **121** (1989), no. 1, 143–146. MR985619

[17] O. E. Lanford III, *Time evolution of large classical systems*, Dynamical systems, theory and applications (Rencontres, Battelle Res. Inst., Seattle, Wash., 1974), Springer, Berlin, 1975, pp. 1–111. Lecture Notes in Phys., Vol. 38. MR0479206

[18] J. L. Lebowitz and H. Spohn, *Steady state self-diffusion at low density*, J. Statist. Phys. **29** (1982), no. 1, 39–55, DOI 10.1007/BF01008247. MR676928

[19] S. Mischler and C. Mouhot, *Kac's program in kinetic theory*, Invent. Math. **193** (2013), no. 1, 1–147, DOI 10.1007/s00222-012-0422-3. MR3069113

[20] S. Olla, S. R. S. Varadhan, and H.-T. Yau, *Hydrodynamical limit for a Hamiltonian system with weak noise*, Comm. Math. Phys. **155** (1993), no. 3, 523–560. MR1231642

[21] M. Pulvirenti, C. Saffirio, and S. Simonella, *On the validity of the Boltzmann equation for short range potentials*, Rev. Math. Phys. **26** (2014), no. 2, 1450001, 64, DOI 10.1142/S0129055X14500019. MR3190204

[22] M. Pulvirenti and S. Simonella, *On the evolution of the empirical measure for the hard-sphere dynamics*, Bull. Inst. Math. Acad. Sin. (N.S.) **10** (2015), no. 2, 171–204. MR3409818

[23] L. Saint-Raymond, *Hydrodynamic limits of the Boltzmann equation*, Lecture Notes in Mathematics, vol. 1971, Springer-Verlag, Berlin, 2009. MR2683475

[24] S. Simonella, *Evolution of correlation functions in the hard sphere dynamics*, J. Stat. Phys. **155** (2014), no. 6, 1191–1221, DOI 10.1007/s10955-013-0905-7. MR3207735

[25] T. Tao, *Nonlinear dispersive equations*, CBMS Regional Conference Series in Mathematics, vol. 106, Published for the Conference Board of the Mathematical Sciences, Washington, DC; by the American Mathematical Society, Providence, RI, 2006. Local and global analysis. MR2233925

[26] H. van Beijeren, O. E. Lanford III, J. L. Lebowitz, and H. Spohn, *Equilibrium time correlation functions in the low-density limit*, J. Statist. Phys. **22** (1980), no. 2, 237–257, DOI 10.1007/BF01008050. MR560556

[27] L. N. Vaserstein, *On systems of particles with finite-range and/or repulsive interactions*, Comm. Math. Phys. **69** (1979), no. 1, 31–56. MR547525

Department of Mathematics, University of Texas at Austin, Austin, Texas 78712

Contemporary Mathematics
Volume **725**, 2019
https://doi.org/10.1090/conm/725/14558

The almost global existence to classical solution for a 3-D wave equation of nematic liquid-crystals

Yi Du, Geng Chen, and Jianli Liu

ABSTRACT. In this paper we study the lifespan of classical solution to Cauchy problem for a nonlinear wave equation, modeling the nematic liquid-crystals in three space dimensions. The almost global existence to classical solution with small initial data will also be presented.

1. Introduction

1.1. Physical background. Liquid crystals are a state of matter that have properties between those of a conventional liquid and those of a solid crystal that are optically anisotropic, even when they are at rest. In this paper, we shall study a variational wave system stem from liquid crystal model.

At the beginning, we introduce the physical derivation by the least action principle for the variational wave system. The mean orientation of the long molecules in a nematic liquid crystal is described by a unit vectors $\mathbf{n} = \mathbf{n}(t, x) \in \mathbb{S}^2$, the unit sphere, where $(t, x) \in \mathbb{R}^+ \times \mathbb{R}^3$, which could be modeled by below Euler-Largrangian equations derived from the least action principle [**1, 30**],

$$\kappa \mathbf{n}_{tt} + \mu \mathbf{n}_t + \frac{\delta \int W(\mathbf{n}, \nabla \mathbf{n})}{\delta \mathbf{n}} = \lambda \mathbf{n}, \qquad \mathbf{n} \cdot \mathbf{n} = 1, \tag{1.1}$$

where the well-known Oseen-Franck potential energy density W is given by

$$W(\mathbf{n}, \nabla \mathbf{n}) = \tfrac{1}{2}\alpha(\nabla \cdot \mathbf{n})^2 + \tfrac{1}{2}\beta(\mathbf{n} \cdot \nabla \times \mathbf{n})^2 + \tfrac{1}{2}\gamma\,|\mathbf{n} \times (\nabla \times \mathbf{n})|^2. \tag{1.2}$$

Here the positive constants α, β and γ are elastic constants of the liquid crystal, corresponding to splay, twist, and bend, respectively. The Lagrangian multiplier λ is determined by the constraint $\mathbf{n} \cdot \mathbf{n} = 1$. The inertia and viscous coefficients κ and μ are two non-negative constants, respectively.

There are many studies on the constrained elliptic system of equations for $\mathbf{n}$, and on the parabolic flow associated with it, where we refer the reader to see [**2, 12, 14, 16, 19, 33**]. In particular, the one-constant elliptic model leads to the equation for harmonic maps taking values in the two-sphere [**12, 14**]. One parabolic

2010 *Mathematics Subject Classification.* Primary 35Q35, 35B44; Secondary 35E15, 35B65.
Key words and phrases. Variational wave equation, liquid-crystal, almost global existence.
The first author was supported in part by NSFC Grant #11471126.
The second author was supported in part by NSFC Grant #11471126 and NSF DMS-1715012.
The corresponding author Jianli Liu was supported in part by NSFC Grant #11401367.

system of equations for $\mathbf{n}$ coupling the compressible Navier-Stokes equations in one space dimension (1-D) has been considered by [**13**].

However, for the complete hyperbolic system (1.1), only the 1-D case has been systematically studied. The main difficulty is the possible gradient blowup. For the 1-D Cauchy problem of (1.1) with $\mu = 0$ and $\kappa = 1$, which describes the model with viscous effects omitted, the global well-posedness of weak solutions have been intensively studied, recently. First, for this case, one example with smooth initial data and singularity formation (gradient blowup) in finite time has been provided in [**15**], and then a lot of results for global weak solutions have been provided: Existence, [**7, 9, 18, 34–37**], Uniqueness [**6, 8**], Lipschitz continuity [**3**] and Generic regularity [**4, 8**]. Furthermore, in [**10**], the singularity formation (gradient blowup) and global weak existence for the 1-D Cauchy problem for the complete system (1.1) (κ and μ are non-negative constants) with $\alpha = \gamma$ have been established. This result shows that for 1-D Cauchy problem we should essentially expect the similar existence and regularity results for the complete system (1.1) and the extreme case with $\mu = 0$ and $\kappa = 1$.

However, there are only very few studies on the multi-d solutions of (1.1). One paper in this approach written by Li, Witt & Yin [**23**] provides the singularity formation and some analysis on the life-span of the classical solution for a 2-D axisymmetric model. Now, to the limit of our knowledge, the well-posedness of solutions in 2-D and 3-D for (1.1) is still wide open. In this paper, we consider the life-span of the classical solution for the 3-D Cauchy problem of (1.1) with small initial data. Note the initial data in the 1-D singularity formation examples in [**10, 15**] have the large C^1-norm.

1.2. The variational wave equation. In this paper, we focus on the Cauchy problem of system (1.1) in the three dimensions with $\mu = 0$ and $\kappa = 1$ and planar deformation:

$$\frac{\delta}{\delta \mathbf{n}} \int \Big\{ \frac{1}{2} |\partial_t \mathbf{n}|^2 - W(\mathbf{n}, \nabla \mathbf{n}) \Big\}\, dx\, dt = 0, \quad |\mathbf{n}| = 1, \tag{1.3}$$

where the planar deformation means that

$$\mathbf{n} = (n_1, n_2, n_3) = (\cos u(t,x), \sin u(t,x), 0) \tag{1.4}$$

with the angle $u = u(t,x)$, $(t,x) \in \mathbb{R}^+ \times \mathbb{R}^3$.

Now, the variation (1.3) can be reformulated as

$$\frac{\delta}{\delta \mathbf{n}} \int \Big\{ \frac{1}{2} |\mathbf{n}_t|^2 - W(\mathbf{n}, \nabla \mathbf{n}) \Big\}\, dx\, dt = 0, \tag{1.5}$$

which gives rise to the Euler-Lagrange equation

$$\partial_{tt} n_i + \partial_{n_i} W(\mathbf{n}, \nabla \mathbf{n}) - \sum_{j=1}^{3} \partial_{x_j} \big(\partial_{\partial_j n_i} W(\mathbf{n}, \nabla \mathbf{n}) \big) = 0. \tag{1.6}$$

Multiplying three equations in (1.6) by $-\sin u$ and $\cos u$, then adding them up, we can get a variational wave equation, by (1.4) and (1.2),

$$u_{tt} - c_1^2(u)\, \partial_{x_1}^2 u - c_2^2(u)\, \partial_{x_2}^2 u - \beta\, \partial_{x_3}^2 u + (\alpha - \gamma) \sin 2u\, \partial_{x_1} \partial_{x_2} u = F(u, \partial u), \tag{1.7}$$

where

$$c_1^2(u) = \alpha \sin^2 u + \gamma \cos^2 u, \quad c_2^2(u) = \alpha \cos^2 u + \gamma \sin^2 u,$$

Moreover, there exists a universal constant $C_2 > 0$ such that if $\|\varepsilon(s)\|_2 \le \alpha$ for all $s \ge 0$, where $\alpha < \alpha_1$, then

$$\left|\frac{\lambda_s}{\lambda}\right| + \left|\frac{x_s}{\lambda} - 1\right| \le C_2\|\varepsilon(s)\|_2. \tag{2.10}$$

PROOF. See [**11**, Lemmas 4 and 12]. □

We are now ready to discuss the approach of monotonicity.

3. Monotonicity

As it was remarked in [**16**], the solution $u(x,t)$ has the mass around its center $x(t)$ versus the mass to the right of the soliton being a time decreasing function. The solution to the right of the soliton is small in L^2 sense. To demonstrate that, we first introduce the L^2 localization functional I, see (3.1), and then show the exponential decay of L^2 norm to the right of the soliton. For $M \ge 4$, denote

$$\psi(x) = \frac{2}{\pi}\arctan\left(e^{\frac{x}{M}}\right).$$

It is easy to check the following properties:

(1) $\psi(0) = \dfrac{1}{2}$,

(2) $\lim_{x\to-\infty}\psi(x) = 0$ and $\lim_{x\to+\infty}\psi(x) = 1$,

(3) $\psi'(x) = \left(\pi M\cosh\left(\dfrac{x}{M}\right)\right)^{-1}$,

(4) $|\psi'''(x)| \le \dfrac{1}{M^2}\psi'(x) \le \dfrac{1}{16}\psi'(x)$.

Let $x(t) \in C^1(\mathbb{R},\mathbb{R})$, and for $x_0, t_0 > 0$ and $t \in [0,t_0]$ define the functional

$$I_{x_0,t_0}(t) = \int u^2(t,x)\,\psi\left(x - x(t_0) + \frac{1}{2}(t_0 - t) - x_0\right)dx, \tag{3.1}$$

where $u \in C(\mathbb{R}, H^1(\mathbb{R}))$ is a solution of the critical gKdV equation (1.1), satisfying

$$\|u(\cdot + x(t)) - Q\|_{H^1} \le \alpha \quad \text{for some} \quad \alpha > 0. \tag{3.2}$$

While the next property is basically known, we provide it here for completeness.

LEMMA 3.1 (Almost Monotonicity). *Let $M \ge 4$ be fixed and assume that $x(t)$ is an increasing function satisfying $x(t_0) - x(t) \ge \frac{3}{4}(t_0 - t)$ for every $t_0, t \ge 0$ with $t \in [0,t_0]$. Then there exist $\alpha_0 > 0$ and $\theta = \theta(M) > 0$ such that if $u \in C(\mathbb{R}, H^1(\mathbb{R}))$ verify (3.2) with $\alpha < \alpha_0$, then for all $x_0 > 0$ and $t_0, t \ge 0$ with $t \in [0,t_0]$, we have*

$$I_{x_0,t_0}(t_0) - I_{x_0,t_0}(t) \le \theta e^{-\frac{x_0}{M}}.$$

PROOF. Using the equation and the fact that $|\psi'''(x)| \le \frac{1}{M^2}\psi'(x) \le \frac{1}{16}\psi'(x)$, we obtain the following bound on the derivative of the functional $I_{x_0,t_0}(t)$:

$$\begin{aligned}
\frac{d}{dt}I_{x_0,t_0}(t) \le& 2\int uu_t\psi - \frac{1}{2}\int u^2\psi' \\
\le& -\int\left(3u_x^2 - \frac{5}{3}u^6\right)\psi' + \int u^2\psi''' - \frac{1}{2}\int u^2\psi' \\
\le& -\int\left(3u_x^2 + \frac{1}{4}u^2\right)\psi' + \frac{5}{3}\int u^6\psi'.
\end{aligned} \tag{3.3}$$

Estimating the last term on the right hand side of the last inequality, we view it in terms of closeness to Q and write

$$\int u^6\psi' = \int Q(\cdot - x(t))u^5\psi' + \int (u - Q(\cdot - x(t)))u^5\psi'. \tag{3.4}$$

Application of the Sobolev Embedding $H^1(\mathbb{R}) \hookrightarrow L^\infty(\mathbb{R})$ in the second term above yields

$$\begin{aligned}\int (u - Q(\cdot - x(t)))u^5\psi' \leq & \|u - Q(\cdot - x(t))\|_\infty \|u\|_\infty^3 \int u^2\psi' \\ \leq & c\alpha \|Q\|_{H^1}^3 \int u^2\psi'. \end{aligned} \tag{3.5}$$

For the first term on the right hand side of (3.4), we divide the integration into two regions $|x - x(t)| \geq R_0$ and $|x - x(t)| < R_0$, where R_0 is a positive number to be chosen later. Therefore, since $Q(x) \leq ce^{-|x|}$, we get

$$\begin{aligned}\int_{|x-x(t)|\geq R_0} Q(\cdot - x(t))u^5\psi' \leq & c\,e^{-R_0}\|u\|_\infty^3 \int u^2\psi' \\ \leq & c\,e^{-R_0}\|Q\|_{H^1}^3 \int u^2\psi'.\end{aligned}$$

When $|x - x(t)| < R_0$, we estimate

$$\begin{aligned}\left| x - x(t_0) + \frac{1}{2}(t_0 - t) - x_0 \right| \geq & (x(t_0) - x(t) + x_0) - \frac{1}{2}(t_0 - t) - |x - x(t)| \\ \geq & \frac{1}{4}(t_0 - t) + x_0 - R_0,\end{aligned}$$

where in the first inequality we used that $x(t)$ is increasing, $t_0 \geq t$ and $x_0 > 0$, to compute the modulus of the first term and in the second line we used the assumption $x(t_0) - x(t) \geq \frac{3}{4}(t_0 - t)$.

Now, since $\psi'(z) \leq c\,e^{-\frac{|z|}{M}}$, we deduce that

$$\begin{aligned}\int_{|x-x(t)|\leq R_0} Q(\cdot - x(t))u^5\psi' \leq & \|u\|_\infty^5 \|Q\|_1 e^{\frac{R_0}{M}} e^{-\frac{\left(\frac{1}{4}(t_0-t)+x_0\right)}{M}} \\ \leq & c\,\|Q\|_{H^1}^5 \|Q\|_1 e^{\frac{R_0}{M}} e^{-\frac{\left(\frac{1}{4}(t_0-t)+x_0\right)}{M}}.\end{aligned} \tag{3.6}$$

Therefore, choosing α such that $c\,\alpha\|Q\|_{H^1}^3 < \frac{3}{5}\cdot\frac{1}{4}$ and R_0 such that $c\,\|Q\|_{H^1}^3 e^{-R_0} < \frac{3}{5}\cdot\frac{1}{4}$, collecting (3.5)-(3.6), we obtain

$$\frac{5}{3}\int u^6\psi' \leq \frac{1}{8}\int u^2\psi' + c\,\|Q\|_{H^1}^5 \|Q\|_1 e^{\frac{R_0}{M}} e^{-\frac{\left(\frac{1}{4}(t_0-t)+x_0\right)}{M}}.$$

Inserting the previous estimate into (3.3), we get that there exists $C > 0$ such that

$$\begin{aligned}\frac{d}{dt} I_{x_0,t_0}(t) \leq & -\int \left(3u_x^2 + \frac{1}{8}u^2\right)\psi' + c\,e^{-\frac{x_0}{M}} \cdot e^{-\frac{1}{4M}(t_0-t)} \\ \leq & c\,e^{-\frac{x_0}{M}} \cdot e^{-\frac{1}{4M}(t_0-t)}.\end{aligned}$$

Finally, integrating on $[t, t_0]$, we obtain the desired inequality for some $\theta = \theta(M) > 0$. □

We now show the exponential decay of the L^2 norm to the right of the soliton.

LEMMA 3.2. *Let $x(t)$ satisfy the assumptions of Lemma 3.1 and assume that $x(t) \geq \frac{1}{2}t$ for all $t \geq 0$. Let $u \in C(\mathbb{R}, H^1(\mathbb{R}))$ be a solution of the gKdV equation (1.1) satisfying (3.2) with $\alpha < \alpha_0$ (with α_0 given in Lemma 3.1) and with initial data u_0 verifying $|u_0(x)| \leq c\,e^{-\delta|x|}$ for some $c > 0$ and $\delta > 0$. Fix $M \geq \max\{4, \frac{1}{\delta}\}$. Then there exists $C = C(M, \delta) > 0$ such that for all $t \geq 0$ and $x_0 > 0$*

$$\int_{x>x_0} u^2(t, x + x(t))\,dx \leq C\,e^{-\frac{x_0}{M}}. \tag{3.7}$$

PROOF. From Lemma 3.1 with $t = 0$ and replacing t_0 by t, we deduce that for all $t \geq 0$

$$I_{x_0,t}(t) - I_{x_0,t}(0) \leq \theta\,e^{-\frac{x_0}{M}}.$$

This is equivalent to

$$\int u^2(t,x)\psi(x - x(t) - x_0)\,dx \leq \int u_0^2(x)\psi(x - x(t) + \frac{1}{2}t - x_0)\,dx + \theta\,e^{-\frac{x_0}{M}}.$$

On the other hand,

$$\begin{aligned}\int u^2(t,x)\psi(x - x(t) - x_0)\,dx &= \int u^2(t, x + x(t))\psi(x - x_0)\,dx \\ &\geq \frac{1}{2}\int_{x>x_0} u^2(t, x + x(t))\,dx,\end{aligned}$$

where in the last inequality we have used the fact that ψ is increasing and $\psi(0) = 1/2$.

Now, since $-x(t) + \frac{1}{2}t \leq 0$ and ψ is increasing, we get

$$\int u_0^2(x)\psi(x - x(t) + \frac{1}{2}t - x_0)\,dx \leq \int u_0^2(x)\psi(x - x_0)\,dx.$$

The assumptions $|u_0(x)| \leq c\,e^{-\delta|x|}$ and $\psi(x) \leq c\,e^{\frac{x}{M}}$ yield

$$\begin{aligned}\int u_0^2(x)\psi(x - x_0)\,dx &\leq c\int e^{-2\delta|x|}e^{\frac{x-x_0}{M}}\,dx \\ &\leq c\,e^{-\frac{x_0}{M}}\int e^{-\left(2\delta|x| - \frac{x}{M}\right)}\,dx \leq \bar{C}(M,\delta)\,e^{-\frac{x_0}{M}},\end{aligned}$$

where in the last inequality we used the fact that

$$2\delta - \frac{1}{M} \geq \delta \Longleftrightarrow M \geq \frac{1}{\delta}.$$

Collecting the above estimates, we obtain the inequality (3.7). □

4. Virial-type estimates

In this section, we define a quantity depending on the ε variable that plays an important role in our instability proof. Indeed, let $\varphi \in C_0^\infty(\mathbb{R})$ be a decreasing function with

$$\varphi(y) = \begin{cases} 1, & \text{if} \quad y \leq 1 \\ 0, & \text{if} \quad y \geq 2. \end{cases}$$

First, for $A \geq 1$ define

$$\varphi_A(y) = \varphi\left(\frac{y}{A}\right).$$

Then $\varphi_A(y) = 1$ for $y \leq A$ and $\varphi_A(y) = 0$ for $y \geq 2A$. Note that

$$(\varphi_A)'(y) = \frac{1}{A}\varphi'\left(\frac{y}{A}\right).$$

Next, define a function

$$F(y) = \int_{-\infty}^{y} \Lambda Q(z)dz. \tag{4.1}$$

From the properties of Q, see (1.4), there exist $c, \delta > 0$ such that

$$|F(y)| \leq c\int_{-\infty}^{y} e^{-\frac{\delta}{2}|z|}dz,$$

and thus,

$$|F(y)| \leq c\, e^{\frac{\delta}{2}y} \quad \text{for} \quad y < 0.$$

Moreover, F is a bounded function on all of $\mathbb{R}$.

Now, we define the functional

$$J_A(s) = \int_{\mathbb{R}} \varepsilon(s)F(y)\varphi_A(y)\, dy. \tag{4.2}$$

The nontruncated version of this virial-type functional appeared in [**11**, Section 3]. The strategy in [**11**] was to show that a virial-type functional is well-defined, bounded above, however, its derivative is bounded from below by a positive constant, which all together led to a contradiction for large times. We will use the same approach, however, in this proof we will avoid using pointwise decay estimates as it was done in [**11**]. Truncation helps in a straightforward way to get an upper bound on the functional. The technique of truncation is, of course, not new and has been previously used in other contexts, even in more intricate forms, for example, see [**15**] or [**14**]. The bound on the derivative from below turns out to be impossible to get only via truncation, and we use the monotonicity. On the other hand, if we remove the truncation, then it is possible to control virial-type functional only with monotonicity, which we show in the last section. We include both proofs to illustrate different approaches.

It is clear that $J_A(s)$ is well-defined if $\varepsilon(s) \in L^2(\mathbb{R})$ due to the truncation. Furthermore, J_A is upper bounded by a constant depending on A and $\|\varepsilon\|_2$. Indeed, from the definition of φ_A, we deduce that

$$\begin{aligned}
|J_A(s)| \leq & \int_{y\leq 0} |\varepsilon(s)F(y)|\, dy + \int_0^{2A} |\varepsilon(s)F(y)|\, dy \\
\leq & c\,\|\varepsilon(s)\|_2 \left(\int_{y\leq 0} e^{\delta y}dy\right)^{1/2} + c\,A^{1/2}\|F\|_\infty \left(\int_0^{2A} |\varepsilon(s)|^2\, dy\right)^{1/2} \\
\leq & c\,(1 + A^{1/2})\|\varepsilon(s)\|_2.
\end{aligned} \tag{4.3}$$

In the next lemma we compute the derivative $\frac{d}{ds}J_A(s)$.

LEMMA 4.1. *Suppose that $\varepsilon(s) \in H^1(\mathbb{R})$ for all $s \geq 0$ and $\|\varepsilon(s)\|_{H^1} \leq 1$. Then the function $s \mapsto J_A(s)$ is C^1 and*

$$\frac{d}{ds}J_A = -\frac{\lambda_s}{2\lambda}\left(J_A - \frac{1}{4}\left(\int Q\right)^2\right) + 2\left(1 - \frac{1}{4}\left(\frac{x_s}{\lambda} - 1\right)\right)\int \varepsilon Q + R(\varepsilon, A),$$

where there exists a universal constant $C_3 > 0$ such that for all $A \geq 1$ we have

$$
\begin{aligned}
|R(\varepsilon, A)| \leq & C_3 \Big(\|\varepsilon\|_2^2 + \|\varepsilon\|_2^2 \|\varepsilon\|_{H^1} + A^{-1/2} \|\varepsilon\|_{L^2(y \geq A)} \\
& + \left| \frac{x_s}{\lambda} - 1 \right| (A^{-1} + \|\varepsilon\|_2) + \left| \frac{\lambda_s}{\lambda} \right| (A^{-1} + A^{1/2} \|\varepsilon\|_{L^2(y \geq A)}) \Big) . \qquad (4.4)
\end{aligned}
$$

PROOF. First, from Lemma 2.1, we have,

$$
\begin{aligned}
\frac{d}{ds} J_A &= \int \varepsilon_s F \varphi_A \\
&= \int \left((L\varepsilon)_y + \frac{\lambda_s}{\lambda} \Lambda \varepsilon + \left(\frac{x_s}{\lambda} - 1 \right) \varepsilon_y \right) F \varphi_A \\
&+ \int \left(\frac{\lambda_s}{\lambda} \Lambda Q + \left(\frac{x_s}{\lambda} - 1 \right) Q_y \right) F \varphi_A \\
&- \int R(\varepsilon)_y F \varphi_A \\
&\equiv (I) + (II) + (III),
\end{aligned}
$$

where $R(\varepsilon)$ is given by relation (2.4).

We start with estimating the last term. Since $\|\varphi_A\|_\infty \leq 1$ and $\varphi_y \in L^\infty(\mathbb{R})$, integrating by parts, we obtain

$$
\begin{aligned}
(III) &= \int R(\varepsilon) \Lambda Q \varphi_A + \int R(\varepsilon) F \frac{1}{A} \varphi' \left(\frac{y}{A} \right) \\
&\leq \|\Lambda Q\|_\infty \int |R(\varepsilon)| + \frac{\|F\|_\infty \|\varphi'\|_\infty}{A} \int |R(\varepsilon)| \\
&\leq c_0 \left(\|\Lambda Q\|_\infty + \frac{\|F\|_\infty \|\varphi'\|_\infty}{A} \right) \left(\|\varepsilon\|_2^2 + \|\varepsilon\|_2^2 \|\varepsilon\|_{H^1} \right), \qquad (4.5)
\end{aligned}
$$

where in the last line we have used the Gagliardo-Nirenberg inequality and the fact that $\|\varepsilon(s)\|_\infty \leq \|\varepsilon(s)\|_{H^1} \leq 1$. Moreover, $c_0 > 0$ is independent of $A \geq 1$.

Dealing with the second term (II), we get

$$
\begin{aligned}
(II) &= \frac{\lambda_s}{\lambda} \int \Lambda Q F \varphi_A + \left(\frac{x_s}{\lambda} - 1 \right) \int Q_y F \varphi_A \\
&\equiv \frac{\lambda_s}{\lambda} (II.1) + \left(\frac{x_s}{\lambda} - 1 \right) (II.2),
\end{aligned}
$$

where, due to $F_y = \Lambda Q$, see (4.1), yields

$$
\begin{aligned}
(II.1) &= \frac{1}{2} \int (F^2)_y \varphi_A \\
&= \frac{1}{2} \int (F^2)_y + \frac{1}{2} \int (F^2)_y (\varphi_A - 1) \\
&= \frac{1}{2} \left(\int \Lambda Q \right)^2 + \frac{1}{2} \int (F^2)_y (\varphi_A - 1) \\
&\equiv \frac{1}{2} \left(\int \Lambda Q \right)^2 + R_1(A).
\end{aligned}
$$

Now, integration by parts gives

$$\int \Lambda Q dy = \frac{1}{2}\int Q dy + \int y Q_y dy = -\frac{1}{2}\int Q,$$

and therefore,

$$\frac{1}{2}\left(\int \Lambda Q\right)^2 = \frac{1}{8}\left(\int Q\right)^2 < +\infty.$$

Moreover, since $\|\varphi_A - 1\|_\infty \leq 2$ and

$$\text{supp}(\varphi_A - 1) \subset \{y \in \mathbb{R} : y \geq A\},$$

we have

$$\begin{aligned} |R_1(A)| \leq & \int_{y\geq A} |\Lambda Q F| \leq \|F\|_\infty \int_{y \geq A} |\Lambda Q| \frac{|y|}{|y|} \\ \leq & \frac{\|F\|_\infty \|y\Lambda Q\|_1}{A}. \end{aligned} \tag{4.6}$$

Note that $\|y\Lambda Q\|_1 \leq 3\,\|yQ\|_2 = const.$

On the other hand, since $\Lambda Q \perp Q$ (we are in the critical case)

$$\begin{aligned} (II.2) =& -\int Q\Lambda Q \varphi_A - \int QF\frac{1}{A}\varphi'\left(\frac{y}{A}\right) \\ =& -\int Q\Lambda Q - \int Q\Lambda Q(\varphi_A - 1) - \int QF\frac{1}{A}\varphi'\left(\frac{y}{A}\right) \\ \equiv& R_2(A), \end{aligned}$$

where, since $\|\varphi_A - 1\|_\infty \leq 2$, $\varphi' \in L^\infty$ and $\text{supp}(\varphi_A - 1) \subset \{y \in \mathbb{R} : y \geq A\}$, we obtain that

$$\begin{aligned} |R_2(A)| \leq & \int_{y\geq A} |Q\Lambda Q| + \|F\|_\infty \|\varphi'\|_\infty \frac{1}{A}\int |Q| \\ \leq & \int_{y \geq A} |Q\Lambda Q| \frac{|y|}{|y|} + \frac{\|F\|_\infty \|\varphi'\|_\infty \|Q\|_1}{A} \\ \leq & \frac{1}{A}\left(\|\Lambda Q\|_2 \|yQ\|_2 + \|F\|_\infty \|\varphi'\|_\infty \|Q\|_1\right) \equiv \frac{c}{A}, \end{aligned} \tag{4.7}$$

with the constant c independent of A and ε.

Next we estimate the term (I). Applying integration by parts, we get

$$\begin{aligned} (I) =& -\int (L\varepsilon)\Lambda Q \varphi_A - \int (L\varepsilon) F \frac{1}{A}\varphi'\left(\frac{y}{A}\right) \\ &+ \frac{\lambda_s}{\lambda}\int \Lambda \varepsilon F \varphi_A \\ &- \left(\frac{x_s}{\lambda} - 1\right)\left(\int \varepsilon \Lambda Q \varphi_A + \int \varepsilon F \frac{1}{A}\varphi'\left(\frac{y}{A}\right)\right) \\ \equiv& (I.1) + \frac{\lambda_s}{\lambda}(I.2) - \left(\frac{x_s}{\lambda} - 1\right)(I.3). \end{aligned}$$

Let us first consider the term $(I.3)$. Using the definition (2.2), we have

$$\begin{aligned}(I.3) =& \frac{1}{2}\int \varepsilon Q + \frac{1}{2}\int \varepsilon Q(\varphi_A - 1) + \int \varepsilon y Q_y \varphi_A + \int \varepsilon F \frac{1}{A}\varphi'\left(\frac{y}{A}\right) \\ \equiv& \frac{1}{2}\int \varepsilon Q + R_3(\varepsilon, A),\end{aligned}$$

where

$$\begin{aligned}|R_3(\varepsilon, A)| \leq& \int_{y\geq A} \varepsilon Q \frac{|y|}{|y|} + \|yQ_y\|_2\|\varepsilon\|_2 + \frac{\|F\|_\infty \|\varphi'\|_\infty}{A}\left(\int_{A\leq y\leq 2A} |\varepsilon| dy\right) \\ \leq& c\left(1 + \frac{1}{A} + \frac{1}{A^{1/2}}\right)\|\varepsilon\|_2 \end{aligned} \tag{4.8}$$

with the constant $c > 0$ independent of ε and A.

Next, we turn to the term $(I.2)$. Integration by parts yields

$$\begin{aligned}(I.2) =& \frac{1}{2}\int \varepsilon F \varphi_A + \int y \varepsilon_y F \varphi_A \\ =& \frac{1}{2}\int \varepsilon F \varphi_A - \int \varepsilon F \varphi_A - \int y \varepsilon \Lambda Q \varphi_A - \int y \varepsilon F \frac{1}{A}\varphi'\left(\frac{y}{A}\right) \\ \equiv& -\frac{1}{2} J_A + R_4(\varepsilon, A),\end{aligned}$$

where in the last line we used the definition (4.2). We estimate $R_4(\varepsilon, A)$. Indeed, it is clear that

$$\int y \varepsilon \Lambda Q \varphi_A \leq \|y\Lambda Q\|_2 \|\varepsilon\|_2.$$

Moreover,

$$\begin{aligned}\int y\varepsilon F \frac{1}{A}\varphi'\left(\frac{y}{A}\right) \leq& \frac{1}{A}\|F\|_\infty \|\varphi'\|_\infty \int_{A\leq |y|\leq 2A} |y\varepsilon| \\ \leq& \frac{1}{A}\|F\|_\infty \|\varphi'\|_\infty \|\varepsilon\|_{L^2(y\geq A)} \left(\int_{A\leq |y|\leq 2A} |y|^2\right)^{1/2} \\ \leq& 4A^{1/2}\|F\|_\infty \|\varphi'\|_\infty \|\varepsilon\|_{L^2(y\geq A)}.\end{aligned}$$

Collecting the last two estimates, we deduce

$$|R_4(\varepsilon, A)| \leq c(\|\varepsilon\|_2 + A^{1/2}\|\varepsilon\|_{L^2(y\geq A)}), \tag{4.9}$$

where $c > 0$ is again independent of ε and A.

To estimate $(I.1)$ we recall the definition of the operator L to deduce

$$\begin{aligned}L(fg) =& -(fg)_{xx} + fg - 5Q^4 fg \\ =& -f_{xx}g - 2f_x g_x - f g_{xx} + fg - 5Q^4 fg \\ =& (Lf)g - 2f_x g_x - f g_{xx}.\end{aligned}$$

Therefore,

$$\begin{aligned}L(\Lambda Q \varphi_A) =& (L\Lambda Q)\varphi_A - 2(\Lambda Q)_y \frac{1}{A}\varphi'\left(\frac{y}{A}\right) - \Lambda Q\left(\frac{1}{A^2}\varphi''\left(\frac{y}{A}\right)\right) \\ \equiv& L(\Lambda Q)\varphi_A + G_A\end{aligned}$$

and

$$L\left(F\frac{1}{A}\varphi'\left(\frac{y}{A}\right)\right) = \frac{1}{A}\left[L\left(\varphi'\left(\frac{y}{A}\right)\right)F - 2\Lambda Q\frac{1}{A}\varphi''\left(\frac{y}{A}\right) - (\Lambda Q)_y\varphi'\right]$$
$$\equiv \frac{1}{A}H_A.$$

So using that L is a self-adjoint operator and $L(\Lambda Q) = -2Q$, we get

$$\begin{aligned}(I.1) =& -\int \varepsilon L(\Lambda Q\varphi_A) - \int \varepsilon L\left(F\frac{1}{A}\varphi_y\left(\frac{y}{A}\right)\right)\\ =& 2\int \varepsilon Q\varphi_A - \int \varepsilon\left(G_A + \frac{1}{A}H_A\right)\\ =& 2\int \varepsilon Q + 2\int \varepsilon Q(\varphi_A - 1) - \int \varepsilon\left(G_A + \frac{1}{A}H_A\right)\\ \equiv& 2\int \varepsilon Q + R_5(\varepsilon, A).\end{aligned}$$

We estimate the terms in $R_5(\varepsilon, A)$ separately. First, note that

$$\int \varepsilon Q(\varphi_A - 1) \leq 2\int_{y\geq A} |\varepsilon Q|\frac{|y|}{|y|} \leq \frac{2}{A}\|yQ\|_2\|\varepsilon\|_2.$$

Next,

$$\int \varepsilon G_A \leq \frac{2}{A}\|\varphi'\|_\infty\|(\Lambda Q)_y\|_2\|\varepsilon\|_2 + \frac{1}{A^2}\|\varphi''\|_\infty\|\Lambda Q\|_2\|\varepsilon\|_2.$$

Now observe that $\|H_A\|_\infty \leq c$ (independent of $A\geq 1$) and $\mathrm{supp}(H_A) \subset \{A\leq y\leq 2A\}$, thus,

$$\frac{1}{A}\int \varepsilon H_A = \frac{1}{A}\int_{A\leq y\leq 2A} \varepsilon H_A \leq \frac{1}{A}\|H_A\|_\infty\left(\int_{A\leq y\leq 2A}|\varepsilon|dy\right) \leq \frac{2\|H_A\|_\infty}{A^{1/2}}\|\varepsilon\|_2.$$

Hence, for $A \geq 1$

$$|R_5(\varepsilon, A)| \leq \frac{c}{A^{1/2}}\|\varepsilon\|_{L^2(y\geq A)}, \tag{4.10}$$

where once again $c > 0$ is independent of ε and A.

Collecting all above estimates, we obtain

$$\begin{aligned}\frac{d}{ds}J_A =& 2\int \varepsilon Q + R_5(\varepsilon, A) + \frac{\lambda_s}{\lambda}\left(-\frac{1}{2}J_A + R_4(\varepsilon, A)\right)\\ &- \left(\frac{x_s}{\lambda} - 1\right)\left(\frac{1}{2}\int \varepsilon Q + R_3(\varepsilon, A)\right)\\ &+ \frac{\lambda_s}{\lambda}\left(\frac{1}{8}\left(\int Q\right)^2 + R_1(A)\right) + \left(\frac{x_s}{\lambda} - 1\right)R_2(A)\\ &+ (III)\\ =& -\frac{\lambda_s}{2\lambda}(J_A - \kappa) + 2\left(1 - \frac{1}{4}\left(\frac{x_s}{\lambda} - 1\right)\right)\int \varepsilon Q + R(\varepsilon, A),\end{aligned}$$

where

$$\kappa = \frac{1}{4}\left(\int Q\right)^2 \tag{4.11}$$

and

$$R(\varepsilon,A)=(III)+R_5(\varepsilon,A)+\Big(\frac{x_s}{\lambda}-1\Big)\left(R_2(A)-R_3(\varepsilon,A)\right)+\frac{\lambda_s}{\lambda}\left(R_1(A)+R_4(\varepsilon,A)\right).$$

Finally, there exists a universal constant $C_3>0$ (independent of ε and A) such that, in view of (4.5), (4.6), (4.7), (4.8), (4.9) and (4.10), for all $A\geq 1$ the inequality (4.4) holds. □

LEMMA 4.2 (Comparison between M_0, ε_0 and $\int\varepsilon_0 Q$). *There exists a universal constant $C_4>0$ such that for $\|\varepsilon_0\|_{H^1}\leq 1$, we have*

$$\left|M_0-2\int\varepsilon_0 Q\right|+\left|E_0+\int\varepsilon_0 Q\right|+\left|E_0+\frac{1}{2}M_0\right|\leq C_4\|\varepsilon_0\|_{H^1}^2.$$

PROOF. First, observe that from the definition (2.6), we have

$$M_0-2\int\varepsilon_0 Q=\int\varepsilon_0^2,$$

and so $\left|M_0-2\int\varepsilon_0 Q\right|=\|\varepsilon_0\|_2^2$. Next, from (2.8) we deduce

$$\begin{aligned}E_0=E[Q+\varepsilon_0]=&\frac{1}{2}\left(L\varepsilon_0,\varepsilon_0\right)-\frac{1}{2}M_0\\&-\frac{1}{6}\left(20\int Q^3\varepsilon_0^3+15\int Q^2\varepsilon_0^4+6\int Q\varepsilon_0^5+\int\varepsilon_0^6\right),\end{aligned}$$

which implies, for some universal constant $c>0$, that

$$\left|E_0+\frac{1}{2}M_0\right|\leq c\|\varepsilon_0\|_{H^1}^2,$$

by the definition of L, and the fact that $\|\varepsilon_0\|_\infty\leq\|\varepsilon_0\|_{H^1}\leq 1$.

Finally,

$$\left|E_0+\int\varepsilon_0 Q\right|\leq\left|E_0+\frac{1}{2}M_0\right|+\frac{1}{2}\left|M_0-2\int\varepsilon_0 Q\right|\leq\left(c+\frac{1}{2}\right)\|\varepsilon_0\|_{H^1}^2,$$

and setting $C_4=c+\frac{1}{2}$ concludes the proof. □

LEMMA 4.3 (Control of $\|\varepsilon(s)\|_{H^1}$). *There exists $0<\alpha_2<1$ such that if $\|\varepsilon(s)\|_{H^1}<\alpha$, $|\lambda(s)-1|<\alpha$, where $\alpha<\alpha_2$, and $\varepsilon(s)\perp\{Q_y,Q^3\}$ for all $s\geq 0$, then there exists a universal constant $C_5>0$ such that*

$$(L\varepsilon(s),\varepsilon(s))\leq\|\varepsilon(s)\|_{H^1}^2\leq C_5\left(\alpha\left|\int\varepsilon_0 Q\right|+\|\varepsilon_0\|_{H^1}^2\right).$$

PROOF. This is Lemma 11 in Martel-Merle [**11**], however, we provide a proof here, since our statement differs slightly from the one in [**11**]. From (2.8) we have

$$\begin{aligned}(L\varepsilon(s),\varepsilon(s))=&2E[Q+\varepsilon(s)]+M_0\\&-\frac{1}{6}\left(20\int Q^3\varepsilon^3+15\int Q^2\varepsilon^4+6\int Q\varepsilon^5+\int\varepsilon^6\right).\end{aligned}$$

Since $\|\varepsilon(s)\|_\infty\leq\|\varepsilon(s)\|_{H^1}\leq 1$, there exists a universal constant $c>0$ such that

$$\begin{aligned}(L\varepsilon(s),\varepsilon(s))\leq&2E[Q+\varepsilon(s)]+M_0+c\,\|\varepsilon(s)\|_{H^1}\|\varepsilon(s)\|_2^2\\\leq&2E[Q+\varepsilon(s)]+M_0+c\,\|\varepsilon(s)\|_{H^1}\left(L\varepsilon(s),\varepsilon(s)\right),\end{aligned}\tag{4.12}$$

where in the last line we have used the coercivity of the quadratic form $(L\cdot,\cdot)$, provided $\varepsilon(s)\perp\{Q_y,Q^3\}$, see Lemma 2.3.

Now, there exists $\alpha_2 > 0$ such that if $\|\varepsilon(s)\|_{H^1} < \alpha$ for all $s \geq 0$, where $\alpha < \alpha_2$, then

$$c\,\|\varepsilon(s)\|_{H^1} \leq \frac{1}{2}.$$

Therefore, the last term on the right hand side of (4.12) may be absorbed into the left side, and we obtain

$$\begin{aligned}(L\varepsilon(s), \varepsilon(s)) \leq& 4E[Q + \varepsilon(s)] + 2M_0\\ \leq& 4\lambda^2(s)E_0 + 2M_0,\end{aligned}$$

where in the last line we have used the second relation in (2.7).

Next, we use the last estimate to control the H^1-norm of $\varepsilon(s)$. Indeed, from the definition of L, we have

$$\begin{aligned}\|\varepsilon(s)\|_{H^1}^2 =& \int \varepsilon^2(s) + \int |\varepsilon_y(s)|^2 = (L\varepsilon(s), \varepsilon(s)) + 5\int Q^4\varepsilon^2(s)\\ \leq& (L\varepsilon(s), \varepsilon(s)) + \|5Q^4\|_\infty \|\varepsilon(s)\|_2^2\\ \leq& \left(1 + \|5Q^4\|_\infty\right)(L\varepsilon(s), \varepsilon(s))\\ \leq& \left(1 + \|5Q^4\|_\infty\right)(4\lambda^2(s)E_0 + 2M_0)\\ \leq& 4\left(1 + \|5Q^4\|_\infty\right)\left((\lambda(s)-1)(\lambda(s)+1)|E_0| + \left|E_0 + \frac{1}{2}M_0\right|\right).\end{aligned}$$

Finally, since $|\lambda(s) - 1| < \alpha$, for $\alpha < 1$ we have $|\lambda(s) + 1| \leq 3$, and applying Lemma 4.2, we deduce

$$\begin{aligned}(\lambda(s)-1)(\lambda(s)+1)|E_0| + \left|E_0 + \frac{1}{2}M_0\right| \leq& 3\alpha\left(\left|E_0 + \int \varepsilon_0 Q\right| + \left|\int \varepsilon_0 Q\right|\right) + C_4\|\varepsilon_0\|_{H^1}^2\\ \leq& 3\alpha\left(C_4\|\varepsilon_0\|_{H^1}^2 + \left|\int \varepsilon_0 Q\right|\right) + C_4\|\varepsilon_0\|_{H^1}^2,\end{aligned}$$

which implies, since $\alpha < 1$, the existence of an universal constant $C_5 > 0$ such that

$$(L\varepsilon(s), \varepsilon(s)) \leq \|\varepsilon(s)\|_{H^1}^2 \leq C_5\left(\alpha\left|\int \varepsilon_0 Q\right| + \|\varepsilon_0\|_{H^1}^2\right),$$

which concludes the proof. □

5. The proof of H^1-instability of Q

In this section we prove Theorem 1.3. Let $0 < b_0 < 1$ to be chosen later and set the initial data

$$u_0 = Q + \varepsilon_0$$

with $\|\varepsilon_0\|_{H^1} \leq 1$ satisfying

$$\|\varepsilon_0\|_{H^1}^2 \leq b_0 \int \varepsilon_0 Q \leq b_0\|Q\|^2. \tag{5.1}$$

Moreover, we can also assume that

$$\varepsilon_0 = u_0 - Q \perp \{Q_y, Q^3\} \tag{5.2}$$

and, for all $y \in \mathbb{R}$

$$|\varepsilon_0(y)| \leq \tilde{c}\, e^{-\tilde{\delta}|y|} \text{ for some } \tilde{c} > 0 \text{ and } \tilde{\delta} > 0. \tag{5.3}$$

REMARK 5.1. One simple example of such initial data is (for all $n \in \mathbb{N}$)

$$\varepsilon_0^n = \frac{1}{n}(Q + rQ^3),$$

where $r \in \mathbb{R}$ is chosen such that $\varepsilon_0^n \perp Q^3$, that is, $r = -\frac{\int Q^4}{\int Q^6}$.

Assume, by contradiction, that Q is stable. Therefore, for $\alpha_0 < \overline{\alpha}$, where $\overline{\alpha} > 0$ is given by Proposition 2.5, if b_0 is sufficiently small, we have $u(t) \in U_{\alpha_0}$ (recall (1.5)). Thus, from Definition 2.6, there exist functions $\lambda(t)$ and $x(t)$ such that

$$\varepsilon(t) = \lambda^{1/2}(t)\, u(\lambda(t)y + x(t)) - Q(y) \perp \{Q_y, Q^3\}$$

and also $\lambda(0) = 1$ and $x(0) = 0$ (by (5.2)).

Next rescaling time $t \mapsto s$ by $\frac{ds}{dt} = \frac{1}{\lambda^3}$ and taking $\alpha_0 < \alpha_1$, where $\alpha_1 > 0$ is given by Lemma 2.7, we obtain that $\lambda(s)$ and $x(s)$ are C^1 functions and $\varepsilon(s)$ satisfies the equation (2.1). Moreover, from Proposition 2.5, since $u(t) \in U_{\alpha_0}$, we have

$$\|\varepsilon(s)\|_{H^1} \le C_1\alpha_0 \quad \text{and} \quad |\lambda(s) - 1| \le C_1\alpha_0. \tag{5.4}$$

Furthermore, in view of (2.10), if $\alpha_0 > 0$ is small enough, we deduce that

$$\left|\frac{\lambda_s}{\lambda}\right| + \left|\frac{x_s}{\lambda} - 1\right| \le C_2\|\varepsilon(s)\|_2 \le C_1C_2\alpha_0.$$

Since $x_t = x_s/\lambda^3$, we conclude that

$$\frac{1 - C_1C_2\alpha_0}{(1 + C_1\alpha_0)^2} \le \frac{1 - C_1C_2\alpha_0}{\lambda^2} \le x_t \le \frac{1 + C_1C_2\alpha_0}{\lambda^2} \le \frac{1 + C_1C_2\alpha_0}{(1 - C_1\alpha_0)^2}.$$

Therefore, we can choose $\alpha_0 > 0$ small enough such that

$$\frac{3}{4} \le x_t \le \frac{5}{4}.$$

The last inequality implies that $x(t)$ is increasing and by the Mean Value Theorem

$$x(t_0) - x(t) \ge \frac{3}{4}(t_0 - t)$$

for every $t_0, t \ge 0$ with $t \in [0, t_0]$. Also, recalling $x(0) = 0$, another application of the Mean Value Theorem yields

$$x(t) \ge \frac{1}{2}t$$

for all $t \ge 0$. Finally, by assumption (5.3) and properties of Q, we have

$$|u_0(x)| \le ce^{-\delta|x|}$$

for some $c > 0$ and $\delta > 0$. Hence, from Lemma 3.2, for a fixed $M \ge \max\{4, \frac{1}{\delta}\}$, there exists $C = C(M, \delta) > 0$ such that for all $t \ge 0$ and $x_0 > 0$, we have

$$\int_{x > x_0} u^2(t, x + x(t))dx \le Ce^{-\frac{x_0}{M}}.$$

The next result provides L^2 exponential decay on the right also for $\varepsilon(s)$.

COROLLARY 5.2. *Let $M \geq \max\{4, \frac{1}{\delta}\}$. If $\alpha_0 > 0$ is sufficiently small, then there exists $C = C(M, \delta) > 0$ such that for every $s \geq 0$ and $y_0 > 0$*

$$\int_{y>y_0} \varepsilon^2(s,y)dy \leq Ce^{-\frac{y_0}{2M}}.$$

PROOF. From the definition of $\varepsilon(s)$, we have

$$\frac{1}{\lambda^{1/2}(s)}\varepsilon\left(s, \frac{y}{\lambda(s)}\right) = u(s, y + x(s)) - \frac{1}{\lambda^{1/2}(s)}Q\left(\frac{y}{\lambda(s)}\right).$$

Recalling the property of Q (and that $M \geq 4$), we get

$$\frac{1}{\lambda^{1/2}(s)}Q\left(\frac{y}{\lambda(s)}\right) \leq \frac{c}{\lambda^{1/2}(s)}e^{-\left|\frac{y}{\lambda(s)}\right|} \leq \sqrt{2}ce^{-\frac{2}{3}|x|} \leq ce^{-\frac{|x|}{M}}, \tag{5.5}$$

and if $\alpha_0 < (2C_1)^{-1}$, then $1/2 \leq \lambda(s) \leq 3/2$. Therefore, using Lemma 3.2 and (5.5), we deduce that

$$\begin{aligned}
\int_{y>y_0} \frac{1}{\lambda(s)}\varepsilon^2\left(s, \frac{y}{\lambda(s)}\right) dy \leq& 2\int_{y>y_0} u^2(s, y + x(s))dy + 2\int_{y>y_0} \frac{1}{\lambda(s)}Q^2\left(\frac{y}{\lambda(s)}\right) \\
\leq& 2ce^{-\frac{y_0}{M}} + 2c\int_{y>y_0} e^{-\frac{2y}{M}}dy \\
\leq& ce^{-\frac{y_0}{M}}.
\end{aligned}$$

Finally, by the scaling invariance of the L^2-norm

$$\int_{y>y_0} \varepsilon^2(s,y)dy = \int_{y>\lambda(s)y_0} \frac{1}{\lambda(s)}\varepsilon^2\left(s, \frac{y}{\lambda(s)}\right) dy \leq ce^{-\frac{\lambda(s)y_0}{M}} \leq Ce^{-\frac{y_0}{2M}},$$

since $\lambda(s) \geq 1/2$. □

Next, as in Martel-Merle [**11**] (see also Bona-Souganidis-Strauss [**1**] and Grillakis-Shatah-Strauss [**5**]), we define a rescaled version of J_A, which plays a central role in the proof [1]. Recall the definition of J_A in (4.2) and let

$$K_A(s) = \lambda^{1/2}(s)(J_A(s) - \kappa),$$

where κ is given by (4.11).

Therefore, from (4.3) and (5.4), it is clear that

$$|K_A(s)| \leq \left(\frac{3}{2}\right)^{1/2}\left(c(1 + A^{1/2})\|F\|_\infty\|\varepsilon(s)\|_2 + \kappa\right) < +\infty, \tag{5.6}$$

for all $s \geq 0$.

Moreover, using Lemma 4.1, we also have

$$\begin{aligned}
\frac{d}{ds}K_A =& \frac{1}{2}\frac{\lambda_s}{\lambda^{1/2}}(J_A - \kappa) + \lambda^{1/2}\frac{d}{ds}J_A \\
=& \lambda^{1/2}\left(\frac{d}{ds}J_A + \frac{\lambda_s}{2\lambda}(J_A - \kappa)\right) \\
=& \lambda^{1/2}\left(2\left(1 - \frac{1}{4}\left(\frac{x_s}{\lambda} - 1\right)\right)\int \varepsilon Q + R(\varepsilon, A)\right).
\end{aligned} \tag{5.7}$$

[1] Note that we do not make use of the functional $I(s) = \frac{1}{2}\int y\varepsilon^2(s)$ introduced by Martel-Merle [**11**] in their instability proof.

In the next result we obtain a strictly positive lower bound for $\frac{d}{ds}K_A(s)$ for a certain choice of α_0, b_0 and A.

THEOREM 5.3. *There exist $b_0, \alpha_0 > 0$ sufficiently small and $A \geq 1$ sufficiently large such that*

$$\frac{d}{ds}K_A(s) \geq \frac{1}{2}\int \varepsilon_0 Q > 0 \quad \textit{for all} \quad s \geq 0. \tag{5.8}$$

PROOF. In view of (5.4), let $\alpha_0 < \min\{\alpha_1(C_1)^{-1}, \alpha_2(C_1)^{-1}, (2C_1)^{-1}, (2C_1C_2)^{-1}, 1/2\}$, so that we can apply Lemmas 2.7 and 4.3. From (5.7) and the definition of M_0, see (2.6), we have

$$\frac{d}{ds}K_A(s) = \lambda^{1/2}\left(\left(1 - \frac{1}{4}\left(\frac{x_s}{\lambda} - 1\right)\right)M_0 + \widetilde{R}(\varepsilon, A)\right), \tag{5.9}$$

where

$$\widetilde{R}(\varepsilon, A) = R(\varepsilon, A) - \left(1 - \frac{1}{4}\left(\frac{x_s}{\lambda} - 1\right)\right)\int \varepsilon^2.$$

Since $\alpha_0 < \min\{(2C_1)^{-1}, (2C_1C_2)^{-1}\}$, we have $1/2 \leq \lambda(s) \leq 3/2$, and using (2.10), we deduce

$$\lambda^{1/2}\left(1 - \frac{1}{4}\left(\frac{x_s}{\lambda} - 1\right)\right) \geq \frac{1}{\sqrt{2}} \cdot \frac{3}{4} > \frac{1}{2}.$$

Moreover, from the definition of M_0, we also obtain

$$M_0 = 2\int \varepsilon_0 Q + \int \varepsilon_0^2 \geq 2\int \varepsilon_0 Q.$$

Hence, the first term in (5.9) is at least

$$\lambda^{1/2}\left(1 - \frac{1}{4}\left(\frac{x_s}{\lambda} - 1\right)\right)M_0 \geq \int \varepsilon_0 Q. \tag{5.10}$$

Now we estimate the second term in (5.9). By Lemma 2.7, we have

$$\left|\frac{\lambda_s}{\lambda}\right| + \left|\frac{x_s}{\lambda} - 1\right| \leq C_2\|\varepsilon(s)\|_2.$$

Therefore, using the inequalities (4.4) and (5.4), there exists a universal constant $C_6 > 0$, such that for $A \geq 1$ we can upper bound the second term as follows

$$\lambda^{1/2}\widetilde{R}(\varepsilon, A) \leq C_6\left(\|\varepsilon(s)\|_2^2 + A^{1/2}\|\varepsilon(s)\|_2\|\varepsilon(s)\|_{L^2(y\geq A)} + A^{-1/2}\|\varepsilon(s)\|_2\right). \tag{5.11}$$

Note that we are at the crucial point here, where we can use the smallness of ε and a large value of A to try to make all three terms in (5.11) small. The issue will be really in the middle term: taking a large A will make $A^{1/2}$ big, while the tail of ε in $\|\varepsilon(s)\|_{L^2(y\geq A)}$ has to balance the growth of $A^{1/2}$, which is delicate and only possible to get via monotonicity (also, with pointwise decay estimates as in the original proof in [**11**], but we are trying to avoid that here).

Continuing, by Lemma 4.3 we get

$$\|\varepsilon(s)\|_{H^1}^2 \leq C_5\left(C_1\alpha_0\left|\int \varepsilon_0 Q\right| + \|\varepsilon_0\|_{H^1}^2\right),$$

and thus, the assumption (5.1) yields

$$\begin{aligned}\|\varepsilon(s)\|_{H^1}^2 \leq & C_5\left(C_1\alpha_0 \int \varepsilon_0 Q + b_0 \int \varepsilon_0 Q\right)\\ \leq & C_5(C_1+1)\,(\alpha_0+b_0)\int \varepsilon_0 Q.\end{aligned}$$

From Corollary 5.2, we obtain the control on the tail of ε in L^2, hence, for every $A \geq 1$ we have

$$\|\varepsilon(s)\|_{L^2(y\geq A)}^2 \leq C_7 e^{-\frac{A}{2M}} \tag{5.12}$$

for some constants $C_7 > 0$ and $M > 0$.

Now, collecting estimates (5.11)-(5.12), there exists $C_8 > 0$ such that

$$\begin{aligned}\lambda^{1/2}\widetilde{R}(\varepsilon, A) \leq & C_8\left((\alpha_0+b_0)\int \varepsilon_0 Q + (A^{1/2}e^{-\frac{A}{4M}} + A^{-1/2})(\alpha_0+b_0)^{1/2}\left(\int \varepsilon_0 Q\right)^{1/2}\right)\\ \leq & C_8(\alpha_0+b_0)^{1/2}\left(\int \varepsilon_0 Q + (A^{1/2}e^{-\frac{A}{4M}} + A^{-1/2})\left(\int \varepsilon_0 Q\right)^{1/2}\right),\end{aligned}$$

where in the last inequality we assume that $\alpha_0 + b_0 < 1$, since these numbers will be chosen small enough in the sequel.

Now, let $\alpha_0, b_0 > 0$ sufficiently small such that $C_8(\alpha_0 + b_0)^{1/2} \leq 1/4$, and then (for α_0 and b_0 fixed) let $A \geq 1$ be large enough such that

$$(A^{1/2}e^{-\frac{A}{4M}} + A^{-1/2}) \leq \left(\int \varepsilon_0 Q\right)^{1/2},$$

this is exactly where monotonicity helps to play against the truncation and make the term $A^{1/2}\, e^{-\frac{A}{4M}}$ not just bounded, but small.

Thus,

$$\lambda^{1/2}\widetilde{R}(\varepsilon, A) \leq 2C_8(\alpha_0+b_0)^{1/2}\int \varepsilon_0 Q \leq \frac{1}{2}\int \varepsilon_0 Q,$$

which implies, in view of (5.9) and (5.10), that

$$\frac{d}{ds}K_A(s) \geq \frac{1}{2}\int \varepsilon_0 Q > 0 \quad \text{for all} \quad s \geq 0.$$

□

Finally, we have all the ingredients to obtain the main result.

PROOF OF THEOREM 1.3. Integrating in s variable both sides of inequality (5.8), we obtain

$$K_A(s) \geq \frac{s}{2}\int \varepsilon_0 Q + K_A(0) \quad \text{for all} \quad s \geq 0.$$

Hence,

$$\lim_{s\to\infty} K_A(s) = \infty,$$

which is a contradiction with (5.6), the boundedness of $K_A(s)$ for all $s > 0$. Thus, our original assumption that Q is stable is not valid and this finishes the proof. □

6. An alternative proof without truncation

We define

$$J(s) = \int_{\mathbb{R}} \varepsilon(s)F(y)\,dy.$$

Since we don't have any truncation now, our first goal is to show that $J(s)$ is well-defined and upper bounded[2] for all $s \geq 0$ under a certain choice of initial condition ε_0. To this end we use Monotonicity, see Lemma 3.2. Indeed, assume that ε_0 satisfies the assumptions (5.1)-(5.3). Therefore, we can apply Corollary 5.2 to deduce, for all $s \geq 0$, that

$$\begin{aligned}
|J(s)| \leq & \int_{y\leq 0} |\varepsilon(s)F(y)|\,dy + \int_{y>0} |\varepsilon(s)F(y)|\,dy \\
\leq & c\|\varepsilon(s)\|_2 \left(\int_{y\leq 0} e^{\delta y}\,dy \right)^{1/2} + \|F\|_\infty \sum_{k=0}^{+\infty} \int_k^{k+1} |\varepsilon(s)|\,dy \\
\leq & c\|\varepsilon(s)\|_2 + \|F\|_\infty \sum_{k=0}^{+\infty} \left(\int_k^{+\infty} |\varepsilon(s)|^2\,dy \right)^{1/2} \\
\leq & c\|\varepsilon(s)\|_2 + c\|F\|_\infty \sum_{k=0}^{+\infty} e^{-\frac{k}{4M}} \\
\leq & c\alpha_0^{1/2} + c\|F\|_\infty, \qquad (6.1)
\end{aligned}$$

where in the last inequality we have used (5.4).

Next, we consider the derivative $\frac{d}{ds}J(s)$. Again, from Martel-Merle [**11**, Lemma 6], we have

LEMMA 6.1. *Suppose that $\varepsilon(s) \in H^1(\mathbb{R})$ with $\|\varepsilon(s)\|_{H^1} \leq 1$ for all $s \geq 0$. Then the function $s \mapsto J(s)$ is C^1 and*

$$\frac{d}{ds}J = -\frac{\lambda_s}{2\lambda}\left(J - \frac{1}{4}\left(\int Q \right)^2 \right) + 2\left(1 - \frac{1}{4}\left(\frac{x_s}{\lambda} - 1 \right) \right) \int \varepsilon Q + R(\varepsilon),$$

where there exists a universal constant $c > 0$ such that

$$|R(\varepsilon)| \leq c\left(\|\varepsilon\|_2^2 + \|\varepsilon\|_2^2\|\varepsilon\|_{H^1} + \left(\left|\frac{x_s}{\lambda} - 1\right| + \left|\frac{\lambda_s}{\lambda}\right| \right) \|\varepsilon\|_2 \right). \tag{6.2}$$

Now define

$$K(s) = \lambda^{1/2}(s)(J(s) - \kappa),$$

where κ is given by (4.11).

Therefore, from (6.1) and (5.4), if α_0 is sufficiently small, it is clear that

$$|K(s)| \leq \left(\frac{3}{2}\right)^{1/2} \left(c\alpha_0^{1/2} + c\|F\|_\infty + \kappa \right) < +\infty \tag{6.3}$$

for all $s \geq 0$.

---

[2]Note that with truncation the boundedness of J_A was basically built in. Now we have to prove it, which we do via monotonicity bounds.

Now, if we obtain a strictly positive lower bound for $\frac{d}{ds}K(s)$, we deduce that $\lim_{s\to\infty} K(s) = \infty$ (as in the first proof of Theorem 1.3 in Section 5) and reach a contradiction with (6.3). From Lemma 6.1 we have

$$\begin{aligned}\frac{d}{ds}K =& \frac{1}{2}\frac{\lambda_s}{\lambda^{1/2}}(J-\kappa) + \lambda^{1/2}\frac{d}{ds}J\\ =&\lambda^{1/2}\left(\frac{d}{ds}J + \frac{\lambda_s}{2\lambda}(J-\kappa)\right)\\ =&\lambda^{1/2}\left(2\left(1-\frac{1}{4}\left(\frac{x_s}{\lambda}-1\right)\right)\int \varepsilon Q + R(\varepsilon)\right).\end{aligned}$$

Using the definition of M_0, see (2.6), we have

$$\frac{d}{ds}K(s) = \lambda^{1/2}\left(\left(1-\frac{1}{4}\left(\frac{x_s}{\lambda}-1\right)\right)M_0 + \widetilde{R}(\varepsilon)\right), \tag{6.4}$$

where

$$\widetilde{R}(\varepsilon) = R(\varepsilon) - \left(1-\frac{1}{4}\left(\frac{x_s}{\lambda}-1\right)\right)\int \varepsilon^2.$$

Since $\alpha_0 < \min\{(2C_1)^{-1}, (2C_1C_2)^{-1}\}$, we have $1/2 \le \lambda(s) \le 3/2$, and using (2.10), we deduce

$$\lambda^{1/2}\left(1-\frac{1}{4}\left(\frac{x_s}{\lambda}-1\right)\right) \ge \frac{1}{\sqrt{2}}\cdot\frac{3}{4} > \frac{1}{2}.$$

Moreover, it is easy to see that

$$M_0 = 2\int \varepsilon_0 Q + \int \varepsilon_0^2 \ge 2\int \varepsilon_0 Q.$$

Thus,

$$\lambda^{1/2}\left(1-\frac{1}{4}\left(\frac{x_s}{\lambda}-1\right)\right)M_0 \ge \int \varepsilon_0 Q. \tag{6.5}$$

On the other hand, by Lemma 2.7 we have

$$\left|\frac{\lambda_s}{\lambda}\right| + \left|\frac{x_s}{\lambda}-1\right| \le C_2\|\varepsilon(s)\|_2.$$

Therefore, using the inequality (6.2), there exists a universal constant $C_6 > 0$, such that

$$\lambda^{1/2}\widetilde{R}(\varepsilon) \le C_6\|\varepsilon(s)\|_2^2. \tag{6.6}$$

Now, by Lemma 4.3, we deduce

$$\|\varepsilon(s)\|_{H^1}^2 \le C_5\left(C_1\alpha_0\left|\int \varepsilon_0 Q\right| + \|\varepsilon_0\|_{H^1}^2\right),$$

and so the assumption (5.1) yields

$$\begin{aligned}\|\varepsilon(s)\|_{H^1}^2 \le& C_5\left(C_1\alpha_0\int \varepsilon_0 Q + b_0\int \varepsilon_0 Q\right)\\ \le& C_5(C_1+1)(\alpha_0+b_0)\int \varepsilon_0 Q.\end{aligned} \tag{6.7}$$

Therefore, collecting (6.6)-(6.7) and (5.4), there exists $C_8 > 0$ such that

$$\lambda^{1/2}\widetilde{R}(\varepsilon) \leq C_8\left(\alpha_0 + b_0\right)\int \varepsilon_0 Q.$$

Now, let $\alpha_0, b_0 > 0$ sufficiently small such that $C_8(\alpha_0 + b_0)^{1/2} \leq 1/4$ and then in view of (6.4) and (6.5), we deduce that

$$\frac{d}{ds}K(s) \geq \frac{1}{2}\int \varepsilon_0 Q > 0 \quad \text{for all } s \geq 0,$$

which gives the contradiction for large s and finishes the proof.

References

[1] J. L. Bona, P. E. Souganidis, and W. A. Strauss, *Stability and instability of solitary waves of Korteweg-de Vries type*, Proc. Roy. Soc. London Ser. A **411** (1987), no. 1841, 395–412. MR897729

[2] S.-M. Chang, S. Gustafson, K. Nakanishi, and T.-P. Tsai, *Spectra of linearized operators for NLS solitary waves*, SIAM J. Math. Anal. **39** (2007/08), no. 4, 1070–1111, DOI 10.1137/050648389. MR2368894

[3] V. Combet, *Construction and characterization of solutions converging to solitons for supercritical gKdV equations*, Differential Integral Equations **23** (2010), no. 5-6, 513–568. MR2654248

[4] L.G. Farah, J. Holmer and S. Roudenko, *Instability of solitons - revisited, II: the supercritical Zakharov-Kuznetsov equation*, preprint.

[5] M. Grillakis, J. Shatah, and W. Strauss, *Stability theory of solitary waves in the presence of symmetry. I*, J. Funct. Anal. **74** (1987), no. 1, 160–197, DOI 10.1016/0022-1236(87)90044-9. MR901236

[6] C. E. Kenig, G. Ponce, and L. Vega, *Well-posedness and scattering results for the generalized Korteweg-de Vries equation via the contraction principle*, Comm. Pure Appl. Math. **46** (1993), no. 4, 527–620, DOI 10.1002/cpa.3160460405. MR1211741

[7] M. K. Kwong, *Uniqueness of positive solutions of $\Delta u - u + u^p = 0$ in $\mathbf{R}^n$*, Arch. Rational Mech. Anal. **105** (1989), no. 3, 243–266, DOI 10.1007/BF00251502. MR969899

[8] M. Mariş, *Existence of nonstationary bubbles in higher dimensions* (English, with English and French summaries), J. Math. Pures Appl. (9) **81** (2002), no. 12, 1207–1239, DOI 10.1016/S0021-7824(02)01274-6. MR1952162

[9] Y. Martel and F. Merle, *A Liouville theorem for the critical generalized Korteweg-de Vries equation*, J. Math. Pures Appl. (9) **79** (2000), no. 4, 339–425, DOI 10.1016/S0021-7824(00)00159-8. MR1753061

[10] Y. Martel and F. Merle, *Asymptotic stability of solitons of the subcritical gKdV equations revisited*, Nonlinearity **18** (2005), no. 1, 55–80, DOI 10.1088/0951-7715/18/1/004. MR2109467

[11] Y. Martel and F. Merle, *Instability of solitons for the critical generalized Korteweg-de Vries equation*, Geom. Funct. Anal. **11** (2001), no. 1, 74–123, DOI 10.1007/PL00001673. MR1829643

[12] Y. Martel and F. Merle, *Blow up in finite time and dynamics of blow up solutions for the L^2-critical generalized KdV equation*, J. Amer. Math. Soc. **15** (2002), no. 3, 617–664, DOI 10.1090/S0894-0347-02-00392-2. MR1896235

[13] Y. Martel and F. Merle, *Stability of blow-up profile and lower bounds for blow-up rate for the critical generalized KdV equation*, Ann. of Math. (2) **155** (2002), no. 1, 235–280, DOI 10.2307/3062156. MR1888800

[14] Y. Martel, F. Merle, and P. Raphaël, *Blow up for the critical generalized Korteweg–de Vries equation. I: Dynamics near the soliton*, Acta Math. **212** (2014), no. 1, 59–140, DOI 10.1007/s11511-014-0109-2. MR3179608

[15] Y. Martel and D. Pilod, *Construction of a minimal mass blow up solution of the modified Benjamin-Ono equation*, Math. Ann. **369** (2017), no. 1-2, 153–245, DOI 10.1007/s00208-016-1497-8. MR3694646

[16] F. Merle, *Existence of blow-up solutions in the energy space for the critical generalized Korteweg-de Vries equation*, J. Amer. Math. Soc. 14 (2001), 555–578.

[17] M. I. Weinstein, *Modulational stability of ground states of nonlinear Schrödinger equations*, SIAM J. Math. Anal. **16** (1985), no. 3, 472–491, DOI 10.1137/0516034. MR783974

Department of Mathematics, Universidade Federal de Minas Gerais, Belo Horizonte, 31270-901, Brazil
Email address: lgfarah@gmail.com

Department of Mathematics, Brown University, Providence, Rhode Island 02912
Email address: holmer@math.brown.edu

Department of Mathematics and Statistics, Florida International University, Miami, Florida 33199
Email address: sroudenko@fiu.edu

Contemporary Mathematics
Volume **725**, 2019
https://doi.org/10.1090/conm/725/14547

Instability of solitons–revisited, II: The supercritical Zakharov-Kuznetsov equation

Luiz Gustavo Farah, Justin Holmer, and Svetlana Roudenko

ABSTRACT. We revisit the phenomenon of instability of solitons in the two dimensional generalization of the Korteweg-de Vries equation, the generalized Zakharov-Kuznetsov (ZK) equation, $u_t + \partial_{x_1}(\Delta u + u^p) = 0$, $(x_1, x_2) \in \mathbb{R}^2$. It is known that solitons are unstable in this two dimensional equation for nonlinearities $p > 3$. This was shown by Anne de Bouard [Proc. Roy. Soc. Edinburgh Sect. A **126**, 89–112 (1996)] generalizing the arguments of Bona, Souganidis and Strauss [Proc. Roy. Soc. London **411**, 395–412 (1987)] for the generalized KdV equation. In this paper, we use a different approach to obtain the instability of solitons, namely, combining truncation and monotonicity properties. Not only does this approach simplify the proof, but it can also be useful for studying various other stability questions in the ZK equation as well as other generalizations of the KdV equation.

CONTENTS

1. Introduction
2. Background on the generalized ZK equation
3. The linearized operator L
4. Decomposition of u and Modulation Theory
5. Virial-type estimates and monotonicity
6. H^1-instability of Q for the supercritical gZK
References

1. Introduction

In this paper we consider the two-dimensional generalized Zakharov-Kuznetsov equation (gZK):

$$u_t + \partial_{x_1}\left(\Delta_{(x_1,x_2)} u + u^p\right) = 0, \quad p > 3, \quad (x_1, x_2) \in \mathbb{R}^2, \quad t \in \mathbb{R}. \tag{1.1}$$

The above equation with $p = 2$, besides being the 2d extension of the well-known KdV equation, governs the behavior of weakly nonlinear ion-acoustic waves in plasma comprising of cold ions and hot isothermal electrons in the present of a uniform magnetic field [**19**, **20**] and was originally derived by Zakharov and Kuznetsov

2010 *Mathematics Subject Classification.* Primary 35Q53, 37K40, 37K45, 37K05.
Key words and phrases. Zakharov-Kuznetsov equation, instability of solitons, monotonicity.

to describe weakly magnetized ion-acoustic waves in a strongly magnetized plasma [**24**].

The equation (1.1) is the two-dimensional extension of the well-studied model describing, for example, the weakly nonlinear waves in shallow water, the Korteweg-de Vries (KdV) equation:

$$u_t + (u_{xx} + u^p)_x = 0, \quad p = 2, \quad x \in \mathbb{R}, \quad t \in \mathbb{R}. \tag{1.2}$$

When other integer powers $p \neq 2$ are considered, it is referred to as the *generalized* KdV (gKdV) equation. It is also possible to consider noninteger powers $p > 1$, however, the nonlinearity would have to be modified as $\partial_{x_1}(|u|^{p-1}u)$. For the odd powers, this would produce a slightly different equation, however, most of the theory would remain the same. Despite its apparent universality, the gKdV equation is limited as a spatially one-dimensional model, and thus, various higher dimensional generalizations exist.

During their lifespan, the solutions $u(t, x_1, x_2)$ to the equation (1.1) conserve the mass and energy:

$$M[u(t)] = \int_{\mathbb{R}^2} u^2(t)\, dx_1 dx_2 = M[u(0)] \tag{1.3}$$

and

$$E[u(t)] = \frac{1}{2}\int_{\mathbb{R}^2} |\nabla u(t)|^2\, dx_1 dx_2 - \frac{1}{p+1}\int_{\mathbb{R}^2} u^{p+1}(t)\, dx_1 dx_2 = E[u(0)]. \tag{1.4}$$

Similar to the gKdV equation, for solutions $u(t, x, y)$ decaying at infinity on $\mathbb{R}^2$ the following invariance holds

$$\int_{\mathbb{R}} u(t, x_1, x_2)\, dx_1 = \int_{\mathbb{R}} u(0, x_1, x_2)\, dx_1,$$

which is obtained by integrating the original equation on $\mathbb{R}$ in the first coordinate x_1.

One of the useful symmetries in the evolution equations is *the scaling invariance*, which states that an appropriately rescaled version of the original solution is also a solution of the equation. For the equation (1.1) it is

$$u_\lambda(t, x_1, x_2) = \lambda^{\frac{2}{p-1}} u(\lambda^3 t, \lambda x_1, \lambda x_2).$$

This rescaling makes a specific Sobolev norm $\dot{H}^s$ invariant, i.e.,

$$\|u(0,\cdot,\cdot)\|_{\dot{H}^s} = \lambda^{\frac{2}{p-1}+s-1}\|u_0\|_{\dot{H}^s},$$

and the index s gives rise to the critical-type classification of equations. For the gKdV equation (1.2) the critical index is $s = \frac{1}{2} - \frac{2}{p-1}$, and for the 2d gZK equation (1.1) it is $s = 1 - \frac{2}{p-1}$. When $s > 0$ (in the 2d gZK equation this corresponds to $p > 3$), the equation (1.1) is often referred as the L^2-supercritical equation. (The gZK equation has other invariances such as translation and dilation.)

The gZK equation has a family of travelling waves (or solitary waves, which sometimes are referred even as solitons), and observe that they travel only in x_1 direction

$$u(t, x_1, x_2) = Q_c(x_1 - ct, x_2) \tag{1.5}$$

with $Q_c(x_1, x_2) \to 0$ as $|x| \to +\infty$. Here, Q_c is the dilation of the ground state Q:

$$Q_c(\vec{x}) = c^{1/p-1} Q(c^{1/2}\vec{x}), \quad \vec{x} = (x_1, x_2),$$

with Q being a radial positive solution in $H^1(\mathbb{R}^2)$ of the well-known nonlinear elliptic equation $-\Delta Q + Q - Q^p = 0$. Note that $Q \in C^\infty(\mathbb{R}^2)$, $\partial_r Q(r) < 0$ for any $r = |x| > 0$ and for any multi-index α

$$|\partial^\alpha Q(\vec{x})| \leq c(\alpha) e^{-|\vec{x}|} \quad \text{for any} \quad \vec{x} \in \mathbb{R}^2. \tag{1.6}$$

In this work, we are interested in stability properties of travelling waves in the supercritical gZK equation (1.1), i.e., in the behavior of solutions close to the ground state Q (perhaps, up to translations). We begin with the precise concept of stability and instability used in this paper. For $\alpha > 0$, the neighborhood (or "tube") of radius α around Q (modulo translations) is defined by

$$U_\alpha = \left\{ u \in H^1(\mathbb{R}^2) : \inf_{\vec{y} \in \mathbb{R}^2} \|u(\cdot) - Q(\cdot + \vec{y})\|_{H^1} \leq \alpha \right\}.$$

DEFINITION 1.1 (Stability of Q). We say that Q is stable if for all $\alpha > 0$, there exists $\delta > 0$ such that if $u_0 \in U_\delta$, then the corresponding solution $u(t)$ is defined for all $t \geq 0$ and $u(t) \in U_\alpha$ for all $t \geq 0$.

DEFINITION 1.2 (Instability of Q). We say that Q is unstable if Q is not stable, in other words, there exists $\alpha > 0$ such that for all $\delta > 0$ the following holds: if $u_0 \in U_\delta$, then there exists $t_0 = t_0(u_0)$ such that $u(t_0) \notin U_\alpha$.

The main goal in this paper is to provide a new proof of the instability result, originally obtained by de Bouard [**5**] in her study of dispersive solitary waves in higher dimensions (her result holds in dimensions 2 and 3). She showed [**5**] that the travelling waves of the form (1.5) are stable (in the two dimensional case) for $p < 3$ and unstable for $p > 3$. She followed the ideas of Bona-Souganidis-Strauss [**1**] for the instability, and Grillakis-Shatah-Strauss [**10**] for the stability arguments. Here, we prove the instability of the traveling wave solution of the form (1.5), in a spirit of Combet [**3**], where he revisited the instability phenomenon for the gKdV equation in the supercritical case, $p > 5$. We show an open set of initial data, in particular, an explicit example of a sequence of initial data, which would contradict the stability of Q. To this end consider, for $n \geq 1$

$$u_{0,n}(\vec{x}) = \lambda_n Q(\lambda_n \vec{x}), \quad \text{where} \quad \lambda_n = 1 + \frac{1}{n} \quad \text{and} \quad \vec{x} = (x_1, x_2).$$

Our main motivation is to show the new methods available (monotonicity and truncation) to obtain the following result:

THEOREM 1.3 (H^1-instability of Q for the supercritical gZK). *Let u_n be the solution with initial data at time zero $u_{0,n}$ for each $n \in \mathbb{N}$, then there exists $\alpha > 0$ such that for every $n \in \mathbb{N}$, there exists $T_n = T_n(u_{0,n})$ such that $u_n(T_n) \notin U_\alpha$, or explicitly,*

$$\inf_{\vec{y} \in \mathbb{R}^2} \|u_n(T_n, \cdot) - Q(\cdot - \vec{y})\| \geq \alpha.$$

The paper is organized as follows. In Section 2 we provide the background information on the well-posedness of the generalized ZK equation in 2 dimensions. In Section 3 we discuss the properties of the linearized operator L around the ground state Q. Section 4 contains the canonical decomposition of the solution around Q, the modulation theory and control of parameters coming from such a decomposition. In Section 5 we discuss the virial-type functional and the concept of monotonicity. The next Section 6 contains a new proof of the instability via truncation and monotonicity.

1.1. Acknowledgments. Most of this work was done when the first author was visiting GWU in 2016-17 under the support of the Brazilian National Council for Scientific and Technological Development (CNPq/Brazil), for which all authors are very grateful as it boosted the energy into the research project. S.R. would like to thank IHES and the organizers for the excellent working conditions during the trimester program "Nonlinear dispersive waves - 2017" in May-July 2016. L.G.F. was partially supported by CNPq and FAPEMIG/Brazil. J.H. was partially supported by the NSF grant DMS-1500106. S.R. was partially supported by the NSF CAREER grant DMS-1151618.

2. Background on the generalized ZK equation

In this section we review the known results on the local and global well-posedness of the generalized ZK equation. To follow the notation in the literature, in this section we denote the power of nonlinearity as u^{k+1} (instead of u^p) and consider the Cauchy problem for the gZK equation as follows:

$$\begin{cases} u_t + \partial_{x_1}\Delta u + \partial_{x_1}(u^{k+1}) = 0, & (x_1, x_2) \in \mathbb{R}^2, \ t > 0, \\ u(0, x_1, x_2) = u_0(x_1, x_2) \in H^s(\mathbb{R}^2). \end{cases} \tag{2.1}$$

The first paper to address the local well-posedness of this Cauchy problem for the $k = 1$ case was by Faminskii [**7**], where he considered $s = 1$ (strictly speaking, he obtained the local well-posedness in H^m, for any integer $m \geq 1$.) The current results on the local well-posedness are gathered in the following statement.

THEOREM 2.1. *The local well-posedness in* (2.1) *holds in the following cases:*

- $k = 1$*: for* $s > \frac{1}{2}$*, see Grünrock-Herr* [**11**] *and Molinet-Pilod* [**18**],
- $k = 2$*: for* $s > \frac{1}{4}$*, see Ribaud-Vento* [**21**],
- $k = 3$*: for* $s > \frac{5}{12}$*, see Ribaud-Vento* [**21**],
- $k = 4, 5, 6, 7$*: for* $s > 1 - \frac{2}{k}$*, see Ribaud-Vento* [**21**],
- $k = 8$*,* $s > \frac{3}{4}$*, see Linares-Pastor* [**15**]
- $k > 8$*,* $s > s_k = 1 - 2/k$*, see Farah-Linares-Pastor* [**9**].

Following the approach of Holmer-Roudenko for the L^2-supercritical nonlinear Schrödinger (NLS) equation, see [**12**] and [**6**], the first author together with F. Linares and A. Pastor obtained the following global well-posedness result in [**9**].

THEOREM 2.2 ([**9**]). *Let* $k \geq 3$ *and* $s_k = 1 - 2/k$*. Assume* $u_0 \in H^1(\mathbb{R}^2)$ *and suppose that*

$$E(u_0)^{s_k} M(u_0)^{1-s_k} < E(Q)^{s_k} M(Q)^{1-s_k}, \ \ E(u_0) \geq 0. \tag{2.2}$$

If

$$\|\nabla u_0\|_{L^2}^{s_k} \|u_0\|_{L^2}^{1-s_k} < \|\nabla Q\|_{L^2}^{s_k} \|Q\|_{L^2}^{1-s_k}, \tag{2.3}$$

then for any t from the maximal interval of existence

$$\|\nabla u(t)\|_{L^2}^{s_k} \|u_0\|_{L^2}^{1-s_k} = \|\nabla u(t)\|_{L^2}^{s_k} \|u(t)\|_{L^2}^{1-s_k} < \|\nabla Q\|_{L^2}^{s_k} \|Q\|_{L^2}^{1-s_k},$$

where Q *is the unique positive radial solution of*

$$\Delta Q - Q + Q^{k+1} = 0.$$

In particular, this implies that H^1 *solutions, satisfying* (2.2)-(2.3) *exist globally in time.*

REMARK 2.3. In the limit case $k = 2$ (or $p = 3$, the modified ZK equation), conditions (2.2) and (2.3) reduce to one condition, which is

$$\|u_0\|_{L^2} < \|Q\|_{L^2}.$$

Such a condition was already used in [**14**] and [**15**] to show the existence of global solutions, respectively, in $H^1(\mathbb{R}^2)$ and $H^s(\mathbb{R}^2)$, $s > 53/63$.

We conclude this section with a note that for the purpose of this paper, it is sufficient to have the well-posedness theory in $H^1(\mathbb{R}^2)$.

3. The linearized operator L

The operator L, which is obtained by linearizing around the ground state Q, is defined by

$$L := -\Delta + 1 - p\,Q^{p-1}. \tag{3.1}$$

We first state the properties of this operator L (see Kwong [**13**] for all dimensions, Weinstein [**23**] for dimension 1 and 3, also Maris [**16**] and [**2**]).

THEOREM 3.1 (Properties of L). *The following holds for an operator L defined in* (3.1)

- *L is a self-adjoint operator and $\sigma_{ess}(L) = [\lambda_{ess}, +\infty)$ for some $\lambda_{ess} > 0$*
- $\ker L =$ *span*$\{Q_{x_1}, Q_{x_2}\}$
- *L has a unique single negative eigenvalue $-\lambda_0$ (with $\lambda_0 > 0$) associated to a positive radially symmetric eigenfunction χ_0. Moreover, there exists $\delta > 0$ such that*

$$|\chi_0(x)| \lesssim e^{-\delta|x|} \quad \textit{for all } x \in \mathbb{R}^2. \tag{3.2}$$

We also define the generator Λ of the scaling symmetry

$$\Lambda f = \frac{1}{p-1} f + \frac{1}{2}\vec{x}\cdot\nabla f,\ (x_1, x_2) \in \mathbb{R}^2.$$

The following identities are useful to have

LEMMA 3.2. *The following identities hold*

(1) $L(\Lambda Q) = -Q$
(2) $\int Q\,\Lambda Q = \frac{3-p}{2(p-1)} \int Q^2 < 0$ *for* $p > 3$, *and* $\int Q\,\Lambda Q = 0$ *if* $p = 3$.

PROOF. The first two identities follow directly from the definition of L, Λ and the equation $-\Delta Q + Q - Q^p = 0$. The third identity is also easy to derive, for example

$$\int Q\Lambda Q = \frac{1}{p-1}\int Q^2 - \frac{1}{2}\left(2\int Q^2 + \int x\cdot\nabla Q\right) = \left(\frac{1}{p-1} - \frac{1}{2}\right)\int Q^2,$$

from which (iii) follows. □

In general, the operator L is not positive-definite, however, if we exclude appropriately the zero eigenvalue and negative eigenvalue directions, then one can expect some positivity properties, which we exhibit in the following two lemmas.

LEMMA 3.3 (Orthogonality Conditions I). *Let χ_0 be the positive radially symmetric eigenfunction associated to the unique single negative eigenvalue $-\lambda_0$ (with $\lambda_0 > 0$). Then, there exists σ_0 such that for any $f \in H^1(\mathbb{R}^2)$ satisfying*

$$(f, \chi_0) = (f, Q_{x_j}) = 0, \quad j = 1, 2,$$

one has

$$(Lf, f) \geq \sigma_0 \|f\|_2^2.$$

PROOF. The result follows directly from Schechter [**22**, Chapter 8, Lemma 7.10] (see also [**22**, Chapter 1, Lemma 7.17]) □

LEMMA 3.4 (Orthogonality Conditions II). *There exist $k_1, k_2 > 0$ such that for all $\varepsilon \in H^1(\mathbb{R}^2)$ satisfying $\varepsilon \perp \{Q_{x_1}, Q_{x_2}\}$ one has*

$$(L\varepsilon, \varepsilon) = \int |\nabla \varepsilon|^2 + \int \varepsilon^2 - p \int Q^{p-1}\varepsilon \geq k_1 \|\varepsilon\|_2^2 - k_2|(\varepsilon, \chi_0)|^2.$$

PROOF. If $\varepsilon \perp \chi_0$, the result is true by the previous lemma. Now, if $(\varepsilon, \chi_0) \neq 0$, let $a \in \mathbb{R}$ be such that $\varepsilon_1 = \varepsilon - a\chi_0$ verifies $\varepsilon_1 \perp \chi_0$, in other words,

$$(\varepsilon_1, \chi_0) = (\varepsilon, \chi_0) - a\|\chi_0\|_2^2 = 0 \iff a = (\varepsilon, \chi_0)\|\chi_0\|_2^{-2}. \tag{3.3}$$

Therefore, by Lemma 3.3 we have

$$(L\varepsilon_1, \varepsilon_1) \geq \sigma_0\|\varepsilon_1\|_2^2 = \sigma_0(\|\varepsilon\|_2^2 - 2a(\varepsilon, \chi_0) + a^2\|\chi_0\|_2^2). \tag{3.4}$$

On the other hand,

$$\begin{aligned}(L\varepsilon_1, \varepsilon_1) =&(L\varepsilon, \varepsilon) - 2a(\varepsilon, L\chi_0) + a^2(L\chi_0, \chi_0)\\ =&(L\varepsilon, \varepsilon) - 2a\lambda_0(\varepsilon, \chi_0) + a^2\lambda_0\|\chi_0\|_2^2.\end{aligned} \tag{3.5}$$

Collecting (3.4) and (3.5), we get

$$\begin{aligned}(L\varepsilon, \varepsilon) \geq&\sigma_0\|\varepsilon\|_2^2 - 2a(\sigma_0 + \lambda_0)(\varepsilon, \chi_0) + a^2(\sigma_0 + \lambda_0)\|\chi_0\|_2^2\\ =&\sigma_0\|\varepsilon\|_2^2 - (\sigma_0 + \lambda_0)\|\chi_0\|_2^{-2}|(\varepsilon, \chi_0)|^2,\end{aligned}$$

where we have used (3.3) in the last line. The result follows by taking $k_1 = \sigma_0$ and $k_2 = (\sigma_0 + \lambda_0)\|\chi_0\|_2^{-2}$. □

Following [**3**], we introduce the Lyapunov-type functional, connecting with the linearized operator L and obtain the upper bounds on it.

LEMMA 3.5 (Weinstein's Functional). *Recall* (1.3)-(1.4) *and define*

$$W[u] = E[u] + \frac{1}{2}M[u].$$

Then

$$W[Q + \varepsilon] = W[Q] + \frac{1}{2}(L\varepsilon, \varepsilon) + H[\varepsilon] \tag{3.6}$$

with $H : H^1 \to \mathbb{R}$, and if $\|\varepsilon\|_{H^1} \leq 1$, there exists $C > 0$ such that

$$|H[\varepsilon]| \leq C\|\varepsilon\|_{H^1}\|\varepsilon\|_2^2. \tag{3.7}$$

PROOF. A straighforward calculation reveals

$$\begin{aligned} E[Q+\varepsilon] &= \frac{1}{2}\int |\nabla(Q+\varepsilon)|^2 - \frac{1}{p+1}\int (Q+\varepsilon)^{p+1} \\ &= E[Q] + \frac{1}{2}\int |\nabla\varepsilon|^2 + \int \nabla Q\cdot\nabla\varepsilon \\ &\quad - \frac{1}{p+1}\int\left((p+1)Q^p\varepsilon + \frac{(p+1)p}{2}Q^{p-1}\varepsilon^2 + G[\varepsilon]\right) \\ &= E[Q] + \frac{1}{2}\int |\nabla\varepsilon|^2 - \int(\Delta Q + Q^p)\varepsilon - \frac{p}{2}\int Q^{p-1}\varepsilon^2 - \frac{1}{p+1}G[\varepsilon] \\ &= E[Q] + \frac{1}{2}\int |\nabla\varepsilon|^2 - \int Q\varepsilon - \frac{p}{2}\int Q^{p-1}\varepsilon^2 + H[\varepsilon], \end{aligned} \tag{3.8}$$

where $H[\varepsilon] = -\frac{1}{p+1}\int G[\varepsilon]$ with[1] $G[\varepsilon] = \sum_{k=3}^{p+1}\binom{p+1}{k}Q^{p+1-k}\varepsilon^k$ and we have used $Q = \Delta Q + Q^p$ in the last line.

Since $Q \in L^\infty(\mathbb{R}^2)$, we use the Gagliardo-Nirenberg inequality

$$\|f\|_q^q \le c\|\nabla f\|_2^{q-2}\|f\|_2^2, \quad q \ge 2, \tag{3.9}$$

to deduce that, if $\|\varepsilon\|_{H^1} \le 1$, then

$$|H[\varepsilon]| \le c\sum_{k=3}^{p+1}\|\varepsilon\|_k^k \le C\|\varepsilon\|_{H^1}\|\varepsilon\|_2^2,$$

for some constant $C > 0$. Next, noticing that

$$M[Q+\varepsilon] = \int Q^2 + 2\int Q\varepsilon + \int \varepsilon^2, \tag{3.10}$$

and putting together (3.10) and (3.8), we obtain (3.6). □

4. Decomposition of u and Modulation Theory

We consider the canonical parametrization of the solution $u(t, x_1, x_2)$ close to Q:

$$\varepsilon(t, x_1, x_2) = u(t, x_1 + y_1(t), x_2 + y_2(t)) - Q(x_1, x_2), \tag{4.1}$$

where u is a solution of (1.1) and $y_1(t)$, $y_2(t)$ are two C^1 functions to be determined later. In the next lemma we deduce the equation for $\varepsilon(t, \vec{x})$.

LEMMA 4.1 (Equation for ε). *There exists $C_0 > 0$ such that for all $t \ge 0$, we have*

$$\varepsilon_t - (L\varepsilon)_{x_1} = (y_1'(t) - 1)(Q+\varepsilon)_{x_1} + y_2'(t)(Q+\varepsilon)_{x_2} - R(\varepsilon), \tag{4.2}$$

where

$$|R(\varepsilon)| \le C_0\sum_{k=2}^{p}(|\varepsilon|^k + |\varepsilon_{x_1}||\varepsilon|^{k-1}). \tag{4.3}$$

---

[1] Recall that $\binom{p+1}{k} = \frac{(p+1)!}{(p+1-k)!k!}$.

PROOF. By definition (4.1), we have

$$u(t,x_1,x_2) = Q(x_1 - y_1(t), x_2 - y_2(t)) + \varepsilon(t, x_1 - y_1(t), x_2 - y_2(t)).$$

Since u is a solution of (1.1), we have

$$y_1'(t)(Q+\varepsilon)_{x_1} + y_2'(t)(Q+\varepsilon)_{x_2} - \varepsilon_t = \partial_{x_1}\Delta(Q+\varepsilon) + (Q^p)_{x_1} + p(Q^{p-1}\varepsilon)_{x_1} + R(\varepsilon), \tag{4.4}$$

where

$$R(\varepsilon) = \partial_{x_1}\left(\sum_{k=2}^{p}\binom{p}{k}Q^{p-k}\varepsilon^k\right) = \sum_{k=2}^{p}\binom{p}{k}((p-k)Q_{x_1}Q^{p-k-1}\varepsilon^k + kQ^{p-k}\varepsilon_{x_1}\varepsilon^{k-1}). \tag{4.5}$$

Recalling that $Q, Q_{x_1} \in L^\infty(\mathbb{R}^2)$, there exists a universal constant $C_0 > 0$ such that the estimate (4.3) holds.

Using that $-\Delta Q + Q - Q^p = 0$ and the definition of L from (3.1), we rewrite the equation (4.4) as

$$-\varepsilon_t + (L\varepsilon)_{x_1} - \varepsilon_{x_1} = Q_{x_1} - y_1'(t)(Q+\varepsilon)_{x_1} - y_2'(t)(Q+\varepsilon)_{x_2} + R(\varepsilon),$$

which implies the equation (4.2), concluding the proof. □

Next we recall the modulation theory close to the ground state Q, as in de Bouard [**5**] (see also Bona-Souganidis-Strauss [**1**] for a similar result in the gKdV model).

PROPOSITION 4.2 (Modulation Theory). *There exists $\alpha_1 > 0$, $C_1 > 0$ and a unique C^1 map*

$$(y_1, y_2) : U_{\alpha_1} \to \mathbb{R}^2$$

such that if $u \in U_{\alpha_1}$ and $\varepsilon_{(y_1(u),y_2(u))}$ is given by

$$\varepsilon_{(y_1(u),y_2(u))}(x_1,x_2) = u(x_1 + y_1(u), x_2 + y_2(u)) - Q(x_1,x_2), \tag{4.6}$$

then

$$\varepsilon_{(y_1(u),y_2(u))} \perp Q_{x_j}, \quad j = 1,2.$$

Moreover, if $u \in U_\alpha$ with $0 < \alpha < \alpha_1$, then

$$\|\varepsilon_{(y_1(u),y_2(u))}\|_{H^1} \leq C_1\alpha. \tag{4.7}$$

Furthermore, if u is cylindrically symmetric (i.e., $u(x_1,x_2) = u(x_1,|x_2|)$), then, reducing α_0 if necessary, we can assume $y_2(u) = 0$.

PROOF. See [**5**, Lemma 4.4]. □

Now, assume that $u(t) \in U_\alpha$, with $\alpha < \alpha_1$, for all $t \geq 0$, and define the functions $y_1(t)$ and $y_2(t)$ as follows.

DEFINITION 4.3. For all $t \geq 0$, let $y_1(t)$ and $y_2(t)$ be such that $\varepsilon_{(y_1(t),y_2(t))}$, defined according to the equation (4.6), satisfy

$$\varepsilon_{(y_1(t),y_2(t))} \perp Q_{x_j}, \quad j = 1,2. \tag{4.8}$$

In this case, we set

$$\varepsilon(t) = \varepsilon_{(y_1(t),y_2(t))} = u(x_1 + y_1(t), x_2 + y_2(t)) - Q(x_1,x_2). \tag{4.9}$$

From the estimate (4.7), it is clear that

$$\|\varepsilon(t)\|_{H^1} \leq C_1\alpha. \tag{4.10}$$

The next lemma provides us with the estimates for $|y_1' - 1|$ and $|y_2'|$.

LEMMA 4.4 (Control of the modulation parameters). *There exists $0 < \alpha_2 < \alpha_1$ such that if for all $t \geq 0$, $u(t) \in U_{\alpha_2}$, then $y_1(t)$ and $y_2(t)$ are C^1 functions of t and they satisfy the following equations*

$$(y_1' - 1)[\|Q_{x_1}\|_2^2 - (\varepsilon, Q_{x_1x_1})] - y_2'(\varepsilon, Q_{x_1x_2}) = (L(Q_{x_1x_1}), \varepsilon) + (R(\varepsilon), Q_{x_1})$$

and

$$-(y_1' - 1)(\varepsilon, Q_{x_1x_2}) + y_2'[\|Q_{x_2}\|_2^2 - (\varepsilon, Q_{x_2x_2})] = (L(Q_{x_1x_2}), \varepsilon) + (R(\varepsilon), Q_{x_2}).$$

Moreover, there exists a universal constant $C_2 > 0$ such that if $\|\varepsilon(t)\|_2 \leq \alpha$, for all $t \geq 0$, where $\alpha < \alpha_2$, then

$$|y_1' - 1| + |y_2'| \leq C_2\|\varepsilon(t)\|_2. \tag{4.11}$$

PROOF. Multiplying the equation (4.4) by Q_{x_1} and then by Q_{x_2} and integrating by parts, we deduce

$$\begin{aligned}&\int \varepsilon_t Q_{x_1} + \int L(\varepsilon) Q_{x_1x_1}\\ &= (y_1'(t) - 1)\left[\int Q_{x_1}^2 - \varepsilon Q_{x_1x_1}\right] + y_2'(t)\left[\int Q_{x_1}Q_{x_2} - \varepsilon Q_{x_1x_2}\right] - \int R(\varepsilon)Q_{x_1}\end{aligned}$$

and

$$\begin{aligned}&\int \varepsilon_t Q_{x_2} + \int L(\varepsilon) Q_{x_1x_2}\\ &= (y_1'(t) - 1)\left[\int Q_{x_1}Q_{x_2} - \varepsilon Q_{x_1x_2}\right] + y_2'(t)\left[\int Q_{x_2}^2 - \varepsilon Q_{x_2x_2}\right] - \int R(\varepsilon)Q_{x_2}.\end{aligned}$$

We first note that the first term on the left hand sides of the last two equalities is zero in view of the orthogonality (4.8). Moreover, we also have $Q_{x_1} \perp Q_{x_2}$, thus, vanishing a couple of terms on the right hand sides. Therefore, using that L is a self-adjoint operator, we obtain the following system

$$(y_1' - 1)[\|Q_{x_1}\|_2^2 - (\varepsilon, Q_{x_1x_1})] - y_2'(\varepsilon, Q_{x_1x_2}) = (L(Q_{x_1x_1}), \varepsilon) + (R(\varepsilon), Q_{x_1})$$

and

$$-(y_1' - 1)(\varepsilon, Q_{x_1x_2}) + y_2'[\|Q_{x_2}\|_2^2 - (\varepsilon, Q_{x_2x_2})] = (L(Q_{x_1x_2}), \varepsilon) + (R(\varepsilon), Q_{x_2}).$$

Next, we observe that there exists $\alpha_2 > 0$ such that

$$\begin{vmatrix} \|Q_{x_1}\|_2^2 - (\varepsilon, Q_{x_1x_1}) & -(\varepsilon, Q_{x_1x_2}) \\ -(\varepsilon, Q_{x_1x_2}) & \|Q_{x_2}\|_2^2 - (\varepsilon, Q_{x_2x_2}) \end{vmatrix} \geq \frac{1}{2}\|Q_{x_1}\|_2^2\|Q_{x_2}\|_2^2,$$

if $\|\varepsilon(t)\|_2 \leq \alpha$ for every $t \geq 0$, where $\alpha < \alpha_2$. Finally, we can find $C_2 > 0$ such that

$$\begin{aligned}|y_1'(t) - 1| \leq &\big[|(L(Q_{x_1x_1}), \varepsilon) + (R(\varepsilon), Q_{x_1})| \cdot |\|Q_{x_2}\|_2^2 - (\varepsilon, Q_{x_2x_2})|\\ &+ |(L(Q_{x_1x_2}), \varepsilon) + (R(\varepsilon), Q_{x_2})| \cdot |(\varepsilon, Q_{x_1x_2})|\big]\left[\frac{1}{2}\|Q_{x_1}\|_2^2\|Q_{x_2}\|_2^2\right]^{-1}\\ \leq &C_2\|\varepsilon(t)\|_2.\end{aligned}$$

Similarly, we obtain $|y_2'(t)| \leq C_2\|\varepsilon(t)\|_2$, concluding the proof. □

5. Virial-type estimates and monotonicity

Our next step is to produce a virial-type functional which will help us study the stability properties of the solutions close to Q. We first define a quantity depending on the ε variable, which incorporates the scaling generator Λ and the eigenfunction of L for the negative eigenvalue. For the gKdV version, compare with Combet [**3**, §2.3]. This turns out to play an important role in our instability proof, and what is absolutely crucial here is that we can find $\beta \neq 0$. We note that this does not work in the critical case ($p = 3$), nor in the critical gKdV equation (with $p = 5$ nonlinearity), since β becomes zero. We now let

$$F(x_1, x_2) = \int_{-\infty}^{x_1} (\Lambda Q(z, x_2) + \beta \chi_0(z, x_2))\, dz,$$

where $\beta \in \mathbb{R}$ is a constant to be chosen later.

From the properties of Q (see (1.6)) and χ_0 (see (3.2)), there exist $c, \delta > 0$ such that

$$|F(x_1, x_2)| \leq c e^{-\frac{\delta}{2}|x_2|} \int_{-\infty}^{x_1} e^{-\frac{\delta}{2}|z|} dz, \tag{5.1}$$

and therefore, $|F(x_1, x_2)| = o(e^{\frac{\delta}{2}x_1})$ when $x_1 \to -\infty$ for every $x_2 \in \mathbb{R}$ fixed. We also note that F is a bounded function in $\mathbb{R}^2$, that is, $F \in L^\infty(\mathbb{R}^2)$.

We next define the virial-type functional

$$J(t) = \int_{\mathbb{R}^2} \varepsilon(t) F(x_1, x_2)\, dx_1 dx_2, \tag{5.2}$$

and would like to show that it is well-defined. The first observation is that $J(t)$ is well-defined if $\varepsilon(t) \in H^{3+}(\mathbb{R}^2)$ for all $t \geq 0$. Indeed, this is a consequence of [**5**, Theorem 2.3], and following the argument on pages 103-104 of [**5**], we find a universal constant $C_3 > 0$ independent of t such that

$$|J(t)| \leq C_3(t^{-3/4} + t^{1/2}). \tag{5.3}$$

We remark that while our functional J differs from the one used in [**5**], the estimate above is the same, see (4.2) in [**5**].

Our next task is to show that $J(t)$ is well-defined only assuming $\varepsilon(t) \in H^1(\mathbb{R}^2)$ for all $t \geq 0$. To this end, we adapt some monotonicity ideas introduced by Martel-Merle [**17**] for the gKdV equation. Define

$$\psi(x_1) = \frac{2}{\pi} \arctan(e^{\frac{x_1}{M}}),$$

where $M \geq 4$. The following properties hold for ψ

(1) $\psi(0) = \dfrac{1}{2}$,
(2) $\lim_{x_1 \to -\infty} \psi(x_1) = 0$ and $\lim_{x_1 \to +\infty} \psi(x_1) = 1$,
(3) $1 - \psi(x_1) = \psi(-x_1)$,
(4) $\psi'(x_1) = \left(\pi M \cosh\left(\frac{x_1}{M}\right)\right)^{-1}$,
(5) $|\psi'''(x_1)| \leq \dfrac{1}{M^2}\psi'(x_1) \leq \dfrac{1}{16}\psi'(x_1)$.

Let $(y_1(t), y_2(t)) \in C^1(\mathbb{R}, \mathbb{R}^2)$ and for $x_0, t_0 > 0$ and $t \in [0, t_0]$ define

$$I_{x_0,t_0}(t) = \int u^2(t, x_1, x_2)\psi(x_1 - y_1(t_0) + \frac{1}{2}(t_0 - t) - x_0) dx_1 dx_2, \tag{5.4}$$

where $u \in C(\mathbb{R}, H^1(\mathbb{R}^2))$ is a solution of the gZK equation (1.1), satisfying

$$\|u(t, x_1 + y_1(t), x_2 + y_2(t)) - Q(x_1, x_2)\|_{H^1} \leq \alpha, \tag{5.5}$$

for some $\alpha > 0$. While this is a similar concept to study the decay of the mass of the solution to the right of the soliton, as it was done in the gKdV equation in works of Martel-Merle, or for example, see our review of the instability of the critical gKdV case via monotonicity [**8**], we note that the integration is two dimensional in the definition (5.4) (but the function ψ is defined only in one variable x_1). For the ZK equation this type of result was also obtained by Côte-Muñoz-Pilod-Simpson [**4**]. We next study the behavior of I in time and our almost monotonicity result is the following.

LEMMA 5.1 (Almost Monotonicity). *Let $M \geq 4$ fixed and assume that $y_1(t)$ is an increasing function satisfying $y_1(t_0) - y_1(t) \geq \frac{3}{4}(t_0 - t)$ for every $t_0, t \geq 0$ with $t \in [0, t_0]$. Then there exist $\alpha_0 > 0$ and $\theta = \theta(M, p) > 0$ such that if $u \in C(\mathbb{R}, H^1(\mathbb{R}^2))$ verify (5.5) with $\alpha < \alpha_0$, then for all $x_0 > 0$, $t_0, t \geq 0$ with $t \in [0, t_0]$, we have*

$$I_{x_0,t_0}(t_0) - I_{x_0,t_0}(t) \leq \theta e^{-\frac{x_0}{M}}.$$

PROOF. Using the equation and the fact that $|\psi'''(x)| \leq \frac{1}{M^2}\psi'(x) \leq \frac{1}{16}\psi'(x)$, we deduce

$$\begin{aligned}
\frac{d}{dt} I_{x_0,t_0}(t) =& 2\int u u_t \psi - \frac{1}{2}\int u^2 \psi' \\
=& -\int \left(3u_{x_1}^2 + u_{x_2}^2 - \frac{2p}{p+1}u^{p+1}\right)\psi' + \int u^2\psi''' - \frac{1}{2}\int u^2\psi' \\
\leq& -\int \left(3u_{x_1}^2 + u_{x_2}^2 + \frac{1}{4}u^2\right)\psi' + \frac{2p}{p+1}\int u^{p+1}\psi'. \qquad (5.6)
\end{aligned}$$

Now, we estimate the last term on the right hand side of the previous inequality. First, we write

$$\int u^{p+1}\psi' = \int Q(\cdot - \vec{y}(t))u^p\psi' + \int (u - Q(\cdot - \vec{y}(t)))u^p\psi', \tag{5.7}$$

where $\vec{y}(t) = (y_1(t), y_2(t))$. To estimate the second term, we use the Sobolev embedding $H^1(\mathbb{R}^2) \hookrightarrow L^q(\mathbb{R}^2)$, for all $2 \leq q < +\infty$, to get

$$\begin{aligned}
\int (u - Q(\cdot - \vec{y}(t)))u^p\psi' \leq& \|(u - Q(\cdot - \vec{y}(t)))u^{p-2}\|_{4/3}\|u^2\psi'\|_4 \\
\leq& c\,\|u - Q(\cdot - \vec{y}(t))\|_2\|u\|_{4(p-2)}^{p-2}\|u\sqrt{\psi'}\|_8^2 \\
\leq& c\,\alpha\,\|Q\|_{H^1}^{p-2}\int (|\nabla u|^2 + |u|^2)\psi'. \qquad (5.8)
\end{aligned}$$

For the first term on the right hand side of (5.7), we divide the integration into two regions $|\vec{x} - \vec{y}(t)| \geq R_0$ and $|\vec{x} - \vec{y}(t)| < R_0$, where R_0 is a positive number to be chosen later. Therefore, since $|Q(\vec{x})| \leq c\,e^{-|\vec{x}|}$, we obtain

$$\begin{aligned}
\int_{|\vec{x} - \vec{y}(t)| \geq R_0} Q(\cdot - \vec{y}(t))u^p\psi' \leq& ce^{-R_0}\|u^{p-2}\|_3\|u\sqrt{\psi'}\|_3^2 \\
\leq& ce^{-R_0}\|Q\|_{H^1}^{p-2}\int (|\nabla u|^2 + |u|^2)\psi'.
\end{aligned}$$

Next, when $|\vec{x}-\vec{y}(t)|\leq R_0$, we have that

$$\left|x_1-y_1(t_0)+\frac{1}{2}(t_0-t)-x_0\right| \geq (y_1(t_0)-y_1(t)+x_0)-\frac{1}{2}(t_0-t)-|x_1-y_1(t)|$$
$$\geq \frac{1}{4}(t_0-t)+x_0-R_0,$$

where in the first inequality we have used that $y_1(t)$ is increasing, $t_0\geq t$ and $x_0>0$ to compute the modulus of the first term, and in the second line we have used the assumption $y_1(t_0)-y_1(t)\geq \frac{3}{4}(t_0-t)$.

Since $\psi'(z)\leq c\,e^{-\frac{|z|}{M}}$, we use again the Sobolev embedding $H^1(\mathbb{R}^2)\hookrightarrow L^q(\mathbb{R}^2)$, for all $2\leq q<+\infty$, to deduce that

$$\int_{|\vec{x}-\vec{y}(t)|\leq R_0} Q(\cdot-\vec{y}(t))u^p\psi' \leq c\|Q\|_\infty e^{\frac{R_0}{M}}e^{-\frac{\left(\frac{1}{4}(t_0-t)+x_0\right)}{M}}\|u\|^p_{H^1}$$
$$\leq c\|Q\|_\infty\|Q\|^p_{H^1}e^{\frac{R_0}{M}}e^{-\frac{\left(\frac{1}{4}(t_0-t)+x_0\right)}{M}}. \tag{5.9}$$

Therefore, choosing α such that $c\,\alpha\,\|Q\|^{p-2}_{H^1}<\frac{p+1}{2p}\cdot\frac{1}{4}$ and R_0 such that $c\,e^{-R_0}\|Q\|^{p-2}_{H^1}<\frac{p+1}{2p}\cdot\frac{1}{4}$, collecting (5.8)-(5.9), we obtain

$$\frac{2p}{p+1}\int u^{p+1}\psi'\leq\frac{1}{8}\int(|\nabla u|^2+|u|^2)\psi'+c\|Q\|_\infty\|Q\|^p_{H^1}e^{\frac{R_0}{M}}e^{-\frac{\left(\frac{1}{4}(t_0-t)+x_0\right)}{M}}.$$

Inserting the previous estimate in (5.6), there exists $C>0$ such that

$$\frac{d}{dt}I_{x_0,t_0}(t)\leq-\int\left(\frac{3}{2}u^2_{x_1}+\frac{1}{2}u^2_{x_2}+\frac{1}{8}u^2\right)\psi'+c\,e^{-\frac{x_0}{M}}\cdot e^{-\frac{1}{4M}(t_0-t)}$$
$$\leq c\,e^{-\frac{x_0}{M}}\cdot e^{-\frac{1}{4M}(t_0-t)}$$

Finally, integrating in time on $[t,t_0]$, we obtain the desired inequality for some $\theta=\theta(M,p)>0$. □

The next lemma is the main tool to obtain the upper bound for $|J(t)|$ independent of $t\geq 0$.

LEMMA 5.2. *Let $y_1(t)$ satisfying the assumptions of Lemma 5.1. Also assume that $y_1(t)\geq\frac{1}{2}t$ and $y_2(t)=0$ for all $t\geq 0$. Let $u\in C(\mathbb{R},H^1(\mathbb{R}^2))$ a solution of the gZK equation (1.1) satisfying (5.5) with $\alpha<\alpha_0$ (with α_0 given in Lemma 5.1) and with initial data u_0 verifying $\int|u_0(x_1,x_2)|^2\,dx_2\leq c\,e^{-\delta|x_1|}$ for some $c>0$ and $\delta>0$. Fix $M\geq\max\{4,\frac{2}{\delta}\}$. Then there exists $C=C(M,\delta,p)>0$ such that for all $t\geq 0$ and $x_0>0$*

$$\int_{\mathbb{R}}\int_{x_1>x_0}u^2(t,x_1+y_1(t),x_2)\,dx_1dx_2\leq C\,e^{-\frac{x_0}{M}}.$$

PROOF. From Lemma 5.1 with $t=0$ and replacing t_0 by t, we deduce for all $t\geq 0$

$$I_{x_0,t}(t)-I_{x_0,t}(0)\leq\theta e^{-\frac{x_0}{M}}.$$

This is equivalent to

$$\int u^2(t,x_1,x_2)\psi(x_1-y_1(t)-x_0)dx_1dx_2$$
$$\leq \int u_0^2(x_1,x_2)\psi(x_1-y_1(t)+\frac{1}{2}t-x_0)dx_1dx_2+\theta e^{-\frac{x_0}{M}}.$$

On the other hand,

$$\int u^2(t,x_1,x_2)\psi(x_1-y_1(t)-x_0)dx_1dx_2=\int u^2(t,x_1+y_1(t),x_2)\psi(x_1-x_0)dx_1dx_2$$
$$\geq \frac{1}{2}\int_{\mathbb{R}}\int_{x_1>x_0} u^2(t,x_1+y_1(t),x_2)dx_1dx_2,$$

where in the last inequality we have used the fact that ψ is increasing and $\psi(0)=1/2$.

Now, since $-y_1(t)+\frac{1}{2}t\leq 0$ and ψ is increasing, we have

$$\int u_0^2(x_1,x_2)\psi(x_1-y_1(t)+\frac{1}{2}t-x_0)dx_1dx_2\leq \int u_0^2(x_1,x_2)\psi(x_1-x_0)dx_1dx_2.$$

The assumptions $\int |u_0(x_1,x_2)|^2\,dx_2\leq c\,e^{-\delta|x_1|}$ and $\psi(x_1)\leq ce^{\frac{x_1}{M}}$ yield

$$\int u_0^2(x_1,x_2)\psi(x_1-x_0)dx_1dx_2 \leq c\int e^{-\delta|x_1|}\,e^{\frac{x_1-x_0}{M}}\,dx_1$$
$$\leq c\,e^{-\frac{x_0}{M}}\int e^{-(\delta-\frac{1}{M})|x_1|}dx_1$$
$$\leq \bar{C}(M,\delta)\,e^{-\frac{x_0}{M}},$$

where in the last inequality we have used the fact that

$$\delta-\frac{1}{M}\geq\frac{\delta}{2}\iff M\geq\frac{2}{\delta}.$$

Collecting the above estimate, we obtain the desired inequality. □

Next, we compute the derivative of $J(t)$.

LEMMA 5.3. *Suppose that $\varepsilon(t)\in H^1(\mathbb{R}^2)$, for all $t\geq 0$, and $y_1(t)$, $y_2(t)$ are two C^1 functions. Then the function $t\mapsto J(t)$ is C^1 and*

$$\frac{d}{dt}J=\beta\lambda_0\int\varepsilon\chi_0+K(\varepsilon), \tag{5.10}$$

where, setting

$$\beta=-\frac{\int Q\Lambda Q}{\int Q\chi_0} \tag{5.11}$$

and recalling the definition (4.5), we have

$$K(\varepsilon)=\int Q\varepsilon-(y_1'-1)\int\varepsilon(\Lambda Q+\beta\chi_0)-y_2'\int\varepsilon F_{x_2}-y_2'\int QF_{x_2}-\int R(\varepsilon)F. \tag{5.12}$$

REMARK 5.4. Since Q and χ_0 are both positive functions, we have $\int Q\chi_0>0$, and thus, β in (5.11) is well-defined. By Lemma 3.2, part (iii), we deduce that $\beta=-\dfrac{\int Q\Lambda Q}{\int Q\chi_0}>0$, if $p>3$. (Note that in the critical case, $p=3$, such a constant β would be zero, since $\int Q\Lambda Q=0$.)

PROOF. From the equation for ε (4.2), we have

$$\begin{aligned}\frac{d}{dt}J &= \int \varepsilon_t F \\ &= \int (L\varepsilon)_{x_1} F + (y_1'(t)-1)\int (Q+\varepsilon)_{x_1} F + y_2'(t)\int (Q+\varepsilon)_{x_2} F - \int R(\varepsilon) F \\ &= -\int \varepsilon L(F_{x_1}) - (y_1'(t)-1)\int (Q+\varepsilon) F_{x_1} - y_2'(t)\int (Q+\varepsilon) F_{x_2} - \int R(\varepsilon) F \\ &= -\int \varepsilon L(\Lambda Q + \beta\chi_0) - (y_1'(t)-1)\int (Q+\varepsilon)(\Lambda Q + \beta\chi_0) \\ &\qquad - y_2'(t)\int (Q+\varepsilon) F_{x_2} - \int R(\varepsilon) F.\end{aligned}$$

Using that $L(\Lambda Q) = -Q$ and $L(\chi_0) = -\lambda_0\chi_0$ (see Theorem 3.1 and Lemma 3.2), we obtain

$$\begin{aligned}\frac{d}{dt}J =& \beta\lambda_0 \int \varepsilon\chi_0 - (y_1'(t)-1)\int Q(\Lambda Q + \beta\chi_0) + \int \varepsilon Q \\ & - (y_1'(t)-1)\int \varepsilon(\Lambda Q + \beta\chi_0) - y_2'(t)\int (Q+\varepsilon) F_{x_2} - \int R(\varepsilon) F.\end{aligned}$$

Setting $\beta = -\frac{\int Q\Lambda Q}{\int Q\chi_0} > 0$, the second term in the last expression is zero, and then relation (5.10) holds with $K(\varepsilon)$ given by (5.12). □

6. H^1-instability of Q for the supercritical gZK

In this section we prove Theorem 1.3. For $n \geq 1$ let

$$u_{0,n}(\vec{x}) = \lambda_n Q(\lambda_n \vec{x}), \quad \text{where} \quad \lambda_n = 1 + \frac{1}{n} \quad \text{and} \quad \vec{x} = (x_1, x_2). \tag{6.1}$$

The following proposition exhibits some properties of the sequence $\{u_{0,n}\}$ with respect to Q that will be useful later.

PROPOSITION 6.1. *Let $u_{0,n}$ be given by (6.1), then for every $n \in \mathbb{N}$*

$$\|u_{0,n}\|_2 = \|Q\|_2 \quad \textit{and} \quad E[u_{0,n}] < E[Q]. \tag{6.2}$$

Moreover,

$$\|u_{0,n} - Q\|_{H^1} \to 0, \quad \textit{as} \quad n \to +\infty. \tag{6.3}$$

PROOF. A simple rescaling shows

$$\int |u_{0,n}(\vec{x})|^2 d\vec{x} = \int \lambda_n^2 |Q(\lambda_n \vec{x})|^2 d\vec{x} = \int |Q(\vec{x})|^2 d\vec{x},$$

which proves the equality in (6.2). Moreover, from $u_{0,n}(\vec{x}) \to Q(\vec{x})$ for all $x \in \mathbb{R}^2$ and the exponential decay of Q (1.6), the limit in (6.3) is true by the Dominated Convergence Theorem.

Next, we turn to the energy inequality in (6.2). Indeed,

$$\begin{aligned} E[u_{0,n}] =&\frac{1}{2}\int |\nabla u_{0,n}(\vec{x})|^2 d\vec{x} - \frac{1}{p+1}\int |u_{0,n}(\vec{x})|^{p+1} d\vec{x} \\ =&\frac{1}{2}\int \lambda_n^4 |\nabla Q(\lambda_n \vec{x})|^2 d\vec{x} - \frac{1}{p+1}\int \lambda_n^{p+1} |Q(\lambda_n \vec{x})|^{p+1} d\vec{x} \\ =&\frac{\lambda_n^2}{2}\int |\nabla Q(\vec{x})|^2 d\vec{x} - \frac{\lambda_n^{p-1}}{p+1}\int |Q(\vec{x})|^{p+1} d\vec{x}. \end{aligned}$$

Therefore, using the Pohozaev identity

$$\int |Q|^{p+1} = \frac{p+1}{p-1}\int |\nabla Q|^2$$

and the definition of $\lambda_n = 1 + \frac{1}{n}$, we obtain

$$\begin{aligned} &E[u_{0,n}] - E[Q] \\ &= \left[\frac{p-1}{2}(\lambda_n^2 - 1) - (\lambda_n^{p-1} - 1)\right]\frac{1}{p+1}\int |Q|^{p+1} d\vec{x} \\ &= \left[\frac{p-1}{2}\left(\frac{2}{n} + \frac{1}{n^2}\right) - \frac{p-1}{n} - \binom{p-1}{2}\frac{1}{n^2} - \sum_{k=3}^{p-1}\binom{p-1}{k}\frac{1}{n^k}\right]\frac{1}{p+1}\int |Q|^{p+1} d\vec{x} \\ &= \left[\left[\frac{p-1}{2} - \binom{p-1}{2}\right]\frac{1}{n^2} - \sum_{k=3}^{p-1}\binom{p-1}{k}\frac{1}{n^k}\right]\frac{1}{p+1}\int |Q|^{p+1} d\vec{x}. \end{aligned}$$

Since for $p > 3$ we have

$$\binom{p-1}{2} = \frac{(p-1)!}{(p-3)!2!} = \frac{(p-1)(p-2)}{2} > \frac{p-1}{2},$$

we deduce the desired inequality. □

Now, assume by contradiction that Q is stable. Then for every $\alpha > 0$ there exists $n(\alpha) \in \mathbb{N}$ such that for every $t \geq 0$ we have

$$u_{n(\alpha)}(t) \in U_\alpha,$$

where $u_{n(\alpha)}$ is the solution with initial data $u_{0,n(\alpha)}$. Since $u_{0,n(\alpha)}(\vec{x}) = \lambda_{n(\alpha)} Q(\lambda_{n(\alpha)}\vec{x})$ is cylindrically symmetric, so will be $u_{n(\alpha)}(t)$ for all $t \geq 0$, as the equation is invariant under rotation in $\mathbb{R}_{x_1}$. Select $\alpha_0 < \alpha_2 < \alpha_1$, where $\alpha_1 > 0$ is given by Proposition 4.2 and $\alpha_2 > 0$ by Lemma 4.4. To simplify the notation, we omit the index $n(\alpha_0)$ from now on. Definition 4.3 provides a function

$$\varepsilon(t) = \varepsilon_{(y_1(t), y_2(t))} = u(x_1 + y_1(t), x_2 + y_2(t)) - Q(x_1, x_2), \tag{6.4}$$

satisfying (4.8), and, from the second conclusion in Proposition 4.2, we also have $y_2(t) = 0$ for all $t \geq 0$. From (6.1) and (6.3), we deduce that $y_1(0) = 0$.

Since $u(t) \in U_{\alpha_0}$, from Proposition 4.2 and Lemma 4.4 we get

$$\|\varepsilon(t)\|_{H^1} \leq C_1\alpha_0 \quad \text{and} \quad |y_1'(t) - 1| \leq C_2 C_1 \alpha_0, \tag{6.5}$$

so taking $\alpha_0 < \{(2C_1)^{-1}, (4C_1C_2)^{-1}\}$, we obtain

$$\|\varepsilon(t)\|_{H^1} \leq 1 \quad \text{and} \quad \frac{3}{4} \leq y_1'(t) \leq \frac{5}{4} \quad \text{for all} \quad t \geq 0.$$

The last inequality implies that $y_1(t)$ is increasing and by the Mean Value Theorem

$$y_1(t_0) - y_1(t) \geq \frac{3}{4}(t_0 - t),$$

for every $t_0, t \geq 0$ with $t \in [0, t_0]$. Also, recalling $y_1(0) = 0$, another application of the Mean Value Theorem yields

$$y_1(t) \geq \frac{1}{2}t,$$

for all $t \geq 0$. Finally, using the two facts that $Q(\vec{x}) \leq c\, e^{-|\vec{x}|}$ for some $c > 0$ and $\lambda_n \geq 1$ for all $n \in \mathbb{N}$ by assumption (6.1), we obtain

$$|u_0(\vec{x})| \leq 2c\, e^{-|\vec{x}|}.$$

Now, from Lemma 5.2, we can deduce the following L^2 exponential decay on the right for $\varepsilon(t)$.

COROLLARY 6.2. *Let $M \geq \max\{4, \frac{2}{\delta}\}$. If $\alpha_0 > 0$ is sufficiently small, there exists $C = C(M, \delta, p) > 0$ such that for every $t \geq 0$ and $x_0 > 0$*

$$\int_{\mathbb{R}} \int_{x_1 > x_0} \varepsilon^2(t, x_1, x_2) dx_1 dx_2 \leq C e^{-\frac{x_0}{2M}}.$$

PROOF. Applying Lemma 5.2, for a fixed $M \geq \max\{4, \frac{2}{\delta}\}$, there exists $C = C(M, \delta, p) > 0$ such that for all $t \geq 0$ and $x_0 > 0$, we have

$$\int_{\mathbb{R}} \int_{x_1 > x_0} u^2(t, x_1 + y_1(t), x_2) dx_1 dx_2 \leq C e^{-\frac{x_0}{M}}. \tag{6.6}$$

From the definition of $\varepsilon(t)$ (see (6.4)), recalling that $y_2(t) = 0$ for all $t \geq 0$, we have that

$$\varepsilon\,(t, x_1, x_2) = u(t, x_1 + y_1(t), x_2) - Q\,(x_1, x_2)\,.$$

Moreover, since $Q(\vec{x}) \leq c\, e^{-|\vec{x}|}$, we obtain

$$\begin{aligned}
\int_{\mathbb{R}} \int_{x_1 > x_0} Q^2\,(x_1, x_2)\, dx_1 dx_2 \leq & c \int_{\mathbb{R}} \int_{x_1 > x_0} e^{-2|\vec{x}|} dx_1 dx_2 \\
\leq & c \left(\int_{\mathbb{R}} e^{-|x_2|} dx_2 \right) \left(\int_{x_1 > x_0} e^{-x_1} dx_1 \right) \\
\leq & c\, e^{-x_0} \leq c\, e^{-\frac{x_0}{M}}, \qquad (6.7)
\end{aligned}$$

where in the last inequality we used the definition of M. Finally, collecting (6.6)-(6.7), we deduce the desired result. □

In the next proposition, we obtain an upper bound for $|J(t)|$ independent of $t \geq 0$ (improving the bound (5.3) previously obtained by De Bouard [**5**]).

PROPOSITION 6.3. *If $\alpha_0 > 0$ is sufficiently small, then there exists a constant $M_0 > 0$ such that*

$$|J(t)| \leq M_0 \quad \textit{for all} \quad t \geq 0. \tag{6.8}$$

PROOF. From the definition of $J(t)$ (see (5.2)), we have that for all $t \geq 0$

$$|J(t)| \leq \int_{\mathbb{R}} \int_{x_1 \leq 0} |\varepsilon(t, x_1, x_2) F(x_1, x_2)|\, dx_1 dx_2 + \int_{\mathbb{R}} \int_{x_1 > 0} |\varepsilon(t, x_1, x_2) F(x_1, x_2)|\, dx_1 dx_2.$$

Note that $F\chi_{\{x_1<0\}} \in L^2(\mathbb{R}^2)$ by (5.1), therefore, the first integral on the right hand side of the last inequality can be bounded as

$$\int_{\mathbb{R}}\int_{x_1\leq 0} |\varepsilon(t,x_1,x_2)F(x_1,x_2)|\,dx_1dx_2 \leq \|\varepsilon(t)\|_2\|F\chi_{\{x_1<0\}}\|_2 \leq cC_1\alpha_0,$$

where in the last inequality we used (6.5).

On the other hand, again from (5.1), for every $x_2 \in \mathbb{R}$ we have

$$\sup_{x_1\in\mathbb{R}} |F(x_1,x_2)| \leq \frac{4c}{\delta} e^{-\frac{\delta}{2}|x_2|}.$$

Therefore,

$$\begin{aligned}
&\int_{\mathbb{R}}\int_{x_1>0} |\varepsilon(t,x_1,x_2)F(x_1,x_2)|\,dx_1dx_2 \\
&\quad\leq \int_{\mathbb{R}} \sup_{x_1\in\mathbb{R}} |F(x_1,x_2)| \left(\int_{x_1>0} |\varepsilon(t,x_1,x_2)|\,dx_1\right) dx_2 \\
&\quad\leq \frac{4c}{\delta}\int_{\mathbb{R}} e^{-\frac{\delta}{2}|x_2|} \left(\int_{x_1>0} |\varepsilon(t,x_1,x_2)|\,dx_1\right) dx_2.
\end{aligned}$$

Furthermore, using the Cauchy-Schwarz inequality in the x_1-variable, we get

$$\begin{aligned}
\int_{x_1>0} |\varepsilon(t,x_1,x_2)|\,dx_1 &= \sum_{k=0}^{+\infty}\int_k^{k+1} |\varepsilon(t,x_1,x_2)|\,dx_1 \\
&\leq \sum_{k=0}^{+\infty}\left(\int_k^{k+1} \varepsilon^2(t,x_1,x_2)dx_1\right)^{1/2}.
\end{aligned}$$

Now, collecting the last two inequalities and using the Cauchy-Schwarz inequality in the x_2-variable, we obtain

$$\begin{aligned}
&\int_{\mathbb{R}}\int_{x_1>0} |\varepsilon(t,x_1,x_2)F(x_1,x_2)|\,dx_1dx_2 \\
&\quad\leq \frac{4c}{\delta}\sum_{k=0}^{+\infty}\int_{\mathbb{R}} e^{-\frac{\delta}{2}|x_2|}\left(\int_k^{k+1} \varepsilon^2(t,x_1,x_2)dx_1\right)^{1/2} dx_2 \\
&\quad\leq \frac{4c}{\delta}\sum_{k=0}^{+\infty}\left(\int_{\mathbb{R}} e^{-\delta|x_2|}dx_2\right)^{1/2}\left(\int_{\mathbb{R}}\int_k^{+\infty} \varepsilon^2(t,x_1,x_2)dx_1dx_2\right)^{1/2} \\
&\quad\leq \frac{4\sqrt{2}c}{\delta^{3/2}}C^{1/2}\sum_{k=0}^{+\infty} e^{-\frac{k}{4M}},
\end{aligned}$$

assuming $\alpha_0 > 0$ is sufficiently small so that we can apply Corollary 6.2 in the last inequality. To complete the proof we take

$$M_0 = cC_1\alpha_0 + \frac{4\sqrt{2}c}{\delta^{3/2}}C^{1/2}\sum_{k=0}^{+\infty} e^{-\frac{k}{4M}} < +\infty.$$

□

The next theorem provides a strictly positive lower bound for $\left|\frac{d}{dt}J(t)\right|$.

THEOREM 6.4. *If $\alpha_0 > 0$ is sufficiently small, then there exists a constant $a_0 > 0$ such that*

$$\left|\frac{d}{dt}J(t)\right| \geq a_0 > 0 \quad \text{for all} \quad t \geq 0.$$

PROOF. Let $\alpha_0 < \min\{\alpha_1, \alpha_2\}$, so that we can apply Lemmas 4.2 and 4.4. In view of (5.10), we have

$$\frac{d}{dt}J = \beta\lambda_0 \int \varepsilon\chi_0 + K(\varepsilon), \tag{6.9}$$

where, since $y_2(t) = 0$ for all $t \geq 0$,

$$K(\varepsilon) = \int Q\varepsilon - (y_1' - 1)\int \varepsilon(\Lambda Q + \beta\chi_0) - \int R(\varepsilon)F$$

and $R(\varepsilon)$ satisfy (4.3). We estimate the terms in $K(\varepsilon)$ separately. First, observe that $\|\varepsilon(t)\| \leq 1$ by (4.10), if $\alpha_0 < (2C_1)^{-1}$. Hence, from the Gagliardo-Nirenberg inequality (3.9), we deduce

$$\begin{aligned}
\|R(\varepsilon)\|_1 \leq & C_0 \sum_{k=2}^{p} (\|\varepsilon\|_k^k + \|\varepsilon_{x_1}\|_2 \|\varepsilon\|_{2(k-1)}^{k-1}) \\
\leq & cC_0 \sum_{k=2}^{p} (\|\nabla\varepsilon\|_2^{k-2}\|\varepsilon\|_2^2 + \|\varepsilon_{x_1}\|_2\|\nabla\varepsilon\|_2^{k-2}\|\varepsilon\|_2) \\
\leq & C_5\|\varepsilon\|_2\|\varepsilon\|_{H^1},
\end{aligned}$$

where $C_5 = 2pcC_0$. Thus,

$$\left|\int R(\varepsilon)F\right| \leq C_5\|F\|_\infty\|\varepsilon\|_2\|\varepsilon\|_{H^1}. \tag{6.10}$$

On the other hand, from (4.11) and Cauchy-Schwarz inequality

$$\left|(y_1' - 1)\int \varepsilon(\Lambda Q + \beta\chi_0)\right| \leq C_2\|\varepsilon\|_2 \left|\int \varepsilon(\Lambda Q + \beta\chi_0)\right| \leq C_2\|\Lambda Q + \beta\chi_0\|_2\|\varepsilon\|_2\|\varepsilon\|_{H^1}.$$

Finally, the mass conservation (1.3) for the solution u, the definition of ε (4.9) and the relation (6.2) imply

$$\int Q^2 = \int u_0 = \int u(t) = \int Q^2 + 2\int \varepsilon Q + \int \varepsilon^2,$$

and hence,

$$\left|\int \varepsilon Q\right| \leq \frac{1}{2}\|\varepsilon\|_2^2. \tag{6.11}$$

Collecting (6.10)-(6.11), we obtain that there exists a universal constant (depending only on p) $C_6 > 0$ such that

$$|K(\varepsilon)| \leq C_6\|\varepsilon\|_2\|\varepsilon\|_{H^1}. \tag{6.12}$$

Now, let

$$\theta(t) = \int \varepsilon(t)\chi_0.$$

From Lemma 3.4 we deduce

$$\theta^2(t) \geq \frac{k_1}{k_2}\|\varepsilon(t)\|_2^2 - \frac{1}{k_2}(L\varepsilon(t), \varepsilon(t)).$$

In order to estimate $(L\varepsilon(t), \varepsilon(t))$, we invoke Lemma 3.5 to get

$$W[u_0] = W[u(t)] = W[Q + \varepsilon(t)] = W[Q] + \frac{1}{2}(L\varepsilon(t), \varepsilon(t)) + H[\varepsilon(t)]$$

and then from the definition of the Weinstein's functional W, we get

$$(L\varepsilon(t), \varepsilon(t)) = 2(E[u_0] - E[Q]) + (M[u_0] - M[Q]) - 2H[\varepsilon(t)] = -\delta_0 - 2H[\varepsilon(t)],$$

where for the first two terms we can find $\delta_0 > 0$ (by the construction of $\{u_{0,n}\}$ in Proposition 6.1). Thus,

$$\begin{aligned} \theta^2(t) \geq& \frac{k_1}{k_2}\|\varepsilon(t)\|_2^2 + \frac{1}{k_2}(\delta_0 + 2H[\varepsilon(t)]) \\ \geq& \frac{k_1}{2k_2}\|\varepsilon(t)\|_2^2 + \frac{\delta_0}{k_2}, \end{aligned} \tag{6.13}$$

where in the last line we chose $\alpha_0 < k_1(4C'C_1k_2)^{-1}$ in order to use inequalities (3.7) and (4.10).

Therefore, $\theta^2(t)$ is a strictly positive number for all $t \in \mathbb{R}$, and hence, the sign of $\theta(t)$ remains the same during the evolution. First, assume that $\theta(t)$ is positive for all $t \in \mathbb{R}$, then (6.13) implies

$$\theta(t) \geq \sqrt{\frac{k_1}{2k_2}\|\varepsilon(t)\|_2^2 + \frac{\delta_0}{k_2}} \geq c\left(\sqrt{\frac{k_1}{2k_2}}\|\varepsilon(t)\|_2 + \sqrt{\frac{\delta_0}{k_2}}\right).$$

Plugging the last inequality in (6.9), we obtain

$$\frac{d}{dt}J \geq c\beta\lambda_0\sqrt{\frac{k_1}{2k_2}}\|\varepsilon(t)\|_2 + c\beta\lambda_0\sqrt{\frac{\delta_0}{k_2}} + K(\varepsilon).$$

Finally, from (6.12) we can choose $\alpha_0 > 0$, sufficiently small, such that

$$\frac{d}{dt}J \geq \frac{c\beta\lambda_0}{2}\sqrt{\frac{k_1}{2k_2}}\|\varepsilon(t)\|_2 + c\beta\lambda_0\sqrt{\frac{\delta_0}{k_2}} \geq c\beta\lambda_0\sqrt{\frac{\delta_0}{k_2}} > 0.$$

If $\theta(t)$ is negative, then arguing as above, we can show that for $\alpha_0 > 0$, sufficiently small, there exists $a_0 > 0$ such that $\frac{d}{dt}J(t) \leq -a_0 < 0$, concluding the proof of Theorem 6.4. □

Proof of Theorem 1.3. If $\frac{d}{dt}J(t) \geq a_0 > 0$, then integrating in t variable both sides, we get

$$J(t) \geq a_0 t + J(0) \quad \text{for all} \quad t \geq 0,$$

which is a contradiction with (6.8), the boundedness of $J(t)$ from above. The case when $J'(t) \leq -a_0 < 0$ treats similarly. □

References

[1] J. L. Bona, P. E. Souganidis, and W. A. Strauss, *Stability and instability of solitary waves of Korteweg-de Vries type*, Proc. Roy. Soc. London Ser. A **411** (1987), no. 1841, 395–412. MR897729

[2] S.-M. Chang, S. Gustafson, K. Nakanishi, and T.-P. Tsai, *Spectra of linearized operators for NLS solitary waves*, SIAM J. Math. Anal. **39** (2007/08), no. 4, 1070–1111, DOI 10.1137/050648389. MR2368894

[3] V. Combet, *Construction and characterization of solutions converging to solitons for supercritical gKdV equations*, Differential Integral Equations **23** (2010), no. 5-6, 513–568. MR2654248

[4] R. Côte, C. Muñoz, D. Pilod, and G. Simpson, *Asymptotic stability of high-dimensional Zakharov-Kuznetsov solitons*, Arch. Ration. Mech. Anal. **220** (2016), no. 2, 639–710, DOI 10.1007/s00205-015-0939-x. MR3461359

[5] A. de Bouard, *Stability and instability of some nonlinear dispersive solitary waves in higher dimension*, Proc. Roy. Soc. Edinburgh Sect. A **126** (1996), no. 1, 89–112, DOI 10.1017/S0308210500030614. MR1378834

[6] T. Duyckaerts, J. Holmer, and S. Roudenko, *Scattering for the non-radial 3D cubic nonlinear Schrödinger equation*, Math. Res. Lett. **15** (2008), no. 6, 1233–1250, DOI 10.4310/MRL.2008.v15.n6.a13. MR2470397

[7] A. V. Faminskiĭ, *The Cauchy problem for the Zakharov-Kuznetsov equation* (Russian, with Russian summary), Differentsial′nye Uravneniya **31** (1995), no. 6, 1070–1081, 1103; English transl., Differential Equations **31** (1995), no. 6, 1002–1012. MR1383936

[8] L. G. Farah, J. Holmer, and S. Roudenko, *Instability of solitons–revisited, I: The critical generalized KdV equation*, Nonlinear Dispersive Waves and Fluids, Contemp. Math., vol. 725, Amer. Math. Soc., Providence, RI, 2019.

[9] L. G. Farah, F. Linares, and A. Pastor, *A note on the 2D generalized Zakharov-Kuznetsov equation: local, global, and scattering results*, J. Differential Equations **253** (2012), no. 8, 2558–2571, DOI 10.1016/j.jde.2012.05.019. MR2950463

[10] M. Grillakis, J. Shatah, and W. Strauss, *Stability theory of solitary waves in the presence of symmetry. I*, J. Funct. Anal. **74** (1987), no. 1, 160–197, DOI 10.1016/0022-1236(87)90044-9. MR901236

[11] A. Grünrock and S. Herr, *The Fourier restriction norm method for the Zakharov-Kuznetsov equation*, Discrete Contin. Dyn. Syst. **34** (2014), no. 5, 2061–2068, DOI 10.3934/dcds.2014.34.2061. MR3124726

[12] J. Holmer and S. Roudenko, *A sharp condition for scattering of the radial 3D cubic nonlinear Schrödinger equation*, Comm. Math. Phys. **282** (2008), no. 2, 435–467, DOI 10.1007/s00220-008-0529-y. MR2421484

[13] M. K. Kwong, *Uniqueness of positive solutions of $\Delta u - u + u^p = 0$ in $\mathbf{R}^n$*, Arch. Rational Mech. Anal. **105** (1989), no. 3, 243–266, DOI 10.1007/BF00251502. MR969899

[14] F. Linares and A. Pastor, *Well-posedness for the two-dimensional modified Zakharov-Kuznetsov equation*, SIAM J. Math. Anal. **41** (2009), no. 4, 1323–1339, DOI 10.1137/080739173. MR2540268

[15] F. Linares and A. Pastor, *Local and global well-posedness for the 2D generalized Zakharov-Kuznetsov equation*, J. Funct. Anal. **260** (2011), no. 4, 1060–1085, DOI 10.1016/j.jfa.2010.11.005. MR2747014

[16] M. Mariş, *Existence of nonstationary bubbles in higher dimensions* (English, with English and French summaries), J. Math. Pures Appl. (9) **81** (2002), no. 12, 1207–1239, DOI 10.1016/S0021-7824(02)01274-6. MR1952162

[17] Y. Martel and F. Merle, *Asymptotic stability of solitons of the subcritical gKdV equations revisited*, Nonlinearity **18** (2005), no. 1, 55–80, DOI 10.1088/0951-7715/18/1/004. MR2109467

[18] L. Molinet and D. Pilod, *Bilinear Strichartz estimates for the Zakharov-Kuznetsov equation and applications*, Ann. Inst. H. Poincaré Anal. Non Linéaire **32** (2015), no. 2, 347–371, DOI 10.1016/j.anihpc.2013.12.003. MR3325241

[19] S. Monro and E. J. Parkes, *The derivation of a modified Zakharov-Kuznetsov equation and the stability of its solutions*, J. Plasma Phys. 62 (3) (1999), 305–317.

[20] S. Monro and E. J. Parkes, *Stability of solitary-wave solutions to a modified Zakharov-Kuznetsov equation*, J. Plasma Phys. 64 (3) (2000), 411–426.

[21] F. Ribaud and S. Vento, *A note on the Cauchy problem for the 2D generalized Zakharov-Kuznetsov equations* (English, with English and French summaries), C. R. Math. Acad. Sci. Paris **350** (2012), no. 9-10, 499–503, DOI 10.1016/j.crma.2012.05.007. MR2929056

[22] M. Schechter, *Spectra of partial differential operators*, 2nd ed., North-Holland Series in Applied Mathematics and Mechanics, vol. 14, North-Holland Publishing Co., Amsterdam, 1986. MR869254

[23] M. I. Weinstein, *Modulational stability of ground states of nonlinear Schrödinger equations*, SIAM J. Math. Anal. **16** (1985), no. 3, 472–491, DOI 10.1137/0516034. MR783974

[24] V. E. Zakharov and E. A. Kuznetsov, *On three dimensional solitons*, Zhurnal Eksp. Teoret. Fiz, 66, 594–597 [in russian]; Sov. Phys JETP, vol. 39, no. 2 (1974), 285–286.

Department of Mathematics, Universidade Federal de Minas Gerais, Belo Horizonte, 31270-901, Brazil
Email address: lgfarah@gmail.com

Department of Mathematics, Brown University, Providence, Rhode Island 02912
Email address: holmer@math.brown.edu

Department of Mathematics and Statistics, Florida International University, Miami, Florida 33199
Email address: sroudenko@fiu.edu

Contemporary Mathematics
Volume **725**, 2019
https://doi.org/10.1090/conm/725/14548

Stabilization of dispersion-generalized Benjamin-Ono

Cynthia Flores, Seungly Oh, and Derek Smith

Abstract. In this article, we examine L^2 well-posedness and stabilization property of the dispersion-generalized Benjamin-Ono equation with periodic boundary conditions. The main ingredient of our proof is a development of dissipation-normalized Bourgain space, which gains smoothing properties simultaneously from dissipation and dispersion within the equation. We will establish a bilinear estimate for the derivative nonlinearity using this space and prove the linear observability inequality leading to small-data stabilization.

1. Introduction

We study dispersion generalized Benjamin-Ono (DGBO) equations with periodic boundary conditions given by

$$\partial_t u + D^\alpha \partial_x u + u\partial_x u = 0, \qquad t \in \mathbb{R}, x \in \mathbb{T}, \tag{1.1}$$

where $\alpha \in (1,2)$ and $\mathbb{T} = [-\pi, \pi]$ is a torus. The endpoint $\alpha = 1$ corresponds to the periodic Benjamin-Ono equation, and $\alpha = 2$ corresponds to the well-known periodic Korteweg-de Vries equation. In this sense, (1.1) defines a continuum of equations of dispersive strength intermediate to two celebrated models. The classical KdV and BO equations arise as models in one-dimensional wave propagation; the former modeling surface waves in a shallow, narrow canal and the second modeling the propagation of internal waves in a stratified fluid of infinite depth.

One difficulty with DGBO models is the strength of the nonlinearity relative to dispersion. This hinders bilinear Strichartz estimates, which are used to establish local well-posedness via contraction. In case of $\alpha = 2$ [**2, 12**] or $\alpha > 2$ [**16**], dispersion is sufficient to establish the bilinear Strichartz estimates, but the analogue fails for any $\alpha < 2$. In fact, for DGBO (1.1) on $\mathbb{R}$, Molinet, Saut and Tzvetkov [**23**] showed that the solution map fails to be C^2, indicating that perturbative methods of directly establishing local well-posedness via bilinear estimates would fail.

Extensive work has been completed on the DGBO on $\mathbb{R}$. To work around the problem of lack of bilinear estimates, energy methods or modifications of initial function spaces have been used in the following. Ginibre and Velo [**4**] proved the existence of global weak solutions. Colliander, Kenig and Staffilani [**3**] proved a well-posedness statement using weighted spaces; and Herr [**7**] showed a local well-posedness in a Sobolev space by imposing a low-frequency restriction and using the contraction principle. Kenig, Ponce and Vega [**11**] used energy method to show that

2010 *Mathematics Subject Classification.* Primary 35Q53; Secondary 93D15.

(1.1) is locally well-posed on $H^s(\mathbb{R})$ for $s \geq \frac{3}{4}(3-\alpha)$. Guo [**6**] improved this range to $s > 2-\alpha$; and Herr, Ionescu, Kenig and Koch [**8**] improved this to $s \geq 0$ using a para-differential renormalization technique. This last result naturally extended to the global well-posedness since L^2 norm is conserved during DGBO evolution.

The well-posedness theory on the periodic domain is not as complete. Molinet and Vento [**24**] worked with equations where $D^\alpha \partial_x$ in (1.1) is replaced by a general class of dispersive symbols. The authors proved that this class of equation is locally well-posed in H^s for $s \geq 1 - \frac{\alpha}{2}$ both on the real line and the torus.

We are interested in stabilization of the periodic DGBO equations as well as well-posedness properties. For the periodic KdV equation, stabilization was first proved by Komornik, Russell and Zhang [**13**]. Russell, Zhang [**25, 26**]; Laurent, Rosier and Zhang [**15**] extended this result and also proved controllability of the equation. For the Benjamin-Ono equation, stabilization is more difficult due to a lack of the bilinear Strichartz estimate. In [**17**], Linares and Rosier proved a local stabilization in $H^s(\mathbb{T})$ for $s > \frac{1}{2}$ and semi-global stabilization in $L^2(\mathbb{T})$ inserting dissipation into the damping term and utilizing propagation of regularity to obtain a smoothing effect on the whole domain. In [**14**], Laurent, Linares and Rosier extended this result to a global L^2 stabilization without dissipation by using Tao's gauge transform [**29**] and bilinear estimates proved by Molinet and Pilod [**19**].

In this article, we investigate the stabilization of a locally-damped variant of (1.1) given by

$$\begin{cases} \partial_t v + D^\alpha \partial_x v + GD^\beta G v = \partial_x(v^2), & x \in \mathbb{T}, t > 0, \\ v|_{t=0} = v_0 \in H^s(\mathbb{T}) \end{cases} \tag{1.2}$$

where $\alpha \in (1,2]$, $\beta \in (2-\alpha, \alpha)$. The definition

$$(Gh)(x,t) := g(x)\left(h(x,t) - \int_{\mathbb{T}} g(y)h(y,t)dy\right) \tag{1.3}$$

ensures that the equation conserves volume. We consider the fixed function $g \in C^\infty(\mathbb{T})$, which is nonnegative and satisfies $2\pi[g] = \displaystyle\int_{\mathbb{T}} g(x)dx = 1$, to be a localizing function supported on an arbitrary interval on $\mathbb{T}$. Note that $\partial_t \int_{\mathbb{T}} v\, dx = 0$ for any smooth solution v of (1.2). Further, from the self-adjoint property of G, L^2 norm of any smooth solution is non-increasing due to the identy

$$\partial_t \|v\|^2_{L^2_x(\mathbb{T})} = -\int_{\mathbb{T}} \left| D^{\frac{\beta}{2}} G v \right|^2 dx.$$

To place our work in context, we closely examine two previously mentioned works on KdV [**15**] and Benjamin-Ono [**17**] equations. Although a better result for Benjamin-Ono exists in [**14**], we do not relate our article to this result due to the difficulty in applying gauge transformation in DGBO contexts. In both articles, stabilization arguments heavily relied on the recovery of one-derivative in the nonlinearity v^2. For the Benjamin-Ono equation, this was achieved by adding a dissipative derivative GDG within the localized damping term. This dissipative derivative was used to recover one derivative in the nonlinearity. On the other hand, a bilinear Strichartz estimate in Bourgain space [**2**] was sufficient to recover one derivative for KdV.

For the intermediate range $1 < \alpha < 2$, it would be possible to obtain the local results of [**17**] with the dissipative GDG control term by adapting the smoothing property from [**4**] to the periodic setting. However, our approach applies to L^2

solutions, as well as bridges the transition between the Benjamin-Ono and the KdV results in the sense that the dissipative derivative is reduced to zero as $\alpha \uparrow 2$. To this end, we develop a functional space with a Bourgain-type weight, yet with a normalized dissipative derivative built in.

Earlier, Molinet and Ribaud [**20–22**] developed a dissipative Bourgain norm to handle multilinear estimates for semilinear equations with mixed dispersion and dissipation. Here, the authors introduced complex factors to the Bourgain weight to account for dissipative smoothing. Our approach to defining the dissipative Bourgain weight will go through a normalization, which appears to be very effective in this case. Also, our setting is distinct from earlier cited references because the dissipation in (1.2) is localized, making a Fourier-analytic approach more challenging.

We will prove the bilinear estimate in Section 5. In doing so, we will establish the global well-posedness of (1.2) in L^2 with mean-zero condition with intermediate dissipation $GD^\beta G$ with $\beta > 2 - \alpha$.

THEOREM 1. *Let $\alpha \in (1, 2]$ and $2 - \alpha < \beta < \alpha$. Then (1.2) is locally and globally well-posed for initial data $v_0 \in L^2$ given $\int_{\mathbb{T}} v_0 = 0$. More specifically, given any $T > 0$, there exists a unique solution $v \in Z_T^{\frac{1}{2}+} \hookrightarrow C_t^0([0, T]; L^2(\mathbb{T}))$. Further, the solution map $v_0 \in L^2(\mathbb{T}) \mapsto v(t) \in C_t^0([0, T]; L^2(\mathbb{T}))$ is uniformly continuous within a bounded set in L^2.*

REMARK. The above well-posedness result is established via a contraction in Z^b, which will be defined in Section 2. For this article, we restrict our attention to the class of initial values satisfying the mean-zero property, but these statements can be easily extended via the transformation: $v(t, x) \mapsto v(t, x - ct)$ for an appropriate constant c.

Further, we have a small-data stabilization theorem in the following:

THEOREM 2. *Let $\alpha \in (1, 2]$ and $2 - \alpha < \beta < \alpha$. Then there exists $0 < \delta \ll 1$ and $\lambda > 0$ such that, if $v_0 \in L^2$ with $\int_{\mathbb{T}} v_0 = 0$ and $\|v_0\|_{L^2} < \delta$, then the solution v of (1.2) satisfies*

$$\|v(t)\|_{L_x^2} \leq e^{-\lambda t} \|v_0\|_{L_x^2}.$$

REMARK. The proof of this theorem follows from stabilization of the associated linear equation combined with the contraction principle. An application of Ingham's inequality yields a linear unique continuation property (Proposition 7), which yields linear stabilization via an observability argument. The existence of an observability inequality for the nonlinear equation would remove the small-data condition above.

Our discussion is organized as follows. In Section 2, we will define necessary notations and the functional space to be used. In Section 3, we will prove key estimates for the new functional space. Sections 4 and 5 focus on proving necessary linear and bilinear estimates to control the right side of the equation. In Section 6, we will prove Theorem 1. Section 7 will deal with linear stabilization by proving the key observability inequality, and Section 8 will provide the proof of Theorem 2. Lastly, Appendix A contains a brief discussion of A_p weight theories, which is used in Section 3 within proofs.

2. Preliminaries

2.1. Notations. We adopt the standard notation in approximate inequalities as follows: By $A \lesssim B$, we mean that there exists an absolute constant $C > 0$ with

$A \le CB$. $A \ll B$ means that the implicit constant is taken to be a *sufficiently* large positive number. For any number of quantities $\alpha_1, \dots, \alpha_k$, $A \lesssim_{\alpha_1,\dots,\alpha_k} B$ means that the implicit constant depends only on $\alpha_1, \dots, \alpha_k$. Finally, by $A \sim B$, we mean $A \lesssim B$ and $B \lesssim A$.

We indicate by η a smooth time cut-off function which is supported on $[-2, 2]$ and equals 1 on $[-1, 1]$. Notations here will be relaxed, since the exact expression of η will not influence the outcome.

For any normed space $\mathcal{Y}$, we denote the quantity $\|\cdot\|_{\mathcal{Y}_T}$ by the expression

$$\|u\|_{\mathcal{Y}_T} = \inf\{\|v\|_{\mathcal{Y}} : v(t) \equiv u(t), \text{ for } t \in [0, T]\}.$$

Fourier coefficients in x and Fourier transform in t are denoted as follows:

$$\mathcal{F}[f]_k = f_k := \int_{\mathbf{T}} f(x)e^{-ikx}\,dx, \qquad \widehat{u}(\tau) = \int_{\mathbf{R}} u(t)e^{-it\tau}\,dt.$$

Also, we define $\langle k \rangle := (1 + |k|^2)^{\frac{1}{2}}$ and denote $L^2_0(\mathbb{T}) := \{u \in L^2(\mathbb{T}) : \int_{\mathbb{T}} u = 0\}$.

2.2. Functional space. In this section, we develop a new type of Fourier-restriction space denoted Z^b. This space is largely motivated by the class of functional space developed by Bourgain [**1, 2**]. Conventionally, Bourgain space is used to gain smoothing via dispersion. On the other hand, the space introduced in this section contains a factor of normalized dissipation so that smoothing can be gained from both dissipation and dispersion simultaneously. This is the main novelty of our method.

Given a linear dispersive symbol L_k, define the norm

$$\|u\|_{Z^b} := \begin{cases} \left\| \langle k \rangle^{b\beta} \left\langle \dfrac{\tau - L_k}{\langle k \rangle^{\beta}} \right\rangle^{b} \widehat{u}_k(\tau) \right\|_{L^2_\tau l^2_k(\mathbb{R}\times\mathbb{Z}^*)} & \text{if } b \in \left(-\frac{1}{2}, \frac{1}{2}\right), \\ \left\| \langle k \rangle^{\operatorname{sgn}(b)\frac{\beta}{2}} \left\langle \dfrac{\tau - L_k}{\langle k \rangle^{\beta}} \right\rangle^{b} \widehat{u}_k(\tau) \right\|_{L^2_\tau l^2_k(\mathbb{R}\times\mathbb{Z}^*)} & \text{otherwise.} \end{cases}$$

where $\mathbb{Z}^* = \mathbb{Z} \setminus \{0\}$. In context of DGBO, the symbol L_k is $k|k|^\alpha$.

By construction, the dual of this space is given as $(Z^b)^* = Z^{-b}$ if L_k is odd in k. Otherwise, $(Z^b)^* = \overline{Z^{-b}}$.

2.3. Equation set-up. We begin our set-up by examining the localized damping $GD^\beta G$. Simple computations show

$$\begin{aligned} GD^\beta Gv &= g\left(D^\beta Gv - \int_{\mathbb{T}} g(y) D^\beta Gv(y)\,dy\right) \\ &= \left(g\,D^\beta(g\,v) - \int g\,D^\beta(g\,v)\right) + \mathcal{R}[v] \end{aligned}$$

where $\mathcal{R}$ is a bounded operator in $L^2(\mathbb{T})$. Now, we examine the main dissipative term $g\, D^\beta[g\, v]$. For $k \neq 0$,

$$\begin{aligned}\mathcal{F}[g(x)D^\beta[g(x)v(x)]]_k &= \sum_{m,n} |m|^\beta\, g_{k-m}\, g_{m-n}\, v_n \\ &= \sum_m |m|^\beta\, g_{k-m} g_{m-k}\, v_k + \sum_m \sum_{n\neq k} |m|^\beta\, g_{k-m}\, g_{m-n}\, v_n \\ &=: c_k v_k + \mathcal{N}_1[v]_k.\end{aligned}$$

From the expression above, the main dissipation occurs when $n = k$, i.e. the diagonal frequency. All off-diagonal frequencies ($n \neq k$) will be treated as a perturbation and estimated on the RHS of the equation. In that spirit, we will denote the second term on the RHS above by $\mathcal{N}_1$.

In the following, we derive a very useful property of c_k.

CLAIM 1. For $k \neq 0$, $c_k \sim_{g,\beta} \langle k\rangle^\beta$.

PROOF. We will use the fact that $g_0 = \int_{\mathbb{T}} g \neq 0$ and also that g is a real-valued non-constant function.

Note $g_{k-m} = \overline{g_{m-k}}$ since g is real-valued. Thus $c_k = \sum_m |m|^\beta |g_{m-k}|^2$. First, by picking $m = k$, we can show the lower bound $c_k \geq |k|^\beta |g_0|^2$ which is non-zero because $g_0 = [g] \neq 0$.

To prove the upper bound,

$$c_k \leq \sum_m (|m-k|^\beta + |k|^\beta)|g_{m-k}|^2 \sim \|g\|_{H^\beta}^2 + |k|^\beta \|g\|_{L^2}^2.$$

This proves the claim. □

To summarize, we can write

$$GD^\beta G v = \widetilde{D^\beta} v + \mathcal{N}_1[v] + \mathcal{R}[v]$$

where $\widetilde{D^\beta}$ is defined via the multiplication of Fourier coefficients by c_k. In light of Claim 1, this approximately acts as a derivative of order β.

Thus, we can re-write (1.2) as

$$\partial_t v + D^\alpha \partial_x v + \widetilde{D^\beta} v = -\mathcal{N}_1[v] - \partial_x(v^2) - \mathcal{R}[v].$$

Taking Fourier-coefficients of the equation above, we can reformulate (1.2) using the variation of parameters,

$$\begin{aligned} v_k(t) =& e^{-(ik|k|^\alpha + c_k)t}(v_0)_k \\ &- \int_0^t e^{-(ik|k|^\alpha + c_k)(t-s)} \left(\mathcal{N}_1[v]_k(s) + (\partial_x(v^2))_k(s) + \mathcal{R}[v]_k(s)\right) ds.\end{aligned}$$

To prevent a backward parabolic propagation, we place absolute values around time variables associated with the dissipative coefficients c_k.

$$\begin{aligned} v_k(t) =& e^{-ik|k|^\alpha t - c_k|t|}(v_0)_k \\ &- \int_0^t e^{-ik|k|^\alpha (t-s) - c_k|t-s|} \left(\mathcal{N}_1[v]_k(s) + (\partial_x(v^2))_k(s) + \mathcal{R}[v]_k(s)\right) ds.\end{aligned} \tag{2.1}$$

3. Bourgain space estimates

In this section, we establish key linear estimates for Z^b.

PROPOSITION 1. *We have following continuous embedding properties:*

$$Z^b \hookrightarrow Z^{b'} \text{ for } \forall b \geq b', \tag{3.1}$$

$$Z^b \hookrightarrow C^0(\mathbb{R}; L_0^2(\mathbb{T})) \cap L^2(\mathbb{R}; H_0^{\frac{\beta}{2}}(\mathbb{T})) \text{ if } b > \frac{1}{2}. \tag{3.2}$$

REMARK. Using definition, (3.2) can be rephased as $Z_T^b \hookrightarrow C^0([0,T]; L_0^2(\mathbb{T})) \cap L^2([0,T]; H_0^{\frac{\beta}{2}}(\mathbb{T}))$ for any $b > \frac{1}{2}$.

PROOF. First, (3.1) follows from definition of the norm.

Consider, (3.2). The first embedding $Z^b \hookrightarrow L_t^2 H_x^{\frac{\beta}{2}}$ directly follows from definition. To see $Z^b \hookrightarrow C_t^0 L_x^2$ for $b > \frac{1}{2}$, note

$$\|v\|_{C_t^0 L_x^2} \leq \|\widehat{v}_k\|_{l_k^2 L_\tau^1} \leq \sup_k \left\| \langle k \rangle^{-\frac{\beta}{2}} \left\langle \frac{\tau - L_k}{\langle k \rangle^\beta} \right\rangle^{-b} \right\|_{L_\tau^2} \left\| \langle k \rangle^{\frac{\beta}{2}} \left\langle \frac{\tau - L_k}{\langle k \rangle^\beta} \right\rangle^{b} \widehat{v}_k \right\|_{L_\tau^2 l_k^2}.$$

The second expression on RHS is $\|v\|_{Z^b}$. So it suffices so show that the first norm is finite. We have

$$\left\| \langle k \rangle^{-\frac{\beta}{2}} \left\langle \frac{\tau - L_k}{\langle k \rangle^\beta} \right\rangle^{-b} \right\|_{L_\tau^2}^2 = \langle k \rangle^{-\beta} \int_{\mathbb{R}} \left\langle \frac{\tau - L_k}{\langle k \rangle^\beta} \right\rangle^{-2b} d\tau = \int_{\mathbb{R}} \langle \tau \rangle^{-2b} \, d\tau, \tag{3.3}$$

which is finite if $b > \frac{1}{2}$. This completes the proof. □

The next lemma asserts that the free solution is bounded in Z^b. To alleviate notations, introduce the semigroup $S(t)$ defined for any $f \in L^2$ via

$$\mathcal{F}[S(t)f]_k = e^{-iL_k t - c_k |t|} f_k.$$

PROPOSITION 2. *For* $b < \frac{3}{2}$,

$$\|S(t)f\|_{Z^b} \lesssim_b \|f\|_{L_x^2}.$$

REMARK. In purely dispersive settings, a smooth time-cutoff function $\eta(t)$ must be multiplied to the LHS above in order to establish the inequality (c.f. [**30**, Lemma 2.8]). However, the built-in dissipation in the free-solution in this case makes it square integrable as long as b is not too large. This upper restriction in b can be removed by imposing a smooth time-cutoff function.

PROOF. It suffices to prove this statement for $b \in (\frac{1}{2}, \frac{3}{2})$ by the embedding (3.1). Denote φ_k to be the time Fourier transform of $e^{-c_k|t|}$:

$$\varphi_k(\tau) = 2\int_0^\infty e^{-c_k t} \cos(\tau t)\, dt = \operatorname{Re}\left[\frac{1}{c_k + i\tau}\right] = \frac{c_k}{c_k^2 + \tau^2} = \frac{1}{c_k} \frac{1}{1 + \left(\frac{\tau}{c_k}\right)^2}. \tag{3.4}$$

Using Claim 1,

$$\mathcal{F}_t[e^{-iL_k t - c_k|t|}](\tau) = \varphi_k(\tau - L_k) = \frac{1}{c_k} \left\langle \frac{\tau - L_k}{c_k} \right\rangle^{-2} \sim \langle k \rangle^{-\beta} \left\langle \frac{\tau - L_k}{\langle k \rangle^\beta} \right\rangle^{-2}.$$

Then for $b > \frac{1}{2}$, we can write

$$\|S(t)f\|_{Z^b} \sim \left\| \langle k \rangle^{-\frac{\beta}{2}} \left\langle \frac{\tau - L_k}{\langle k \rangle^\beta} \right\rangle^{b-2} f_k \right\|_{L^2_\tau l^2_k} = \sup_k \left\| \langle k \rangle^{-\frac{\beta}{2}} \left\langle \frac{\tau - L_k}{\langle k \rangle^\beta} \right\rangle^{b-2} \right\|_{L^2_\tau} \|f_k\|_{l^2_k}.$$

The second term is $\|f\|_{L^2_x}$. The computation (3.3), where $-b$ is replaced with $b-2$, shows that the first term on the RHS above is finite if and only if $2b - 4 < -1 \iff b < \frac{3}{2}$. This proves the claim. □

The next proposition shows that the Z^b space defined here inherits a special property for dispersive Bourgain spaces which is used to derive a contraction factor of T^ε. It is not immediately obvious that this property should carry through in case of the newly defined space Z^b. Thus, we will carefully prove these results here. In the process, we will need a few items from theories of A_p weights which are listed in Appedix A.

PROPOSITION 3. *Let $\eta \in \mathcal{S}_t$. Then for any $b \in \mathbb{R}$,*

$$\|\eta(t)u\|_{Z^b} \lesssim_{\eta,b} \|u\|_{Z^b}. \tag{3.5}$$

Also, for same η, given $T > 0$ and $-\frac{1}{2} < b' \le b < \frac{1}{2}$, we have

$$\|\eta(t/T)u\|_{Z^{b'}} \lesssim_{\eta,b,b'} T^{b-b'} \|u\|_{Z^b}. \tag{3.6}$$

PROOF. First, consider (3.5).

$$\|\eta(t)u\|_{Z^b} = \left\| \langle k \rangle^{\operatorname{sgn}(b)\beta \min(\frac{1}{2},|b|)} \left\langle \frac{\tau - L_k}{\langle k \rangle^\beta} \right\rangle^b \int_{\mathbb{R}} \widehat{\eta}(\tau - \sigma)\widehat{u}_k(\sigma)\, d\sigma \right\|_{L^2_\tau l^2_k}.$$

Using the algebraic identity $\langle \alpha + \beta \rangle^b \lesssim \langle \alpha \rangle^{|b|} \langle \beta \rangle^b$, we have

$$\left\langle \frac{\tau - L_k}{\langle k \rangle^\beta} \right\rangle^b \lesssim \left\langle \frac{\tau - \sigma}{\langle k \rangle^\beta} \right\rangle^{|b|} \left\langle \frac{\sigma - L_k}{\langle k \rangle^\beta} \right\rangle^b \lesssim_c \langle \tau - \sigma \rangle^{|b|} \left\langle \frac{\sigma - L_k}{\langle k \rangle^\beta} \right\rangle^b.$$

Now $\|\eta(t)u\|_{Z^b}$ is bounded by

$$\left\| \int_{\mathbb{R}} \langle \tau - \sigma \rangle^{|b|} |\widehat{\eta}| (\tau - \sigma) \left(\langle k \rangle^{\operatorname{sgn}(b)\beta \min(\frac{1}{2},|b|)} \left\langle \frac{\sigma - L_k}{\langle k \rangle^\beta} \right\rangle^b |\widehat{u}_k| (\sigma) \right) d\sigma \right\|_{L^2_\tau l^2_k}.$$

Using Young's inequality, above is bounded $\left\| \langle \cdot \rangle^{|b|} \widehat{\eta} \right\|_{L^1_\tau} \|u\|_{Z^b}$. This proves (3.5).

Next, consider (3.6). Say that $0 \le b' \le b < \frac{1}{2}$, then the negative range will follow from duality. This can be seen as follows. Say that $-\frac{1}{2} < b' < b < 0$, then

$$\begin{aligned}
\|\varphi(t/T)u\|_{Z^{b'}} &= \sup_{\|v\|_{Z^{-b'}}=1} \left| \int_{\mathbb{R}} \varphi(t/T)uv\, dt\, dx \right| \\
&\lesssim \sup_{\|v\|_{Z^{-b'}}=1} \|u\|_{Z^b} \|\varphi(t/T)v\|_{Z^{-b}} \\
&\lesssim_{b,b',\varphi} \sup_{\|v\|_{Z^{-b'}}=1} \|u\|_{Z^b} T^{-b'+b} \|v\|_{Z^{-b'}} = T^{b-b'} \|u\|_{Z^b}.
\end{aligned}$$

Also, if $-\frac{1}{2} < b' < 0 < b < \frac{1}{2}$, use $\varphi(t/T) = \varphi(t/T)\varphi(t/2T)$ to write

$$\|\varphi(t/T)u\|_{Z^{b'}} = \|\varphi(t/T)\varphi(t/2T)u\|_{Z^{0,b'}} \lesssim T^{-b'}\|\varphi(t/2T)u\|_{Z^0} \lesssim T^{b-b'}\|u\|_{Z^b}.$$

By above arguments, it suffices to assume $0 \le b' \le b < \frac{1}{2}$. Given $b < \frac{1}{2}$, we will show (3.6) by interpolating the following two inequalities:

$$\|\eta(t/T)v\|_{Z^b} \lesssim \|v\|_{Z^b}, \tag{3.7}$$

$$\|\eta(t/T)v\|_{Z^0} \lesssim T^b\|v\|_{Z^b}. \tag{3.8}$$

First, (3.7) is proved using A_p weights from Appendix A. By definition of the maximal function defined in Appendix A, note

$$\left\| \langle k\rangle^{\beta b}\left\langle \frac{\tau - L_k}{\langle k\rangle^{\beta}}\right\rangle^{b} \left(T\widehat{\eta}(T\cdot) * \widehat{u_k}\right)(\tau)\right\|_{l_k^2 L_\tau^2} \le \left\| \langle k\rangle^{\beta b}\left\langle \frac{\tau - L_k}{\langle k\rangle^{\beta}}\right\rangle^{b} \left(M\widehat{u_k}\right)(\tau)\right\|_{l_k^2 L_\tau^2}$$

where M is the maximal function defined in Lemma 8. Writing out this expression, note that we need to bound M in $L^2(W)$ where $W = W(\tau) = \left\langle \frac{\tau - L_k}{\langle k\rangle^{\beta}}\right\rangle^{2b}$. By Lemma 6,we have

$$\left[\left\langle \frac{\tau - L_k}{\langle k\rangle^{\beta}}\right\rangle^{2b}\right]_{A_2} = \left[\langle \tau\rangle^{2b}\right]_{A_2},$$

where Claim 8 gives that this expression is finite when $b \in (-\frac{1}{2}, \frac{1}{2})$. Finally, Lemma 8 gives

$$\left\| \langle k\rangle^{\beta b}\left\langle \frac{\tau - L_k}{\langle k\rangle^{\beta}}\right\rangle^{b} \left(M\widehat{u_k}\right)(\tau)\right\|_{l_k^2 L_\tau^2} \lesssim \left\| \langle k\rangle^{\beta b}\left\langle \frac{\tau - L_k}{\langle k\rangle^{\beta}}\right\rangle^{b} |\widehat{u_k}|(\tau)\right\|_{l_k^2 L_\tau^2},$$

which leads to (3.7).

To prove (3.8), we follow as decomposition of Fourier support similar to the proof given in [**30**, Lemma 2.11]. Let $v = v^1 + v^2$ so that the Fourier transform of v^1 is supported in the region $\langle k\rangle^{\beta}\left\langle \frac{\tau - L_k}{\langle k\rangle^{\beta}}\right\rangle \ge \frac{1}{T}$. Then the estimate (3.8) for v^1 directly follows. To obtain (3.8) for v^2, note

$$\|\eta(t/T)v^2\|_{Z^0} = \|\eta(t/T)\|_{L_t^2}\|v^2\|_{L_t^\infty L_x^2} \lesssim T^{\frac{1}{2}}\left\|\widehat{v^2}_k(\tau)\right\|_{l_k^2 l_\tau^1}.$$

We now estimate $\left\|\widehat{v^2}_k(\tau)\right\|_{l_k^2 l_\tau^1}$. Denote $\mathcal{A}$ to be the support of $\widehat{v^2}$, we have

$$\left\|\widehat{v^2}\right\|_{l_k^2 L_\tau^1} \lesssim \sup_k \left(\int_{\mathcal{A}} \langle k\rangle^{-\beta}\left\langle \frac{\tau - L_k}{\langle k\rangle^{\beta}}\right\rangle^{-2b} d\tau\right)^{\frac{1}{2}} \|\widehat{v}\|_{Z^b}.$$

Note $\langle k\rangle^{\beta}\left\langle \frac{\tau - L_k}{\langle k\rangle^{\beta}}\right\rangle \ge |\tau - L_k|$, so we have

$$\mathcal{A} = \left\{\tau : \langle k\rangle^{\beta}\left\langle \frac{\tau - L_k}{\langle k\rangle^{\beta}}\right\rangle \le \frac{1}{T}\right\} \subset \left\{\tau : |\tau - L_k| \le \frac{1}{T}\right\}.$$

Thus,

$$\int_{\mathcal{A}} \langle k \rangle^{-\beta} \left\langle \frac{\tau - L_k}{\langle k \rangle^{\beta}} \right\rangle^{-2b} d\tau \leq \int_{\tau:|\tau - L_k| < \frac{1}{T}} \frac{1}{|\tau - L_k|^{2b}} \, d\tau = 2\left(\frac{1}{T}\right)^{1-2b} = 2T^{2b-1}$$

as long as $1 - 2b > 0 \iff b < \frac{1}{2}$. Plugging this into the previous computations, we have

$$\left\|\eta(t/T)v^2\right\|_{Z^0} \lesssim T^{\frac{1}{2}+\frac{1}{2}(2b-1)} \|v\|_{Z^b},$$

which leads to (3.8). This completes the proof. □

The next proposition is the main tool used to establish smoothing of the nonlinear term. Here, we can see an effect of dissipative smoothing that takes place in Z^b, which is unique to our case.

PROPOSITION 4. *Let f be smooth and rapidly decaying. For any $b \in \left(\frac{1}{2}, \frac{3}{2}\right)$,*

$$\left\| \int_0^t S(t-s)f(s)\, ds \right\|_{Z^b} \lesssim \left\| D^{-\beta(b-\frac{1}{2})} f \right\|_{Z^{b-1}}.$$

REMARK. If we consider $b \approx \frac{1}{2}$, then the norm on the RHS above is approximately $\|f\|_{Z^{-\frac{1}{2}}}$. By definition of this norm, it carries $\|f\|_{Z^{-\frac{1}{2}}} \lesssim \|f\|_{L_t^2 H_x^{-\frac{\beta}{2}}}$. Thus, this proposition shows a smoothing of order $\beta/2$. We remark that this is analogous to the $\frac{1}{2}$ derivative gain [**17**, Proposition 2.16] achieved for the proof of stabilization of Benjamin-Ono equation using the operator GD^1G. However, the method used in [**17**] cannot take advantage of dispersive estimates which we will establish for our case. Thus, we are able to use a smaller dissipation $\beta < 1$ and still acquire enough smoothing to overcome the full nonlinear derivative $\partial_x(v^2)$ in (1.2).

PROOF. First, note the identity $\chi_{(0,t)}(s) = \frac{1}{2}(\operatorname{sgn}(s) + \operatorname{sgn}(t-s))$ for any $t > 0$ and $s \in \mathbb{R}$. Then, we can write the integral on the LHS of the statement as

$$\int_{\mathbb{R}} e^{-i(t-s)L_k - c_k|t-s|} \operatorname{sgn}(s) f_k(s)\, ds + \int_{\mathbb{R}} \operatorname{sgn}(t-s) e^{-i(t-s)L_k - c_k|t-s|} f_k(s)\, ds \tag{3.9}$$

where we have omitted the factor of $\frac{1}{2}$. Note that both integrals are convolutions. We establish the following claim.

CLAIM 2. The following hold,

$$\varphi_k(\tau) := \mathcal{F}_t[e^{-c_k|t|}](\tau) = \frac{1}{c_k} \left\langle \frac{\tau}{c_k} \right\rangle^{-2} \sim \langle k \rangle^{-\beta} \left\langle \frac{\tau}{\langle k \rangle^{\beta}} \right\rangle^{-2}, \tag{3.10}$$

$$\varphi_k^a(\tau) := \mathcal{F}_t[\operatorname{sgn}[t] e^{-c_k|t|}](\tau) = \frac{1}{c_k} \frac{\tau}{c_k} \left\langle \frac{\tau}{c_k} \right\rangle^{-2} \lesssim \langle k \rangle^{-\beta} \left\langle \frac{\tau}{\langle k \rangle^{\beta}} \right\rangle^{-1}. \tag{3.11}$$

PROOF. First, recall that (3.10) was computed (3.4). To compute (3.11), note

$$\varphi_k^a(\tau) = 2\int_0^\infty e^{-c_k t} \sin(\tau t)\, dt = \operatorname{Im}\left[\frac{1}{c_k - i\tau}\right] = \frac{\tau}{c_k^2 + \tau^2} = \frac{1}{c_k} \frac{\frac{\tau}{c_k}}{1 + \left(\frac{\tau}{c_k}\right)^2}.$$

Using Claim 1, this leads to (3.11). □

Using the same notations as in Claim 2, the Fourier transform in time of (3.9) can be written as

$$-i\varphi_k(\tau - L_k)\widehat{\mathcal{H}_\tau f_k}(\tau) + \varphi_k^a(\tau - L_k)\widehat{f_k}(\tau),$$

where $\mathcal{H}_\tau$ is the Hilbert transform in τ. Using Claim 2, the Z^b norm of the Duhamel term is bounded by

$$\left\| \langle k\rangle^{-\frac{\beta}{2}} \left\langle \frac{\tau - L_k}{\langle k\rangle^\beta} \right\rangle^{b-1} \widehat{\mathcal{H}_\tau f_k} \right\|_{L^2_\tau l^2_k} + \left\| \langle k\rangle^{-\frac{\beta}{2}} \left\langle \frac{\tau - L_k}{\langle k\rangle^\beta} \right\rangle^{b-1} \widehat{f_k} \right\|_{L^2_\tau l^2_k} \tag{3.12}$$

where we have given up one power of $\left\langle \frac{\tau - L_k}{\langle k\rangle^\beta} \right\rangle$ in the first norm. The second term in (3.12) is more than sufficiently bounded, so we need to prove the bound for the first term. Here again, we use properties of A_p weights.

We need to show that $\mathcal{H}_\tau$ is bounded in $L^2(W)$ where $W = \left\langle \frac{\tau - L_k}{\langle k\rangle^\beta} \right\rangle^{2b-2}$. As before, we use properties given in Lemma 6 and Claim 8 to conclude that $W \in A_2$ if $2b - 2 \in (-1, 1) \iff b \in (\frac{1}{2}, \frac{3}{2})$. Finally, we use Lemma 7 to conclude that $\mathcal{H}_\tau$ is bounded in $L^2(W)$ uniformly in k.

Then (3.12) $\lesssim \left\| \langle k\rangle^{-\frac{\beta}{2}} \left\langle \frac{\tau - L_k}{\langle k\rangle^\beta} \right\rangle^{b-1} \widehat{f_k} \right\|_{L^2_{\tau,k}}$. Noting $-\frac{\beta}{2} = \beta(b-1) - \beta(b-\frac{1}{2})$, we obtain the claim. □

4. Estimate of linear terms

In this section, we establish that $\mathcal{N}_1$ and $\mathcal{R}$ from (2.1) are bounded. Estimating $\mathcal{R}$ is very simple, but $\mathcal{N}_1$ will be dealt with more carefully. We begin with estimate of $\mathcal{N}_1$.

LEMMA 1. *Let $\alpha > \beta$ be fixed. Given $0 < T \ll 1$ and $b \in \left(\frac{1}{2}, \frac{\alpha}{\alpha+\beta}\right)$, there exists $\varepsilon > 0$ such that*

$$\left\| D^{-\beta(b-\frac{1}{2})} \mathcal{N}_1[v] \right\|_{Z_T^{b-1}} \lesssim_{\varepsilon, b} T^\varepsilon \|v\|_{Z_T^b}.$$

PROOF. Let $u \in Z^b$ such that $u(t) \equiv v(t)$ on $[0, T]$, and $\|u\|_{Z^b} \leq 2\|v\|_{Z_T^b}$. Then the LHS of the desired estimate is bounded by $\left\| D^{-\beta(b-\frac{1}{2})} \eta(t/T) \mathcal{N}_1[u] \right\|_{Z^{b-1}}$.

Denote $\varepsilon > 0$ to be a constant to be chosen later. Note that $b - 1 \in (-\frac{1}{2}, \frac{1}{2})$. Then, by Proposition 3,

$$\left\| D^{-\beta(b-\frac{1}{2})} \eta(t/T) \mathcal{N}_1[u] \right\|_{Z^{b-1}} \lesssim_{\eta, b, \varepsilon} T^\varepsilon \left\| D^{-\beta(b-\frac{1}{2})} \mathcal{N}_1[u] \right\|_{Z^{b-1+\varepsilon}}.$$

The norm on the RHS above can be written as

$$\left\| |k|^{-\frac{\beta}{2}} \left\langle \frac{\tau - L_k}{\langle k\rangle^\beta} \right\rangle^{b-1+\varepsilon} \sum_m \sum_{n \neq k} |m|^\beta g_{k-m} g_{m-n} \widehat{u_n}(\tau) \right\|_{L^2_\tau l^2_k}. \tag{4.1}$$

Denote $f_n(\tau) := \langle n\rangle^{\frac{\beta}{2}} \left\langle \frac{\tau - L_n}{\langle n\rangle^\beta} \right\rangle^b |\widehat{u_n}|(\tau)$. Then

$$(4.1) \lesssim \left\| \sum_{m,n\neq k} |g_{k-m}|\, |g_{m-n}|\, [\mathcal{M}_{n,m,k}(\tau)]\, f_n(\tau) \right\|_{L^2_\tau l^2_k}$$

where

$$\mathcal{M} = \mathcal{M}_{n,m,k}(\tau) := |m|^\beta |k|^{-\frac{\beta}{2}} \langle n\rangle^{-\frac{\beta}{2}+\varepsilon} \left\langle \frac{\tau - L_k}{\langle k\rangle^\beta} \right\rangle^{b-1+\varepsilon} \left\langle \frac{\tau - L_n}{\langle n\rangle^\beta} \right\rangle^{-b}.$$

We focus on this term $\mathcal{M}$. Our goal here is to deal with the β derivative in m frequency, and also to assure that it is summable in m, n.

First, note that if $k \not\sim m$, then we can use the decay in $|g_{k-m}|$ to produce weights $|k-m|^{-N} \lesssim_N \max(|k|, |m|)^{-N}$ for any $N \gg 1$. Same goes for the case when $n \not\sim m$ since we have $|g_{m-n}|$. Thus, we only need to deal with the case when $k \sim m \sim n$. In this case,

$$|m|^\beta |k|^{-\frac{\beta}{2}} \langle n\rangle^{-\frac{\beta}{2}+\varepsilon} \sim \langle k\rangle^\varepsilon, \tag{4.2}$$

so we must recover ε derivatives from the remaining terms.

CLAIM 3. Let $\alpha > 0$. For all $n \neq k$,

$$\max\left(\left\langle \frac{\tau - k|k|^\alpha}{\langle k\rangle^\beta} \right\rangle, \left\langle \frac{\tau - n|n|^\alpha}{\langle n\rangle^\beta} \right\rangle \right) \gtrsim \max(\langle n\rangle, \langle k\rangle)^{\alpha-\beta}.$$

PROOF. Without loss of generality, assume $|n| \geq |k|$,

$$\left(\frac{\tau - k|k|^\alpha}{\langle k\rangle^\beta} \right) - \left(\frac{\tau - n|n|^\alpha}{\langle n\rangle^\beta} \right) = (\tau - k|k|^\alpha) \left(\frac{1}{\langle k\rangle^\beta} - \frac{1}{\langle n\rangle^\beta} \right) + \frac{n|n|^\alpha - k|k|^\alpha}{\langle n\rangle^\beta}$$

Since $\langle k\rangle^{-\beta} \geq \langle n\rangle^{-\beta}$, the first term is at most size of $2\left\langle \frac{\tau - k|k|^\alpha}{\langle k\rangle^\beta} \right\rangle$.

Next, consider the second term above. This expression is apparently larger when n and k have different signs. Also, if $|n| \gg |k|$, this expression has order $\langle n\rangle^{\alpha+1-\beta}$, which is more than sufficient for our desired estimate. So it suffices to bound this expression from below when $n \sim k$. Thus,

$$\left| \frac{n|n|^\alpha - k|k|^\alpha}{\langle n\rangle^\beta} \right| = \frac{|n|^{\alpha+1} - |k|^{\alpha+1}}{\langle n\rangle^\beta} = \frac{(\alpha+1)|k^*|^\alpha\, |n-k|}{\langle n\rangle^\beta},$$

where $k^* \in (k, n)$. Since $|n-k| \geq 1$ and $n \sim k$, the RHS above has the size $\langle n\rangle^{\alpha-\beta}$. This gives our claim. □

Using Claim 3, we have

$$\left\langle \frac{\tau - L_k}{\langle k\rangle^\beta} \right\rangle^{b-1+\varepsilon} \left\langle \frac{\tau - L_n}{\langle n\rangle^\beta} \right\rangle^{-b} \lesssim \langle k\rangle^{(\alpha-\beta)(b-1+\varepsilon)}. \tag{4.3}$$

Collecting estimates (4.2) and (4.3), we have

$$\mathcal{M} \lesssim \langle k\rangle^{\varepsilon+(\alpha-\beta)(b-1+\varepsilon)}$$

when $k \sim m \sim n$. To make this exponent non-positive, we need

$$0 < \varepsilon < \frac{(\alpha - \beta)(1-b)}{\alpha - \beta + 1}.$$

Now, with this condition satisfied, we apply Young's inequality to bound (4.1) by

$$\left\| \sum_m \sum_{n \neq k} |g_{k-m}^N||g_{m-n}^N| \mathcal{M} f_n(\tau) \right\|_{L^2_\tau l^2_k} \leq \|g_k\|^2_{l^1_k} \|f\|_{L^2_{\tau,k}} \lesssim_g \|u\|_{Z^b} \leq 2\|v\|_{Z^b_T}$$

where $g_k^N := \langle k \rangle^{-N} g_k$ for some $N \gg 1$. □

The next lemma deals with the bounded operator $\mathcal{R}$.

LEMMA 2. *Let $b \in (\frac{1}{2}, 1)$. Then, for some $0 < \varepsilon \ll 1$,*

$$\left\| D^{-\beta(b-\frac{1}{2})} \mathcal{R}[v] \right\|_{Z_T^{b-1}} \lesssim_\varepsilon T^{\frac{3}{2}-b-\varepsilon} \|v\|_{Z^b_T}.$$

PROOF. Let $u \in Z^b$ such that $u(t) \equiv v(t)$ on $[0, T]$, and $\|u\|_{Z^b} \leq 2\|v\|_{Z^b_T}$. Then LHS of the desired estimate is bounded by $\|\eta(t/T)\mathcal{R}[u]\|_{Z^{b-1}}$. Then using Propostion 3, (3.1) and the fact that $\mathcal{R}$ is bounded in L^2_x,

$$\begin{aligned} \left\| \eta\left(\frac{t}{T}\right) \mathcal{R}[u] \right\|_{Z^{b-1}} &\lesssim T^{1-b} \left\| \eta\left(\frac{t}{2T}\right) \mathcal{R}[u] \right\|_{L^2_t L^2_x} \lesssim T^{1-b} \left\| \eta\left(\frac{t}{2T}\right) u \right\|_{L^2_t L^2_x} \\ &\lesssim_\varepsilon T^{1-b+\frac{1}{2}-\varepsilon} \|u\|_{Z^{\frac{1}{2}-\varepsilon}} \lesssim T^{\frac{3}{2}-b-\varepsilon} \|u\|_{Z^b}. \end{aligned}$$

Recalling $\|u\|_{Z^b} \leq 2\|v\|_{Z^b_T}$, we complete the proof. □

5. Bilinear estimate

In this section, we perform the bilinear estimate necessary for our argument. This is the key estimate that orchestrates the whole proof. In Proposition 4, we saw a dissipative smoothing effect of order $\beta/2$. In the following lemma, we observe a dispersive gain which compensates for the nonlinear derivative $\partial_x(v^2)$.

LEMMA 3. *Let $\alpha > 1$ and $2 - \alpha < \beta \leq 1$. Given $0 < T \ll 1$ and $s \geq 0$, there exists $\varepsilon > 0$ and $b > \frac{1}{2}$ such that*

$$\left\| D^{-\beta(b-\frac{1}{2})} \partial_x(uv) \right\|_{Z_T^{b-1}} \lesssim_{\varepsilon,b} T^\varepsilon \|u\|_{Z^b_T} \|v\|_{Z^b_T}.$$

REMARK. From [**12**, Lemma 6.1], it is widely known that such dispersive bilinear estimate for KdV (i.e. $\alpha = 2$ and $\beta = 0$) cannot be established for $b > \frac{1}{2}$. On the other hand, by adding a slight localized dissipation ($\beta > 0$), this bilinear estimate can be established even for the periodic KdV.

PROOF. Let $\varepsilon > 0$ be a small number to be chosen later. Note that $b - 1 > -\frac{1}{2}$. So using (3.1) and (3.6), we write

$$\left\| D^{-\beta(b-\frac{1}{2})} \partial_x(uv) \right\|_{Z_T^{b-1}} \lesssim_b T^\varepsilon \left\| D^{-\beta(b-\frac{1}{2})} \partial_x(uv) \right\|_{Z^{b-1+\varepsilon}}.$$

Now we estimate the norm on the RHS above. Note the dual of $(Z^{b-1+\varepsilon})^* = Z^{1-b-\varepsilon}$. Using duality, the norm on the LHS can be written as

$$\sup_{\|w\|_{Z^{1-b-\varepsilon}}=1} \left| \int_{\mathbb{R}} \int_{\mathbb{T}} u(t,x)v(t,x)\, D^{-\beta(b-\frac{1}{2})} \partial_x w(t,x)\, dx\, dt \right|.$$

Using Plancherel and neglecting the complex conjugate on w (due to the fact $\overline{Z^b} = Z^b$ in our case), we can write the integral as

$$(5.1) \qquad \sum_{k_1+k_2+k_3=0} \int_{\tau_1+\tau_2+\tau_3=0} \widehat{u}_{k_1}(\tau_1)\, \widehat{v}_{k_2}(\tau_2)\, k_3|k_3|^{-\beta(b-\frac{1}{2})} \widehat{w}_{k_3}(\tau_3)\, d\sigma,$$

where $d\sigma$ is the inherited measure on the plane $\tau_1 + \tau_2 + \tau_3 = 0$.

For simplification, we introduce a few notations developed by Tao [**28**]:

For $j = 1, 2, 3$, denote N_j and L_j to be dyadic indices such that $|k_j| \sim N_j$ and $\left\langle \frac{\tau_j - L_{k_j}}{\langle k_j \rangle^\beta} \right\rangle \sim L_j$. Denote $N_{\max} := \max\{N_1, N_2, N_3\}$, and analogously for notations N_{med}, $N_{\min}$, $L_{\max}$, L_{med}, $L_{\min}$.

Following are key algebraic lemmas, which will be used to prove our desired estimate.

CLAIM 4. Let $k_1 + k_2 + k_3 = 0$ and $k_1 k_2 k_3 \neq 0$. Then, for $\alpha \geq 1$,

$$\left| \sum_{j=1}^{3} k_j |k_j|^\alpha \right| \gtrsim N_{\max}^\alpha N_{\min}.$$

REMARK. Similar arithmetic estimates are shown in [**16, 24**].

PROOF. Without loss of generality, assume $|k_1| \geq |k_2| \geq |k_3|$. Note that the restriction $k_1 + k_2 + k_3 = 0$ forces the identity that both k_2 and k_3 share a sign that is opposite to k_1. Further, this leads to the identity: $|k_1| - |k_2| = |k_3|$. We will use this in our computation in the following.

We can split into two cases. First is when $|k_1| \sim |k_2| \gg |k_3|$, and the second is when $|k_1| \sim |k_2| \sim |k_3|$.

In the first case, the third term $k_3|k_3|^\alpha$ can be ignored. Then, since k_1 and k_2 have opposite signs,

$$\sum_{j=1}^{3} k_j|k_j|^\alpha \sim |k_1|^{\alpha+1} - |k_2|^{\alpha+1} = (|k_1| - |k_2|)(\alpha + 1)|k^*|^\alpha$$

for some $k^* \in (|k_2|, |k_1|) \sim N_{\max}$. Finally, note that $|k_1| - |k_2| = |k_3| \sim N_{\min}$. This proves the claim for the case $N_1 \sim N_2 \gg N_3$.

Next, consider the second case. By writing $k_2 = -(k_1 + k_3)$, we can write

$$\sum_{j=1}^{3} k_j|k_j|^\alpha = k_1(|k_1|^\alpha - |k_2|^\alpha) - k_3(|k_2|^\alpha - |k_3|^\alpha).$$

Since k_1 and k_3 have opposite signs and both parenthesized terms are positive, it suffices to take the first term to estimate the sum. Again, using MVT,

$$\sum_{j=1}^{3} k_j|k_j|^\alpha \sim k_1(|k_1|^\alpha - |k_2|^\alpha) = k_1(|k_1| - |k_2|)\alpha|k^*|^{\alpha-1}$$

for some $k^* \in (|k_2|, |k_1|) \sim N_{\max}$. Again, using $|k_1| - |k_2| = |k_3|$, we conclude the proof. □

The following claim will give us the dispersive gain via our new Bourgain space.

CLAIM 5. Let $\tau_1 + \tau_2 + \tau_3 = 0$, $k_1 + k_2 + k_3 = 0$, and $k_1 k_2 k_3 \neq 0$. Then $L_{\max} \gtrsim N_{\max}^{\alpha-\beta} N_{\min}$.

Further, consider the special case when $L_{\max}$ occurs at the same index as $N_{\min}$: Let $j_0 \in \{1, 2, 3\}$ satisfy $N_{j_0} = N_{\min}$. If $L_{j_0} = L_{\max}$, then $L_{\max} \gtrsim N_{\max}^{\alpha} N_{\min}^{1-\beta}$.

PROOF. Again, we assume without loss of generality that $|k_1| \geq |k_2| \geq |k_3|$. Then note

$$\frac{\tau_1 + k_1|k_1|^\alpha}{\langle k_1 \rangle^\beta} + \frac{\tau_2 + k_2|k_2|^\alpha}{\langle k_1 \rangle^\beta} + \frac{\tau_3 + k_j|k_3|^\alpha}{\langle k_1 \rangle^\beta} = \frac{\sum_{j=1}^3 k_j|k_j|^\alpha}{\langle k_1 \rangle^\beta}.$$

Since $N_1 \sim N_2$, three terms on the LHS above are respectively of order L_1, L_2, $L_3 N_{\min}^{\beta} N_{\max}^{-\beta}$. By Claim 4, the RHS is of order $N_{\max}^{\alpha-\beta} N_{\min}$. So it must be the case that

$$\max\left(L_1, L_2, L_3 N_3^\beta N_1^{-\beta}\right) \gtrsim N_{\max}^{\alpha-\beta} N_{\min}.$$

Noting $N_3^\beta N_1^{-\beta} \leq 1$, we obtain $L_{\max} \gtrsim N_{\max}^{\alpha-\beta} N_{\min}$. Also, the special case in the claim directly follows from the expression above. This proves our claim. $\square$

We return our attention now to the estimate of (5.1). First, we will localize spatial and modulational frequencies by N_j and L_j using partition of unity and write

$$\sum_{N_j, L_j \geq 1} \int_\Gamma \widehat{u_{k_1}}(\tau_1) \widehat{v_{k_2}}(\tau_2) k_3 |k_3|^{-\beta(b-\frac{1}{2})} \widehat{w_{k_3}}(\tau_3)\, d\sigma, \tag{5.2}$$

where $\Gamma := \{(\tau_1, \tau_2, \tau_3; k_1, k_2, k_3) : \tau_1 + \tau_2 + \tau_3 = 0, k_1 + k_2 + k_3 = 0\}$. Now, note that the integrand can be written as

$$\frac{N_3^{1-\frac{\beta}{2}+\varepsilon\beta}}{N_1^{\frac{\beta}{2}} N_2^{\frac{\beta}{2}} L_1^b L_2^b L_3^{1-b-\varepsilon}} \left(N_1^{\frac{\beta}{2}} L_1^b \widehat{u_{k_1}}(\tau_1)\right) \left(N_2^{\frac{\beta}{2}} L_2^b \widehat{v_{k_2}}(\tau_2)\right) \left(N_3^{1-b-\varepsilon} L_3^{1-b-\varepsilon} \widehat{w_{k_3}}(\tau_3)\right).$$

Denote the parenthesized functions above repectively by $f_{k_1}(\tau_1)$, $g_{k_2}(\tau_2)$, $h_{k_3}(\tau_3)$.

To estimate this expression, we need the following main claim:

CLAIM 6. Let $\alpha + \beta > 2$. Then there exists $\varepsilon > 0$ and $b > \frac{1}{2}$ and $j_1 \in \{1, 2, 3\}$ satisfying

$$\frac{N_3^{1-\frac{\beta}{2}+\varepsilon\beta}}{N_1^{\frac{\beta}{2}} N_2^{\frac{\beta}{2}} L_1^b L_2^b L_3^{1-b-\varepsilon}} \lesssim L_{\max}^{-\varepsilon} N_{min}^{-\frac{1}{2}-\varepsilon} N_{j_1}^{-\frac{\beta}{2}} L_{j_1}^{-b}.$$

PROOF. First, note that $\max(N_1, N_2) \sim N_{\max}$. So

$$\frac{N_3^{1-\frac{\beta}{2}+\varepsilon\beta}}{N_1^{\frac{\beta}{2}} N_2^{\frac{\beta}{2}} L_1^b L_2^b L_3^{1-b-\varepsilon}} \lesssim \frac{N_{\max}^{1-\beta+\varepsilon\beta}}{N_{\min}^{\frac{\beta}{2}} L_{\max}^{1-b-\varepsilon} L_{\text{med}}^b L_{\min}^b}.$$

First, we will show the estimate using $L_{\max} \gtrsim N_{\max}^{\alpha-\beta} N_{\min}$ from Claim 5. This will cover all cases except the special case mentioned in Claim 5 Consider

$$\frac{N_{\max}^{1-\beta+\varepsilon\beta}}{N_{\min}^{\frac{\beta}{2}} L_{\max}^{1-b-\varepsilon} L_{\text{med}}^b L_{\min}^b} \lesssim \frac{N_{\max}^{1-\beta+\varepsilon\beta-(\alpha-\beta)(1-b-2\varepsilon)}}{N_{\min}^{\frac{\beta}{2}+(1-b-2\varepsilon)} L_{\max}^{\varepsilon} L_{\text{med}}^b L_{\min}^b}. \tag{5.3}$$

First, we need to ensure that the exponent of $N_{\max}$ is negative. For this, it is sufficient to establish is $(1-\beta) < (1-b)(\alpha-\beta) \implies \dfrac{\alpha-1}{\alpha-\beta} > b$. In order to ensure that this is compatible with $b > \frac{1}{2}$, we need

$$\frac{\alpha-1}{\alpha-\beta} > \frac{1}{2} \iff \alpha+\beta > 2 \qquad \text{assuming } \alpha > \beta.$$

Next, we need to make the exponent of $N_{\min}$ in the denominator to equal $\frac{\beta}{2} + \frac{1}{2} + \varepsilon$. For this, we need a contribution of $N_{\min}^{-(b-\frac{1}{2}+3\varepsilon)}$, which will come from the left-over gain of $N_{\max}$. To ensure that this is possible, we need to establish

$$(1-\beta) - (\alpha-\beta)(1-b) + \left(b - \frac{1}{2}\right) < 0 \implies \frac{\alpha - \frac{1}{2}}{\alpha-\beta+1} > b.$$

But again, simple calculations show that

$$\alpha+\beta > 2 \iff \frac{\alpha-\frac{1}{2}}{\alpha-\beta+1} > \frac{1}{2}.$$

To conclude the numerology, given $\alpha > \beta$ satisfying $\alpha+\beta > 2$, we can choose b to satisfy

$$\frac{1}{2} < b < \min\left(\frac{\alpha-1}{\alpha-\beta}, \frac{\alpha-\frac{1}{2}}{\alpha-\beta+1}\right)$$

and we can choose $\varepsilon > 0$ so that

$$\frac{N_{\max}^{1-\beta+\varepsilon\beta-(\alpha-\beta)(1-b-2\varepsilon)}}{N_{\min}^{\frac{\beta}{2}+(1-b-2\varepsilon)} L_{\max}^{\varepsilon} L_{\text{med}}^{b} L_{\min}^{b}} \lesssim \frac{1}{N_{\min}^{\frac{\beta}{2}+(\frac{1}{2}+\varepsilon)} L_{\max}^{\varepsilon} L_{\text{med}}^{b} L_{\min}^{b}}.$$

Note that this gives the desired statement except when we are in the special case mentioned in Claim 5. Now, consider the special case when $N_{j_0} = N_{\min}$ and $L_{\max} = L_{j_0}$. In that case, the estimate for (5.3) is slightly modified as follows. Since $L_{\max} \gtrsim N_{\max}^{\alpha} N_{\min}$, instead of (5.3), we get

$$\frac{N_{\max}^{1-\beta+\varepsilon\beta}}{N_{\min}^{\frac{\beta}{2}} L_{\max}^{1-b-\varepsilon} L_{\text{med}}^{b} L_{\min}^{b}} \lesssim \frac{N_{\max}^{1-\beta+\varepsilon\beta-\alpha(1-b-2\varepsilon)}}{N_{\min}^{\frac{\beta}{2}+(1-\beta)(1-b-2\varepsilon)} L_{\max}^{\varepsilon} L_{\text{med}}^{b} L_{\min}^{b}}.$$

Let $j_1 \in \{1,2,3\} \setminus \{j_0\}$. Since $L_{j_1} \neq L_{\max}$, either $L_{j_1} = L_{\text{med}}$ or $L_{\min}$. We borrow $\beta/2$ power of $N_{\max}$ to write

$$\frac{N_{\max}^{1-\beta+\varepsilon\beta-\alpha(1-b-2\varepsilon)}}{N_{\min}^{\frac{\beta}{2}+(1-\beta)(1-b-2\varepsilon)} L_{\max}^{\varepsilon} L_{\text{med}}^{b} L_{\min}^{b}} \leq \frac{N_{\max}^{1-\frac{\beta}{2}+\varepsilon\beta-\alpha(1-b-2\varepsilon)}}{N_{\min}^{\frac{\beta}{2}+(1-\beta)(1-b-2\varepsilon)} L_{\max}^{\varepsilon} N_{j_1}^{\frac{\beta}{2}} L_{j_1}^{b}}.$$

We proceed as before: First, to ensure that the exponent of $N_{\max}$ is negative,

$$1 - \frac{\beta}{2} - \alpha(1-b) < 0 \iff 1 - \frac{1-\beta/2}{\alpha} > b.$$

This is compatible with $b > \frac{1}{2}$ if and only if $\alpha+\beta > 2$.

Next, to ensure that we can borrow enough remaining derivative from $N_{\max}$ to contribute to $N_{\min}$, we need

$$1 - \frac{\beta}{2} - \alpha(1-b) - \frac{\beta}{2} - (1-\beta)(1-b) < -\frac{1}{2} \iff 1 - \frac{\frac{3}{2}-\beta}{\alpha-\beta+1} > b$$

which is compatible with $b > \frac{1}{2}$ if and only of $\alpha+\beta > 2$.

Thus, to conclude the special case, given $\alpha > \beta$ with $\alpha + \beta > 2$, we can choose b satisfying

$$\frac{1}{2} < b < \min\left(1 - \frac{1-\beta/2}{\alpha}, 1 - \frac{\frac{3}{2}-\beta}{\alpha-\beta+1}\right)$$

and we can choose $\varepsilon > 0$ so that

$$\frac{N_{\max}^{1-\frac{\beta}{2}+\varepsilon\beta-\alpha(1-b-2\varepsilon)}}{N_{\min}^{\frac{\beta}{2}+(1-\beta)(1-b-2\varepsilon)} L_{\max}^{\varepsilon} N_{j_1}^{\frac{\beta}{2}} L_{j_1}^{b}} \lesssim \frac{1}{L_{\max}^{\varepsilon} N_{\min}^{\frac{1}{2}+\varepsilon} N_{j_1}^{\frac{\beta}{2}} L_{j_1}^{b}}.$$

This proves the claim. □

Now we return to the proof of the Lemma 3. It follows from Claim 6 that

$$(5.2) \lesssim \sum_{N_j, L_j \geq 1} L_{\max}^{-\varepsilon} N_{\min}^{-\frac{1}{2}-\varepsilon} N_{j_1}^{-\frac{\beta}{2}} L_{j_1}^{-b} \int_{\Gamma} f_{k_1}(\tau_1) g_{k_2}(\tau_2) h_{k_3}(\tau_3)\, d\sigma,$$

where $j_1 \in \{1, 2, 3\}$. To close this estimate, we will need a following, very rough, multilinear L^2 convolution-type estimates:

Claim 7. The following holds

$$N_{\min}^{-\frac{1}{2}-\varepsilon} N_{j_1}^{-\frac{\beta}{2}} L_{j_1}^{-b} \int_{\Gamma} f_{k_1}(\tau_1) g_{k_2}(\tau_2) h_{k_3}(\tau_3)\, d\sigma \lesssim \|f\|_{L^2_{\tau,k}} \|g\|_{L^2_{\tau,k}} \|h\|_{L^2_{\tau,k}}.$$

Proof. Without loss of generality, assume that $N_1 = N_{\min}$ and $j_1 = 2$. The case $j_0 = j_1$ will be simpler, so we omit this case. The LHS of the claim is bounded by

$$\int_{\Gamma} \left(\langle k_1\rangle^{-\frac{1}{2}-\varepsilon} f_{k_1}(\tau_1)\right) \left(\langle k_2\rangle^{-\frac{\beta}{2}} \left\langle \frac{\tau_2 - L_{k_2}}{\langle k_2\rangle^{\beta}} \right\rangle^{-b} g_{k_2}\right)(\tau_2) h_{k_3}(\tau_3)\, d\sigma.$$

We estimate the above integral via Cauchy-Schwarz and Young's inequality in $l^1_k L^2_\tau \times l^2_k L^1_\tau \times L^2_{\tau,k}$,

$$(5.4) \qquad \left\|\langle k\rangle^{-\frac{1}{2}-\varepsilon} f_k(\tau)\right\|_{l^1_k L^2_\tau} \left\|\langle k\rangle^{\frac{-\beta}{2}} \left\langle \frac{\tau - L_k}{\langle k\rangle^{\beta}} \right\rangle^{-b} g_k(\tau)\right\|_{l^2_k L^1_\tau} \|h\|_{L^2_\tau l^2_k}.$$

The first term in (5.4) is bounded by Cauchy-Schwarz in k:

$$\left\|\langle k\rangle^{-\frac{1}{2}-\varepsilon} f_k(\tau)\right\|_{l^1_k L^2_\tau} = \left\|\langle k\rangle^{-\frac{1}{2}-\varepsilon} \|f_k(\tau)\|_{L^2_\tau}\right\|_{l^1_k} \leq \left\|\langle k\rangle^{-\frac{1}{2}-\varepsilon}\right\|_{l^2_k} \|f_k(\tau)\|_{L^2_\tau l^2_k}$$

which is bounded as long as $\varepsilon > 0$. The second term in (5.4) is bounded by Cauchy-Swartz in τ:

$$\left\|\langle k\rangle^{\frac{-\beta}{2}} \left\langle \frac{\tau - L_k}{\langle k\rangle^{\beta}} \right\rangle^{-b} g_k(\tau)\right\|_{l^2_k L^1_\tau} \leq \sup_k \left\|\langle k\rangle^{-\frac{\beta}{2}} \left\langle \frac{\tau - L_k}{\langle k\rangle^{\beta}} \right\rangle^{-b}\right\|_{L^2_\tau} \|g\|_{L^2_\tau l^2_k}.$$

The first term on the RHS above is bounded by (3.3) if and only if $b > \frac{1}{2}$. This concludes the proof. □

Using this claim, we have

$$(5.2) \lesssim \sum_{N_j, L_j \geq 1} L_{\max}^{-\varepsilon} \|f\|_{L^2} \|g\|_{L^2} \|h\|_{L^2}.$$

Recalling the definition of f, g, h

$$\|f\|_{L^2}\|g\|_{L^2}\|h\|_{L^2} \sim \|u\|_{Z^b}\|v\|_{Z^b}\|w\|_{Z^{1-b-\varepsilon}}.$$

Finally, since we are summing over dyadic indices and $L_{\max}$ dominates all other dyadic index, the factor of $L_{\max}^{-\varepsilon}$ makes the summations converge.

This proves the desired bilinear estimate. □

6. Proof of Theorem 1

In this section, we will prove Theorem 1 using contraction in Z^b. We construct a contraction argument in Z_T^b for some $b > \frac{1}{2}$ which satisfies Lemma 1, 2 and 3.

Let $0 < T \ll 1$ be a constant to be determined later. We want to prove that, for $T > 0$ sufficiently small, the operator defined on the RHS of (2.1) is a contraction in a ball in Z_T^b centered at the free solution $S(t)v_0$. Define

$$\Gamma_1(v) := \int_0^t S(t-s)\mathcal{N}_1[v](s)\, ds,$$

$$\Gamma_2(v) := \int_0^t S(t-s)\mathcal{R}[v](s)ds,$$

$$\Gamma_3(u,v) := \int_0^t S(t-s)\partial_x(u\,v)\, ds.$$

Our claim is that, given $v_0 \in L_0^2$, we can find $T > 0$ sufficiently small such that $\Gamma(v) := S(t)v_0 + \Gamma_1(v) + \Gamma_2(v) + \Gamma_3(v,v)$ is a contraction in Z_T^b norm for all v in a ball of radius R (to be determined) centered at the free solution. We need to prove two statements: given $B_R := \{u : \|u - S(t)v_0\|_{Z_T^b} \leq R\}$,

$$\|\Gamma(v) - S(t)v_0\|_{Z_T^b} \leq R \qquad \text{for } \forall v \in B_R, \tag{6.1}$$

$$\|\Gamma(u) - \Gamma(v)\|_{Z_T^b} \leq (1-\delta)\|u-v\|_{Z_T^b} \qquad \text{for } \forall u, v \in B_R \text{ and some } \delta > 0. \tag{6.2}$$

We select $R \ll \|S(t)v_0\|_{Z_T^b} \lesssim_b \|v_0\|_{L_x^2}$ so that $\|v\|_{Z_T^b} \sim \|S(t)v_0\|_{Z_T^b}$ for any $v \in B_R$. By Proposition 2, if $v \in B_R$, then

$$\|v\|_{Z_T^b} \sim \|S(t)v_0\|_{Z_T^b} \lesssim_b \|v_0\|_{L_x^2}.$$

To show (6.1), we use Proposition 4 and Lemma 1, 2 and 3 to write

$$\begin{aligned}\|\Gamma_1(v)\|_{Z_T^b} + \|\Gamma_2(v)\|_{Z_T^b} + \|\Gamma_3(v,v)\|_{Z_T^b} &\lesssim_{\varepsilon,b} T^\varepsilon \left(\|v\|_{Z_T^b} + \|v\|_{Z_T^b}^2\right) \\ &\lesssim_b T^\varepsilon \left(\|v_0\|_{L_x^2} + \|v_0\|_{L_x^2}^2\right).\end{aligned}$$

So, we will have (6.1) as long as T satisfies

$$T^\varepsilon < \frac{R}{\|v_0\|_{L_x^2} + \|v_0\|_{L_x^2}^2} \ll \frac{1}{1 + \|v_0\|_{L_x^2}}.$$

Next, we prove (6.2). Let $u, v \in B_R \subset Z_T^b$ be arbitrary functions. By triangular inequality, suffices to show that $\|\Gamma_j(u) - \Gamma_j(v)\|_{Z_T^b} \leq \frac{1}{4}\|u-v\|_{Z_T^b}$ for $j = 1, 2$ and $\|\Gamma_3(u,u) - \Gamma_3(v,v)\|_{Z_T^b} \leq \frac{1}{4}\|u-v\|_{Z_T^b}$.

Using Proposition 4, Lemma 1, 2 and linearity of Γ_1, we can induce for $j = 1, 2$,

$$\|\Gamma_j(u) - \Gamma_j(v)\|_{Z_T^b} \lesssim_{b,\varepsilon} T^\varepsilon \|u-v\|_{Z_T^b}$$

For Γ_3, we use Proposition 4 and Lemma 3. Adding and subtracting by $\Gamma_3(u,v)$ and using bilinearity of Γ_3,

$$\begin{aligned}\|\Gamma_3(u,u)-\Gamma_3(v,v)\|_{Z_T^b} &= \|\Gamma_3(u+v,u-v)\|_{Z_T^b}\\ &\lesssim_{b,\varepsilon} T^{\varepsilon}\|u+v\|_{Z_T^b}\|u-v\|_{Z_T^b}\\ &\le T^{\varepsilon}\left(\|u\|_{Z_T^b}+\|v\|_{Z_T^b}\right)\|u-v\|_{Z_T^b}\\ &\lesssim_b T^{\varepsilon}\|v_0\|_{L_x^2}\|u-v\|_{Z_T^b}.\end{aligned}$$

Thus, (6.2) can be satisfied as long as

$$T^{\varepsilon} \ll \frac{1}{1+\|v_0\|_{H_x^s}}.$$

This shows that $\Gamma : B_R \to B_R$ is a contraction map for T sufficiently small with respect to the initial data, which proves the local well-posedness statement of Theorem 1. Further, using the energy estimate $\|v(t)\|_{L_x^2} \le \|v_0\|_{L_x^2}$ for any solution v of (1.2), we can patch the uniform-size local time intervals to prove global well-posedness.

To show uniform continuity, let u_0 and v_0 be mean-zero initial data for (1.2) and u, v be the respective solutions. Then by (3.2),

$$\|u-v\|_{C_T^0 H_x^s} \lesssim \|u-v\|_{Z_T^b} \le \|S(t)u_0 - S(t)v_0\|_{Z_T^b} + \|\Gamma(u)-\Gamma(v)\|_{Z_T^b}.$$

Then by (6.2) and Proposition 2, we obtain

$$\|u-v\|_{Z_T^b} \lesssim_b \|S(t)u_0 - S(t)v_0\|_{Z_T^b} \lesssim \|u_0-v_0\|_{L_x^2}$$

This proves the uniform continuity.

7. Linear stabilization

In this section we establish the linear stabilization which is needed to prove Theorem 2. Consider the equation

$$\partial_t v + D^{\alpha}\partial_x v + GD^{\beta}Gv = 0, \qquad x\in\mathbb{T}, t\ge 0, \tag{7.1}$$

with $\alpha > 0$ and $\beta \ge 0$. Note that since the operator $GD^{\beta}G$ is positive definite in $L^2(\mathbb{T})$, the operator $A = -(D^{\alpha}\partial_x + GDG)$ is a dissipative perturbation of a dispersive operator. Therefore, A generates a C^0 semigroup on $L^2(\mathbb{T})$ for any $\alpha,\beta\ge 0$. We denote the semigroup generated by A to be

$$W(t) := e^{-(D^{\alpha}\partial_x + GD^{\beta}G)t}.$$

REMARK. If we want this semigroup to act in $H^s(\mathbb{T})$ for $s \ne 0$, then we would need an additional restriction $\beta \le 1$. See for instance [**17**, Claim 1]. Since we only work in $L^2(\mathbb{T})$ where $GD^{\beta}G$ is positive-definite, we do not impose this additional restriction.

We state the main result of this section:

PROPOSITION 5. *Let $\alpha > 1$ and $\beta \ge 0$. Then there exists $\lambda > 0$ and $C > 0$ such that for any $v_0 \in L_0^2(\mathbb{T})$, the associated solution $W(t)v_0$ to (7.1) satisfies*

$$\|W(t)v_0\|_{L_x^2(\mathbb{T})} \le Ce^{-\lambda t}\|v_0\|_{L_x^2(\mathbb{T})} \qquad t\ge 0. \tag{7.2}$$

Note that scaling (7.1) by the solution v results in

$$\partial_t \int_{\mathbb{T}} |v|^2(t)dx = -\int_{\mathbb{T}} \left|D^{\frac{\beta}{2}} Gv(t)\right|^2 dx. \tag{7.3}$$

To prove Proposition 5, it would suffice to have the observability inequality:

$$\|v_0\|^2_{L^2(\mathbb{T})} \lesssim \int_{\mathbb{T}} \left|D^{\frac{\beta}{2}} Gv(t)\right|^2 dx. \tag{7.4}$$

Proof of this observability relies on a unique continuation property which will be given in Proposition 7. First, we introduce a classical tool that will be used to prove the uniquc continuation result. The following is known as a generalized Ingham's lemma.

PROPOSITION 6. **[10, 18]** *Let $\{\lambda_k\}_{k\in\mathbb{Z}}$ be a sequence of real numbers. If there exists $\gamma > 0$, $\gamma_\infty > 0$ and a positive integer N such that*

$$\begin{aligned} \lambda_{n+1} - \lambda_n &\geq \gamma > 0 \qquad \textit{for any } n \in \mathbb{Z}, \text{ and} \\ \lambda_{n+1} - \lambda_n &\geq \gamma_\infty > 0 \qquad \textit{whenever } |n| > N, \end{aligned}$$

then the sequence $\{e^{i\lambda_k t}\}$ is a Riesz-Fischer sequence in $L^2[0,T]$ for any $T > \frac{\pi}{\gamma_\infty}$. More specifically, there exists a sequence of functions $\{q_j\}_{j\in\mathbb{Z}} \in L^2[0,T]$ such that q_j is orthogonal to $e^{i\lambda_k t}$ if and only if $j \neq k$.

The following unique continuation principle leads to the linear stabilization results for (7.1).

PROPOSITION 7. *Let $\alpha > 0$. If, for some $T > 0$ and $a < b$, $v \in C^0_t L^2_x$ satisfy*

$$\partial_t v + D^\alpha \partial_x v = 0, \qquad v = 0 \text{ a.e. in } [0,T] \times [a,b].$$

Then $v \equiv 0$ a.e. on $[0,T] \times \mathbb{T}$.

REMARK. Above Proposition also works when $\alpha = 0$ using the same technique, but with an additional restriction that T must be sufficiently large. This is natural considering that $\alpha = 0$ gives finite speed of propagation where $\alpha > 0$ corresponds to infinite speed of propagation.

PROOF. Let $v(0,x) =: f(x) \in L^2_x(\mathbb{T})$. Then we can write

$$v(t,x) = \sum_{k\in\mathbb{Z}} e^{itk|k|^\alpha + ikx} f_k \qquad \text{where } f_k = \frac{1}{2\pi}\int_{\mathbb{T}} f(x) e^{-ikx}\, dx.$$

Note that the sequence $\{\lambda_k\} = \{k|k|^\alpha\}$ satisfies the conditions of Proposition 6 with γ_∞ sufficiently large as long as $\alpha > 0$. This means that we can use this statement with $T > 0$ as small as desired. Using Ingham's Lemma, select a sequence $\{q_j\}_{j\in\mathbb{Z}}$ which is biorthogonal to $\{e^{itj|j|^\alpha}\}_{j\in\mathbb{Z}}$. Then, for each $j \in \mathbb{Z}$,

$$\langle v(\cdot,x), q_j\rangle_{L^2_T} = \int_0^t \sum_{k\in\mathbb{Z}} e^{itk|k|^\alpha + ikx} f_k\, q_j(t)\, dt = e^{ijx} f_j.$$

But notice that for almost every $x \in [a,b]$, the LHS of above is equal to zero. That implies that for at least one $x_0 \in [a,b]$ (which may depend on j), $e^{ijx_0} f_j = 0$. This implies $f_j = 0$ for each $j \in \mathbb{Z}$. This implies $v \equiv 0$ for a.e. $(t,x) \in [0,T] \times [a,b]$. $\square$

Given this unique continuation property, observability (7.4) can be proved via compactness arguments similar to [**17**]. Proof in this case is a little different because we do not have the smoothing statement as given in [**17**, Proposition 2.16], but we can overcome this via the smoothing given by the following lemma.

LEMMA 4. *For any $f \in L^2_0$ and $T > 0$,*

$$\|W(t)f\|_{Z^b_T} \lesssim_{b,T} \|f\|_{L^2(\mathbb{T})}$$

PROOF. As in the nonlinear equation, we can see that $W(t)f$ satisfies

$$W(t)f = S(t)f - \int_0^t S(t-s)\left(\mathcal{N}_1[W(\cdot)f](s) + \mathcal{R}[W(\cdot)f](s)\right) ds.$$

Using Lemma 1, Lemma 2 and Proposition 2 for some $0 < T_0 \ll 1$,

$$\|W(t)f\|_{Z^b_{T_0}} \lesssim_{T_0} \|S(t)f\|_{Z^b_{T_0}} \lesssim_b \|f\|_{L^2(\mathbb{T})}.$$

Note that this time T_0 does not depend on size of the initial data but is an absolute constant. Iterating the time T_0 gives the claim for an arbitrary $T > 0$. □

In the following, we prove the linear observability.

PROOF OF OBSERVABILITY (7.4). We prove observability via contradiction. Assume to the contrary that there exists a sequence of non-zero functions $\{v_0^n\}_{n\in\mathbb{Z}^+} \subset L^2_0(\mathbb{T})$ and corresponding solutions $\{v^n\}$ of (7.1) such that, after rescaling,

$$1 = \|v_0^n\|^2 > n \int_0^T \|D^{\beta/2}(Gv^n)\|^2 \, d\tau, \tag{7.5}$$

with $v^n = W(t)v_0^n$ denoting solutions to (7.1) corresponding to initial data v_0^n. The smoothing effect of Z^b and Lemma 4 will demonstrate that the limit of the v^n exists and satisfies the hypothesis of Proposition 7.

Setting $\gamma^* = \beta/2 - (1+\alpha)$, observe

$$\|D^\alpha \partial_x v^n\|_{L^2(0,T;H^{\gamma^*}(\mathbb{T}))} \leq \|v^n\|_{L^2(0,T;H^{\beta/2}(\mathbb{T}))}.$$

By (3.2) and Lemma 4

$$\|v^n\|_{L^2(0,T;H^{\beta/2}(\mathbb{T}))} = \|W(t)v_0^n\|_{L^2(0,T;H^{\beta/2}(\mathbb{T}))} \lesssim \|W(t)v_0^n\|_{Z^b_T} \lesssim_{b,T} \|v_0^n\|_{L^2_x} = 1.$$

So these are uniformly bounded. Using commutator estimate, Sobolev embedding and the fact that G is bounded on $H^s_0(\mathbb{T})$,

$$\begin{aligned} \|D^{\gamma^*} G D^\beta G v^n\| &\leq \|G D^{\gamma^*+\beta} G v^n\| + \|[D^{\gamma^*}, G] D^\beta G v^n\| \\ &\lesssim (\|Gv^n\|_{\gamma^*+\beta} + \|Gv^n\|_{\gamma^*+\beta-1}) \\ &\lesssim \|Gv^n\|_{\beta/2}. \end{aligned}$$

Combining this with the energy identity (7.3) applied to v^n, we find

$$\|GD^\beta G v^n\|^2_{L^2(0,T;H^{\gamma^*}(\mathbb{T}))} \lesssim \int_0^T \|D^{\beta/2}(Gv^n)\|^2 \, dt \lesssim \|v_0^n\|^2,$$

which is uniformly bounded. Using the equation implies

$$\|\partial_t v^n\|_{L^2(0,T;H^{\gamma^*}(\mathbb{T}))} \leq \|D^\alpha \partial_x v^n + GD^\beta G v^n\|_{L^2(0,T;H^{\gamma^*}(\mathbb{T}))} \leq C$$

for some C indepdent of n. Additionally, recall from above that the sequence $\{v^n\}_{n\in\mathbb{Z}^+}$ is uniformly bounded in $L^2(0,T;H^{\beta/2}(\mathbb{T}))$. Applying the Banach-Alaoglu theorem and the Aubin-Lions lemma, we extract a subsequence with the following properties:

$$\begin{aligned} v^n &\to v \quad \text{in } L^2(0,T;H^\gamma(\mathbb{T})) && \forall \gamma < \beta/2 \\ v^n &\to v \quad \text{in } L^2(0,T;H^{\beta/2}(\mathbb{T})) \text{ weak} \\ v^n &\to v \quad \text{in } L^\infty(0,T;L^2(\mathbb{T})) \text{ weak}* \end{aligned}$$

for some $v \in L^2(0,T;H_0^\gamma(\mathbb{T})) \cap L^\infty(0,T;L^2(\mathbb{T}))$. Taking $\gamma = 0$ implies

$$(v^n)^2 \to v^2 \qquad \text{in } L^1(\mathbb{T}\times(0,T)).$$

Next, we verify that $\{v_0^n\}_{n\in\mathbb{Z}}$ is Cauchy in $L_0^2(\mathbb{T})$ as a consequence of the choice $\gamma = 0$ above. Scaling the equation (7.1) by $(T-t)v$ yields

$$\frac{T}{2}\|v_0\|^2 + \frac{1}{2}\int_0^T \|v(t)\|^2\,dt + \int_0^T (T-t)\|D^{\beta/2}(Gv)\|^2\,dt = 0.$$

Applying this to the difference of two solutions produces

$$\begin{aligned} \|v_0^n - v_0^m\| &\leq \frac{1}{T}\int_0^T \|v^n - v^m\|^2\,dt + 2\int_0^T \|D^{\beta/2}G(v^n - v^m)\|^2\,dt \\ &\leq \frac{1}{T}\int_0^T \|v^n - v^m\|^2\,dt + 4\left(\frac{1}{n} + \frac{1}{m}\right) \end{aligned}$$

after using (7.5). Thus v_0^n converges strongly to some v_0 in $L_0^2(\mathbb{T})$ and it follows that the solution of (7.1) associated to v_0 agrees with the limit v of the sequence $\{v^n\}_{n\in\mathbb{Z}}$. Thus $v \in C([0,T];L_0^2(\mathbb{T}))$ and $v_0 = v(0)$.

Letting $n \to \infty$ in (7.5) we find that

$$\int_0^T \|D^{\beta/2}(Gv)\|^2\,dt = 0.$$

Hence $Gv = 0$ a.e. $\mathbb{T}\times(0,T)$ and using (1.3) we may write

$$v(x,t) = \int_{\mathbb{T}} g(y)v(y,t)\,dy := c(t) \qquad \text{for all } (x,t) \in \omega \times (0,T),$$

where $\omega = \{x \in \mathbb{T} : g(x) > 0\}$ and $c \in L^\infty(0,T)$. Thus $\partial_x v$ satisfies the hypothesis of Proposition 7 implying that $\partial_x v \equiv v \equiv 0$ (since v has mean value zero). This leads to a contradiction with the fact that $\|v(0)\| = \|v_0^n\| = 1$. □

8. Proof of Theorem 2

Using semigroup $W(t)$, the solution $v \in C_t^0 L_x^2$ of (1.2) satisfies

$$v(t) = W(t)v_0 + \int_0^t W(t-s)\partial_x(v^2)(s)\,ds. \tag{8.1}$$

Given a bilinear estimate in Lemma 3 and semigroup estimate for $W(t)$ in Lemma 4, we can extend the bilinear estimate to the semigroup. Together with linear stabilization given in Proposition 5, this leads directly to the proof of Theorem 2. In the following lemma, we prove the following extension of the bilinear estimate, following the proof scheme given in [**15**, Lemma 4.4].

LEMMA 5. *Let $b > \frac{1}{2}$ and $v \in Z^b$ be the solution of (1.2). Given any $T > 0$,*

$$\left\| \int_0^t W(t-s)\partial_x(v^2)(s)\,ds \right\|_{Z_T^b} \lesssim_T \|v\|_{Z_T^b}^2.$$

PROOF. Here, we justify the second inequality more carefully. Combining (2.1) with (8.1), we can write

$$v(t) = S(t)v_0 + \int_0^t S(t-s)\mathcal{N}_1[W(t)v_0](s)\,ds$$
$$+ \int_0^t S(t-s)\mathcal{N}_1\left[\int_0^s W(s-s')\partial_x(v^2)(s')\,ds'\right](s)\,ds + \int_0^t S(t-s)\partial_x(v^2)(s)\,ds.$$

Here, we have intentionally omitted the third term involving $\mathcal{R}$ to simplify the resulting expression, but it will be evident from the argument that $\mathcal{R}$ does not impose any additional difficulty. Replacing $v(t)$ on the LHS above by (8.1),

$$\int_0^t W(t-s)\partial_x(v^2)(s)\,ds = S(t)v_0 - W(t)v_0 + \int_0^t S(t-s)\mathcal{N}_1[W(t)v_0](s)\,ds$$
$$+ \int_0^t S(t-s)\mathcal{N}_1\left[\int_0^s W(s-s')\partial_x(v^2)(s')\,ds'\right](s)\,ds + \int_0^t S(t-s)\partial_x(v^2)(s)\,ds.$$

Noting $W(t)v_0 = S(t)v_0 + \int_0^t S(t-s)\mathcal{N}_1[W(t)v_0](s)\,ds$, the above simplifies to

$$\int_0^t W(t-s)\partial_x(v^2)(s)\,ds$$
$$= \int_0^t S(t-s)\mathcal{N}_1\left[\int_0^s W(s-s')\partial_x(v^2)(s')\,ds'\right](s)\,ds + \int_0^t S(t-s)\partial_x(v^2)(s)\,ds.$$

Taking $Z^b_{T_0}$ norm of both sides for some $0 < T_0 \ll 1$ and using Lemma 1 and Lemma 3, we obtain

$$\left\|\int_0^t W(t-s)\partial_x(v^2)(s)\,ds\right\|_{Z^b_{T_0}} \lesssim_{T_0} \left\|\int_0^t S(t-s)\partial_x(v^2)(s)\,ds\right\|_{Z^b_{T_0}} \lesssim_b \|v\|^2_{Z^b_{T_0}}.$$

Since T_0 is again an absolute constant as in the proof of Lemma 4, we can iterate this to any arbitrary $T > 0$ to obtain the desired statement. □

We have acquired all tools needed to prove Theorem 2, so we conclude with the following proof of local nonlinear stabilization of (1.2).

PROOF OF THEOREM 2. Using Proposition 5, we can fix a $T \gg 1$ and $0 < \lambda' \ll \lambda$ such that

$$\|W(T)v_0\|_{L^2_x(\mathbb{T})} \le \frac{1}{2}e^{-\lambda' T}\|v_0\|_{L^2_x(\mathbb{T})} \quad \text{for any } v_0 \in L^2_0(\mathbb{T}).$$

Standard contraction arguments give a contraction map of (8.1) on a sufficiently small closed ball B_M in Z^b_T centered around the linear solution $W(t)v_0$. Furthermore, as long as $M \ll \|W(t)v_0\|_{Z^b_T}$, we can have that $\|v\|_{Z^b_T} \sim \|W(t)v_0\|_{Z^b_T} \lesssim_T \|v_0\|_{L^2_x}$. Then, using (8.1), we obtain

$$\|v(T)\|_{L^2_x} \le \|W(T)v_0\|_{L^2_x} + C_{b,T}\left\|\int_0^T W(T-s)\partial_x(v^2)(s)\,ds\right\|_{Z^b_T}$$
$$\le \frac{1}{2}e^{-\lambda' T}\|v_0\|_{L^2_x} + C_{b,T}\|v\|^2_{Z^b_T} \le \frac{1}{2}e^{-\lambda' T}\|v_0\|_{L^2_x} + C_{b,T}\|v_0\|^2_{L^2_x}.$$

Thus, if we choose $\delta > 0$ such that $C_{b,T}\delta < \frac{1}{2}e^{-\lambda' T}$, we have

$$\|v(T)\|_{L^2_x} \le e^{-\lambda' T}\|v_0\|_{L^2_x}$$

for $v_0 \in L^2_0$ satisfying $\|v_0\|_{L^2_x} < \delta$. This proves Theorem 2. □

Appendix A. Theory of A_p weights

We will briefly introduce A_p weights, which will be used to simplify the proofs of Propositions 3 and 4. Our aim is to show that $\left\langle \frac{\tau - L_k}{\langle k\rangle^\beta} \right\rangle^\alpha \in A_2$ for a certain range of α and then incur widely-known properties A_2 weights to aid with technical estimates. In particular, we will use that, if W is an A_2 weight, then Maximal functions and Hilbert transform are bounded in $L^2(W)$.

We begin with a few basic definitions. Given a positive weight, we introduce the weighted norm $L^p(W)$ as follows:

$$\|v\|_{L^p_\tau(W)} := \left(\int_{\mathbb{R}} |v|^p(\tau) W(\tau)\, d\tau \right)^{\frac{1}{p}}.$$

A non-negative function $W = W(\tau)$ belongs to A_p class if and only if

$$\sup_I \left(\frac{1}{|I|} \int_I W(\tau)\, d\tau \right) \left(\frac{1}{|I|} \int_I W^{-\frac{1}{p-1}}(\tau)\, d\tau \right)^{p-1} < \infty \tag{A.1}$$

where I is an arbitrary connected interval on $\mathbb{R}$. For $W \in A_2$, the LHS of (A.1) is denoted $[W]_{A_p}$. We are only interested in the case $p = 2$. The following lemma gives convenient properties of this quantity:

LEMMA 6. [**5**, Proposition 9.1.5]

$$[W(\lambda\,\cdot)]_{A_p} = [W(\cdot)]_{A_p} \text{ for } \lambda > 0, \qquad [W(\cdot - z)]_{A_p} = [W(\cdot)]_{A_p} \text{ for } z \in \mathbb{R}.$$

In addition, for $p = 2$, $[W^{-1}]_{A_2} = [W]_{A_2}$.

In view of above, note that $\left[\left\langle \frac{\tau - L_k}{\langle k\rangle^\beta} \right\rangle^{2(b-1)} \right]_{A_2} = [\langle \tau \rangle^{2(b-1)}]_{A_2}$.

Following are two of the celebrated results in the theory of A_p weights.

LEMMA 7. [**9**] *Let* $p \in (1, \infty)$ *and* W *be non-negative. Then*

$$\mathcal{H} : L^p(W) \to L^p(W) \iff W \in A_p,$$

where $\mathcal{H}$ *denotes the Hilbert transform.*

LEMMA 8. [**27**, Theorem V.3.1] *Let* $p \in (1, \infty)$ *and* $W \in A_p$. *Then*

$$\int_{\mathbb{R}} |Mf|^p(\tau) W(\tau)\, d\tau \lesssim_{[W]_{A_p}} \int_{\mathbb{R}} |f|^p(\tau) W(\tau)\, d\tau$$

where $(Mf)(x) := \sup_{T>0} T\eta(T\,\cdot) * |f|$.

Now it remains to show the following claim:

CLAIM 8. $\langle \tau \rangle^\alpha \in A_2$ if $\alpha \in (-1, 1)$.

PROOF. Due to Lemma 6, it suffices to show the claim for $\alpha \in [0,1)$. More specifically, we must show

$$\left(\frac{1}{|I|}\int_I W(\tau)\,d\tau\right)\left(\frac{1}{|I|}\int_I W(\tau)^{-1}\,d\tau\right) = \frac{1}{|I|^2}\int_{I\times I} \langle\tau\rangle^{-\alpha}\langle\sigma\rangle^{\alpha}\,d\tau\,d\sigma < C$$

where C is independent of the choice of interval I. Given an interval I, denote $M = \sup\{|x| : x \in I\}$.

First, note that the inequality is harmless when $|I| \leq 1$. So we can assume that $|I| \sim \langle |I| \rangle$ for harmful cases. We split into two cases.

CASE 1. $\{-\frac{M}{2}, \frac{M}{2}\} \notin I$. Note that in this case $m := \inf\{|x| : x \in I\} \geq \frac{M}{2}$. Then, for any $\tau, \sigma \in I$, $\langle\tau\rangle^{-\alpha}\langle\sigma\rangle^{\alpha} \lesssim 1$. Then

$$\frac{1}{|I|^2}\int_{I\times I} \langle\tau\rangle^{-\alpha}\langle\sigma\rangle^{\alpha}\,d\tau\,d\sigma \lesssim \frac{1}{|I|^2}\int_{I\times I} 1\,d\tau\,d\sigma \leq 1$$

as long as $\alpha \geq 0$.

CASE 2. $\{-\frac{M}{2}, \frac{M}{2}\} \in I$. In this case, note that $\frac{M}{2} \leq |I| \leq 2M$, so $|I| \sim M$. We can write

$$\frac{1}{|I|^2}\int_{I\times I} \langle\tau\rangle^{-\alpha}\langle\sigma\rangle^{\alpha}\,d\tau\,d\sigma \lesssim \frac{1}{|I|^2}\int_{I\times I} \langle\tau\rangle^{-\alpha}\langle M\rangle^{\alpha}\,d\tau\,d\sigma = \frac{M^{\alpha}}{|I|}\int_I \langle\tau\rangle^{-\alpha}\,d\tau.$$

It remains to estimate the integral above. Note that, $I \subseteq [-M, M]$. So

$$\int_I \langle\tau\rangle^{-\alpha}\,d\tau \leq 2\int_0^M \tau^{-\alpha}\,d\tau = \frac{2}{-\alpha+1}M^{-\alpha+1} \sim |I|^{-\alpha+1}$$

as long as $-\alpha > -1 \implies \alpha < 1$. Combining the above computations, we conclude the proof.

□

References

[1] J. Bourgain, *Fourier transform restriction phenomena for certain lattice subsets and applications to nonlinear evolution equations. I. Schrödinger equations*, Geom. Funct. Anal. **3** (1993), no. 2, 107–156, DOI 10.1007/BF01896020. MR1209299

[2] J. Bourgain, *Fourier transform restriction phenomena for certain lattice subsets and applications to nonlinear evolution equations. II. The KdV-equation*, Geom. Funct. Anal. **3** (1993), no. 3, 209–262, DOI 10.1007/BF01895688. MR1215780

[3] J. Colliander, C. Kenig, and G. Staffilani, *Local well-posedness for dispersion-generalized Benjamin-Ono equations*, Differential Integral Equations **16** (2003), no. 12, 1441–1472. MR2029909

[4] J. Ginibre and G. Velo, *Smoothing properties and existence of solutions for the generalized Benjamin-Ono equation*, J. Differential Equations **93** (1991), no. 1, 150–212, DOI 10.1016/0022-0396(91)90025-5. MR1122309

[5] L. Grafakos, *Modern Fourier analysis*, 2nd ed., Graduate Texts in Mathematics, vol. 250, Springer, New York, 2009. MR2463316

[6] Z. Guo, *Local well-posedness for dispersion generalized Benjamin-Ono equations in Sobolev spaces*, J. Differential Equations **252** (2012), no. 3, 2053–2084, DOI 10.1016/j.jde.2011.10.012. MR2860610

[7] S. Herr, *Well-posedness for equations of Benjamin-Ono type*, Illinois J. Math. **51** (2007), no. 3, 951–976. MR2379733

[8] S. Herr, A. D. Ionescu, C. E. Kenig, and H. Koch, *A para-differential renormalization technique for nonlinear dispersive equations*, Comm. Partial Differential Equations **35** (2010), no. 10, 1827–1875, DOI 10.1080/03605302.2010.487232. MR2754070

[9] R. Hunt, B. Muckenhoupt, and R. Wheeden, *Weighted norm inequalities for the conjugate function and Hilbert transform*, Trans. Amer. Math. Soc. **176** (1973), 227–251, DOI 10.2307/1996205. MR0312139

[10] A. E. Ingham, *Some trigonometrical inequalities with applications to the theory of series*, Math. Z. **41** (1936), no. 1, 367–379, DOI 10.1007/BF01180426. MR1545625

[11] C. E. Kenig, G. Ponce, and L. Vega, *Well-posedness of the initial value problem for the Korteweg-de Vries equation*, J. Amer. Math. Soc. **4** (1991), no. 2, 323–347, DOI 10.2307/2939277. MR1086966

[12] C. E. Kenig, G. Ponce, and L. Vega, *A bilinear estimate with applications to the KdV equation*, J. Amer. Math. Soc. **9** (1996), no. 2, 573–603, DOI 10.1090/S0894-0347-96-00200-7. MR1329387

[13] V. Komornik, D. L. Russell, and B. Y. Zhang, *Stabilisation de l'équation de Korteweg-de Vries* (French, with English summary), C. R. Acad. Sci. Paris Sér. I Math. **312** (1991), no. 11, 841–843. MR1108503

[14] C. Laurent, F. Linares, and L. Rosier, *Control and stabilization of the Benjamin-Ono equation in $L^2(\mathbb{T})$*, Arch. Ration. Mech. Anal. **218** (2015), no. 3, 1531–1575, DOI 10.1007/s00205-015-0887-5. MR3401014

[15] C. Laurent, L. Rosier, and B.-Y. Zhang, *Control and stabilization of the Korteweg-de Vries equation on a periodic domain*, Comm. Partial Differential Equations **35** (2010), no. 4, 707–744, DOI 10.1080/03605300903585336. MR2753618

[16] J. Li and S. Shi, *Local well-posedness for the dispersion generalized periodic KdV equation*, J. Math. Anal. Appl. **379** (2011), no. 2, 706–718, DOI 10.1016/j.jmaa.2011.01.026. MR2784352

[17] F. Linares and L. Rosier, *Control and stabilization of the Benjamin-Ono equation on a periodic domain*, Trans. Amer. Math. Soc. **367** (2015), no. 7, 4595–4626, DOI 10.1090/S0002-9947-2015-06086-3. MR3335395

[18] S. Micu and E. Zuazua, *Boundary controllability of a linear hybrid system arising in the control of noise*, SIAM J. Control Optim. **35** (1997), no. 5, 1614–1637, DOI 10.1137/S0363012996297972. MR1466919

[19] L. Molinet and D. Pilod, *The Cauchy problem for the Benjamin-Ono equation in L^2 revisited*, Anal. PDE **5** (2012), no. 2, 365–395, DOI 10.2140/apde.2012.5.365. MR2970711

[20] L. Molinet and F. Ribaud, *The Cauchy problem for dissipative Korteweg de Vries equations in Sobolev spaces of negative order*, Indiana Univ. Math. J. **50** (2001), no. 4, 1745–1776, DOI 10.1512/iumj.2001.50.2135. MR1889080

[21] L. Molinet and F. Ribaud, *The global Cauchy problem in Bourgain's-type spaces for a dispersive dissipative semilinear equation*, SIAM J. Math. Anal. **33** (2002), no. 6, 1269–1296, DOI 10.1137/S0036141000374634. MR1920630

[22] L. Molinet and F. Ribaud, *On the low regularity of the Korteweg-de Vries-Burgers equation*, Int. Math. Res. Not. **37** (2002), 1979–2005, DOI 10.1155/S1073792802112104. MR1918236

[23] L. Molinet, J. C. Saut, and N. Tzvetkov, *Ill-posedness issues for the Benjamin-Ono and related equations*, SIAM J. Math. Anal. **33** (2001), no. 4, 982–988, DOI 10.1137/S0036141001385307. MR1885293

[24] L. Molinet and S. Vento, *Improvement of the energy method for strongly nonresonant dispersive equations and applications*, Anal. PDE **8** (2015), no. 6, 1455–1495, DOI 10.2140/apde.2015.8.1455. MR3397003

[25] D. L. Russell and B. Y. Zhang, *Controllability and stabilizability of the third-order linear dispersion equation on a periodic domain*, SIAM J. Control Optim. **31** (1993), no. 3, 659–676, DOI 10.1137/0331030. MR1214759

[26] D. L. Russell and B. Y. Zhang, *Exact controllability and stabilizability of the Korteweg-de Vries equation*, Trans. Amer. Math. Soc. **348** (1996), no. 9, 3643–3672, DOI 10.1090/S0002-9947-96-01672-8. MR1360229

[27] E. M. Stein, *Harmonic analysis: real-variable methods, orthogonality, and oscillatory integrals*, Princeton Mathematical Series, vol. 43, Princeton University Press, Princeton, NJ, 1993. With the assistance of Timothy S. Murphy; Monographs in Harmonic Analysis, III. MR1232192

[28] T. Tao, *Multilinear weighted convolution of L^2-functions, and applications to nonlinear dispersive equations*, Amer. J. Math. **123** (2001), no. 5, 839–908. MR1854113

[29] T. Tao, *Global well-posedness of the Benjamin-Ono equation in $H^1(\mathbf{R})$*, J. Hyperbolic Differ. Equ. **1** (2004), no. 1, 27–49, DOI 10.1142/S0219891604000032. MR2052470

[30] T. Tao, *Nonlinear dispersive equations*, CBMS Regional Conference Series in Mathematics, vol. 106, Published for the Conference Board of the Mathematical Sciences, Washington, DC; by the American Mathematical Society, Providence, RI, 2006. Local and global analysis. MR2233925

California State University, Channel Islands, One University Drive, BTE 2762, Camarillo, California 93012

Email address: cynthia.flores@csuci.edu

Western New England University, 1215 Wilbraham Road, Springfield, Massachusetts 01119

Email address: seungly.oh@wne.edu

Independent Researcher, Harrisburg, Pennsylvania 17111

Email address: dereklee smith@gmail.com

Contemporary Mathematics
Volume **725**, 2019
https://doi.org/10.1090/conm/725/14555

Uniqueness of standing-waves for a non-linear Schrödinger equation with three pure-power combinations in dimension one

Daniele Garrisi and Vladimir Georgiev

ABSTRACT. We show that symmetric and positive profiles of ground-state standing-wave of the non-linear Schrödinger equation are non-degenerate and unique up to a translation of the argument and multiplication by complex numbers in the unit sphere. The non-linear term is a combination of two or three pure-powers. The class of non-linearities satisfying the mentioned properties can be extended beyond two or three power combinations. Specifically, it is sufficient that an Euler differential inequality is satisfied and that a certain auxiliary function is such that the first local maximum is also an absolute maximum.

1. The role of the uniqueness and non-degeneracy in the stability

A standing-wave is a function defined as $\phi(t,x) := e^{i\omega t}u(x)$, where ω is a real number, u is a complex-valued function in $H^1(\mathbb{R};\mathbb{C})$ and ϕ is a solution to the non-linear Schrödinger equation

$$i\partial_t\phi(t,x) + \partial^2_{xx}\phi(t,x) - F'(\phi(t,x)) = 0, \tag{1.1}$$

The profile of a standing-wave is just $R(x) := |u(x)|$. The literature is concerned with the existence and the stability of standing-waves whose profiles obey prescribed variational characterizations. The profiles we are interested in are minima of the energy functional

$$E(u) := \frac{1}{2}\int_{-\infty}^{+\infty} |u'(x)|^2 dx + \int_{-\infty}^{+\infty} F(u(x))dx$$

on the constrained defined as $S(\lambda) := \{u \in H^1(\mathbb{R}) \mid \|u\|^2_{L^2} = \lambda\}$ where $\lambda > 0$. As one can easily check, if u is a minimum of the energy functional, then $v(x) := zu(x+y)$

2010 *Mathematics Subject Classification.* Primary 35Q55; Secondary 47J35.

D. Garrisi was supported by INHA UNIVERSITY Research Grant and by the London Mathematical Society through the Research in Pairs Scheme 4, Grant ref. 41753 in the project "Uniqueness and non-degeneracy of normalized standing-waves".

V. Georgiev was supported in part by Project 2017 "Problemi stazionari e di evoluzione nelle equazioni di campo nonlineari" of INDAM, GNAMPA - Gruppo Nazionale per l'Analisi Matematica, la Probabilità e le loro Applicazioni, by Institute of Mathematics and Informatics, Bulgarian Academy of Sciences and Top Global University Project, Waseda University, by the University of Pisa, Project PRA 2018 49 and project "Dinamica di equazioni nonlineari dispersive", "Fondazione di Sardegna", 2016.

belongs to the same constraint and has the same energy. Therefore, it is a new minimum, for every choice of z in S^1 (complex numbers in the unit sphere) and y in $\mathbb{R}$. Then, u is clearly a degenerate critical point of E on the constraint $S(\lambda)$, as the transformations defined above show that u is the limit of a sequence of critical points. Therefore, both uniqueness and non-degeneracy need to be defined. We introduce the notation

$$\mathcal{G}_\lambda := \{u \in S(\lambda) \mid E(u) = \inf_{S(\lambda)} E\}.$$

The set we defined is sometimes called *ground state*, as in [**3**], even if the literature occasionally adopts this term to address more generally positive solutions to semi-linear elliptic equations, [**9**]. We denote by $H^1_r(\mathbb{R})$ the set of real-valued H^1 functions which are radially symmetric with respect to the origin.

DEFINITION 1.1 (Uniqueness and non-degeneracy). A pair (F, λ) satisfies the uniqueness property if given u and v in $\mathcal{G}_\lambda$, there exists (z, y) in $S^1 \times \mathbb{R}$ such that $u(x) = zv(x+y)$ for every x in $\mathbb{R}$. It satisfies the non-degeneracy property if the function E_r obtained as a restriction of E on $S(\lambda) \cap H^1_r(\mathbb{R})$ has non-degenerate minima.

Uniqueness and non-degeneracy are not only interesting features of the energy functional, but also play a role in the orbital stability of standing-wave solutions to (1.1). We say that (1.1) is globally well-posed in $H^1(\mathbb{R};\mathbb{C})$ if, given u_0 in $H^1(\mathbb{R};\mathbb{C})$, there exists a solution

$$\phi\colon [0, +\infty) \times \mathbb{R} \to \mathbb{C}$$

such that $\phi(0, x) = u_0(x)$ and the map

$$U\colon [0, +\infty) \times H^1(\mathbb{R};\mathbb{C}) \to H^1(\mathbb{R};\mathbb{C}), \quad U(t, u_0) := \phi(t, \cdot)$$

is such that

$$U(\cdot, u_0) \in C^1\big([0, +\infty); H^{-1}(\mathbb{R};\mathbb{C})\big) \cap C\big([0, +\infty); H^1(\mathbb{R};\mathbb{C})\big)$$

for every u_0 in $H^1(\mathbb{R};\mathbb{C})$. On the set $H^1(\mathbb{R};\mathbb{C})$ we consider the metric induced by the scalar product

$$(u, w)_{H^1(\mathbb{R};\mathbb{C})} := \operatorname{Re} \int_{\mathbb{R}} u(x)\overline{w}(x)dx + \operatorname{Re} \int_{\mathbb{R}} u'(x) \cdot \overline{w'(x)}dx$$

and denote it by d.

DEFINITION 1.2 (Stability). A subset $\mathcal{G}$ of $H^1(\mathbb{R};\mathbb{C})$ is said stable if for every $\delta > 0$ there exists $\varepsilon > 0$ such that $d(u_0, \mathcal{G}) < \varepsilon \implies d(U_t(u_0), \mathcal{G}) < \delta$ for every $t \geq 0$.

Given u in $\mathcal{G}_\lambda$, we define

$$\mathcal{G}_\lambda(u) := \{zu(\cdot + y) \mid (z, y) \in S^1 \times \mathbb{R}\}. \tag{1.2}$$

In general, if u is a minimum of E, then $\mathcal{G}_\lambda(u)$ is a subset of the ground state $\mathcal{G}_\lambda$. The stability of these two sets is object of interest of the literature since the work of T. Cazenave and P. L. Lions, [**8**], where pure-powers are considered. Results of stability of the ground-state have been extended to more general non-linearities, as in [**4, 21**]. We also mention other references which target the stability of the ground-state in other evolutionary equation, as multi-constraint non-linear Schrödinger systems, [**13, 16**], coupled non-linear Schrödinger systems (NLS

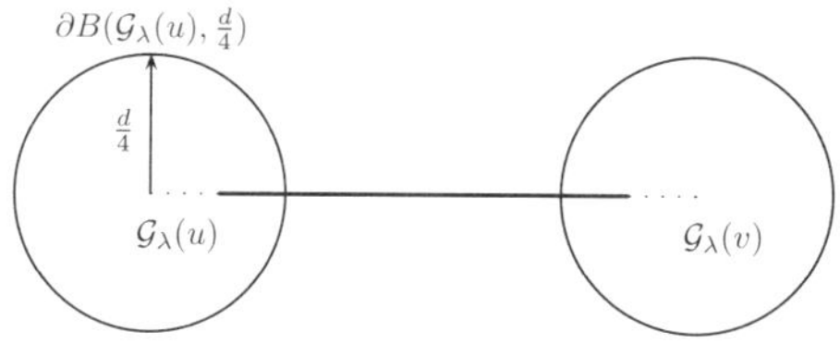

FIGURE 1. Trajectories bridging two different sets

+ NLS), [**12, 18, 19**], coupled non-linear Schrödinger and Kortweg-de Vries equation (NLS + KdV), [**1**], non-linear Klein-Gordon equation (NLKG), [**3**], (NLKG + NLKG), [**10**]. In most cases, the stability of the ground-state is a consequence of the Concentration-Compactness Lemma, [**14, 15**]. Coupled equations present some additional difficulties (rescalings do not work) but they can be worked around with *ad hoc* rescalings, as in [**1**] or with inequalities obtained through symmetric rearrangements for more general non-linearities, as [**10**, Lemma 3.1] and [**6**, Proposition 1.4], or through the *coupled rearrangement* defined in [**20**, §2.2].

The stability of $\mathcal{G}_\lambda(u)$ is more challenging than the stability of $\mathcal{G}_\lambda$: there might be solutions to (1.1) with initial values close to $\mathcal{G}_\lambda(u)$, but intermediate values far from it. Another application of the Concentration-Compactness Lemma and the stability of the ground state implies that these intermediate values are close to another set $\mathcal{G}_\lambda(v)$ (as shown in FIGURE 1). A simple way to rule out the existence of these trajectories is to prove that there is only one $\mathcal{G}_\lambda(u)$, as u varies in $\mathcal{G}_\lambda$. This is the approach followed in [**8**] with the help of a uniqueness result, [**17**], which specifically applies to pure-powers. Therefore, $\mathcal{G}_\lambda = \mathcal{G}_\lambda(u)$, and the second set is stable because the first one is stable. Another way is to show that there are only finitely many of these sets $\mathcal{G}_\lambda(u)$. In this case (see FIGURE 1), trajectories bridging two different sets need to achieve a minimum amount of energy, which is too high if the initial value is too close to $\mathcal{G}_\lambda(u)$, as it follows from [**11**, §4]. Now, from the work of L. Jeanjean and J. Byeon, [**7**], in every set $\mathcal{G}_\lambda(u)$ there exists a unique positive R in H^1_r. Therefore, the problem of the stability of the set (1.2) reduces to showing that $\mathcal{G}_{\lambda,r} := \mathcal{G}_\lambda \cap H^1_r$ is finite, [**11**, Proposition 5]. And this follows straightforwardly from the non-degeneracy of minima of E_r. From [**11**, Corollary 2], the uniqueness holds if $\mathcal{G}^+_{\lambda,r} := \mathcal{G}_\lambda \cap H^1_{r,+}$ is a singleton.

2. Assumptions on F and non-degeneracy

The non-linearity F is a C^2 real valued function defined on $\mathbb{C}$; $F(s) = G(|s|)$ for every s in $\mathbb{C}$. We list our assumptions trying to keep the notation consistent with [**11**]:

$$\text{(G1)} \qquad \exists s_0 > 0 \text{ such that } G(s_0) < 0$$

$$\text{(G2b)} \qquad \exists s_*, p^* \text{ such that } -C|s|^{p^*} \leq G(s), \quad s \geq s_*, \quad 2 < p_* < 6$$

$$\text{(G4')} \qquad G(0) = G'(0) = 0, \quad |G''(s)| \leq C(|s|^{p-2} + |s|^{q-2}), \quad 2 < p < q$$

are satisfied. (G4') makes sure that the energy functional E is $C^2(H^1(\mathbb{R};\mathbb{C}), \mathbb{R})$, which is a consequence of regularity theorems on Nemytski operators proved in [**2**]; (G2b) makes E a coercive functional on $S(\lambda)$ and provides a-priori estimates, and thus the global well-posedness of (1.1). For (G1) we refer to [**4**]; together with (G2b), it ensures that a minimum of E on $S(\lambda)$ exists, if λ is large enough, see [**4**, Theorem 2]. In the quoted reference, a stronger condition than (G2b) is actually used, by setting s_* is to zero. However, their proof applies under the weaker assumption as well. We illustrate how the non-degeneracy of minima of E on $S(\lambda)$ are obtained in [**11**]. We set $S_r(\lambda) := S(\lambda) \cap H^1_r(\mathbb{R};\mathbb{R})$. Suppose that R_0 is a minimum of E_r on $S_r(\lambda)$. Then, there exists ω_0 such that

$$R_0'' - G'(R_0) - \omega_0 R_0 = 0. \tag{2.1}$$

By looking at conserved quantities, as in [**11**, Proposition 3], there holds $\omega_0 > 0$. In order to prove that R_0 is a non-degenerate minimum of E_r, it is enough to consider a function v in H^1_r such that $(v, R_0)_2 = 0$ and show that there exists C not depending on v such that

$$D^2E(R_0)[v,v] \geq C\|v\|^2_{H^1}. \tag{2.2}$$

In order to evaluate the Hessian, it only takes to define a smooth curve α from $(-\varepsilon, \varepsilon)$ to $S_r(\lambda)$ such that $\alpha(0) = R_0$, $\alpha'(0) = v$. Then

$$(E \circ \alpha)''(0) = D^2E(R_0)[v,v] = \int_{-\infty}^{+\infty} \Big(|v'(x)|^2 + (G''(R_0(x)) + \omega_0)v(x)^2\Big)\,dx =: \xi(v).$$

Since the functional above is homogeneous, we only need to show that the infimum of ξ is positive when restricted to the unit sphere $S_r(1)$ in L^2. Then (2.2) will follow from the Banach-Steinhaus theorem. The functional ξ is certainly non-negative, because the fact that R_0 is a minimum is part of our assumptions. It is convenient to prove that ξ does actually achieve its infimum. We use as a reference the proof of [**23**, Proposition 2.9]. In fact, although that deals specifically with pure-powers, it can be applied to more general non-linearities, provided $G''(0) = 0$. Let v_0 be a minimum of ξ. If $\xi(v) = 0$, then there exists β in $\mathbb{R}$ such that

$$L_+(v) := -v'' + G''(R_0)v + \omega_0 v = \beta R_0. \tag{2.3}$$

We construct a one-parameter family of solutions starting from R_0. We premise a few remarks. Firstly, $R_0(0)$ is a solution to the equation

$$V(R_0(0)) = \omega_0, \quad V(s) := -\frac{2G(s)}{s^2} \tag{2.4}$$

which is the auxiliary function mentioned in the abstract of this paper. Secondly, from [**5**, Theorem 5], $R_0(0)$ is the least positive solution to (2.4). Moreover, R_0 is also an even function decreasing on $[0, +\infty)$. Therefore, $R_0''(0) < 0$. Then

$$R_0(0) = \inf\{s > 0 \mid V(s) = \omega_0\}, \quad V'(R_0(0)) > 0.$$

The second inequality is obtained by combining two equalities which in turn can be obtained by multiplying (2.1) by R_0 and R_0', as in [**11**, Proposition 4]. The construction of the one-parameter family is made as follows: the function

$$R_*(\omega) := \inf\{s > 0 \mid V(s) = \omega\}$$

is smooth in a neighborhood of ω_0, because $V'(R_0(0)) > 0$. We define the function R_ω as solution to the initial value problem

$$R_\omega''(x) - G'(R_\omega(x)) - \omega R_\omega(x) = 0 \quad R_\omega'(0) = 0, \quad R_\omega(0) = R_*(\omega). \tag{2.5}$$

We set $S(\omega, x) = \frac{\partial R}{\partial \omega}(\omega, x)$ and define $S(\omega_0, x) := S_0(x)$. Therefore, taking the derivative with respect to ω in (2.5), and evaluating at $\omega = \omega_0$, we obtain

$$L_+(S_0) = -R_0. \tag{2.6}$$

Taking the L^2 scalar product with S_0, we obtain

$$\frac{1}{2}\frac{d}{d\omega}\|R(\omega_0,\cdot)\|_2^2 = -(L^+(S_0), S_0)_2. \tag{2.7}$$

From (2.6) and (2.3) we obtain $\beta S_0 + v = 0$, because S_0 and v are even functions and the kernel of L^+ is generated by R_0', which is odd. Since v is even, $\beta \neq 0$. Then $(L^+(S_0), S_0)_2 = 0$. Therefore, it is worth to investigate the behavior of the derivative of the function $\lambda(\omega) := \|R(\omega,\cdot)\|_2^2$ at the point $\omega = \omega_0$: if $\lambda'(\omega_0) \neq 0$, then (2.7) states the equality between zero and a positive number (in fact, $\lambda'(\omega_0) > 0$ is considered sufficient for the stability of R_0, a criterion named after N. G. Vakhitov and A. A. Kolokolov [**22**]). The calculations made in [**11**, §4] can be summarized as follows: there exists a positive function Ψ such that

$$\frac{d\lambda}{d\omega}(\omega_0) = -\frac{2R_*'(\omega_0)}{R_*(\omega_0)^5}\int_0^1 \frac{\theta^4(K(R_*(\omega_0)) - K(\theta R_*(\omega_0)))}{(\Psi(\theta, R_*(\omega_0), \omega_0))^{3/2}}d\theta, \tag{2.8}$$

where $K(s) = \frac{1}{s^2}(-6G(s) + sG'(s))$. At this point, provided K is a strictly non-decreasing function, we have $\lambda'(\omega_0) > 0$. Since

$$K'(s) = \frac{12sG(s) - 7s^2G'(s) + s^3G''(s)}{s^4},$$

this computation suggests requiring that $12G(s) - 7sG'(s) + s^2G''(s) > 0$ for every s in the interval $(0, R_*(\omega_0))$. In fact, there is no need to have a strict inequality here: since the integrand in (2.8) is non-negative, if $\lambda'(\omega_0)$ vanishes, then $12G(s) - 7sG'(s) + s^2G''(s) = 0$ on $(0, R_*(\omega_0))$, which means that on this interval G is a linear combination of s^2 and s^6. However, the coefficient of s^2 is zero, by (G4'), while the coefficient of s^6 is equal to zero because it is the pure-power critical case where minima of E over $S(\lambda)$ do not exist, see [**11**, Proof of Lemma 3.1]. Since we wish to address all the minima, regardless of the constraint, the set where the requirement holds should apply to the images of all the minima. Actually, it is not always easy to determine which choice of ω_0 in (2.1) give arise to minima of E on $S(\lambda)$ or other types of critical points. Therefore, it is better to address all the positive, symmetric solutions decaying at infinity, which we denote by $\mathcal{S}$. We define

$$\Omega := \bigcup_{R\in\mathcal{S}} \mathrm{Img}(R).$$

By [**11**, Proposition 4], this set can be defined as $(0, +\infty)$ if V is not bounded or V is bounded but $\sup(V)$ is not achieved, or $(0, R_*(\max(V))$, as we did in [**11**]. We require

$$L(s) := 12G(s) - 7sG'(s) + s^2G''(s) \geq 0 \text{ on } \Omega. \tag{G3}$$

There are several non-linearities satisfying the condition above, starting from pure-powers $G(s) = -as^p$ with $a > 0$ and $2 < p \leq 6$. Another example is the combined pure-power $G(s) = -as^p + bs^q$ with $a, b > 0$ and $p < q$; clearly, in the latter case,

$$L(s) = a(p-2)(6-p)s^p - b(q-2)(6-q)s^q$$

might changes sign. However, the function is non-negative on Ω which is a bounded interval for this choice of G. In fact, (G3) is satisfied as shown in [**11**, §5].

3. Uniqueness of standing-waves

The idea of how we obtain the uniqueness of standing-waves is the following: if there are two minima R_0 and R_1 belonging to the same constraint $S(\lambda)$, we consider the corresponding Lagrange multipliers $-\omega_0$ and $-\omega_1$. From (2.8) and (G3), the function λ is injective on $[\omega_0, \omega_1]$, which implies that λ is constant, because achieves the same values at the endpoints. Then $L \equiv 0$ on $(0, R_*(\omega_1))$ which implies that G is a linear combination of s^2 and s^6 and gives a contradiction with the sub-critical assumptions. The only thing we need to take care of is the definition of λ, which is smooth as long as R_* is smooth. In turn R_* is smooth on ω if $V'(R_*(\omega)) \neq 0$. Therefore, critical points of V represent potential discontinuities of the function R_*. However, R_* is continuous everywhere if, for instance, V does not have local maxima or the first local maximum is an absolute maximum. Therefore, we set

$$A := \{s > 0 \mid s \text{ is a local maximum of } V\}.$$

The assumption introduced in [**11**] reads

(G5) $\qquad A = \emptyset$ or $(A \neq \emptyset,\ V$ is bounded and $V(\inf(A)) = \sup(V) < +\infty)$.

When A is non-empty and $\sup(V)$ is achieved we will use the notation $s_V := \inf(A)$. To summarize, condition (G3) allows to state that the set $\mathcal{G}^+_{\lambda,r}$ is finite. If (G5) holds as well, then $\mathcal{G}^+_{\lambda,r}$ is a singleton, [**11**, Theorem 1.4].

THEOREM 3.1 ([**11**, Theorem 1.4]). *If the conditions* (G1), (G2b), (G3), (G4') *and* (G5) *hold, then $\mathcal{G}_\lambda \cap H^1_r$ consists of exactly two functions, R_+ and R_-. The first is positive while $R_- = -R_+$.*

The assumption (G2a) in [**11**] has been omitted here, as it can be replaced by (G4'). This explains the slight difference with the referenced theorem. We consider

$$G(s) = \varepsilon_a a s^p + \varepsilon_b b s^q + \varepsilon_c c s^r,\ \{\varepsilon_a, \varepsilon_b, \varepsilon_c\} \subseteq \{-1, 0, 1\},\ a, b, c > 0,\ 2 < p < q < r$$

and discuss the assumptions mentioned above. In the remainder of the paper we will describe the behavior of the two properties for pure-powers, combined pure-power, and three pure-power combinations. Some cases have already been illustrated in [**11**, §5], but we included them for the sake of completeness. We will leave out the cases $\{\varepsilon_a, \varepsilon_b, \varepsilon_c\} \subseteq \{0, 1\}$ as (G1) is not fulfilled. (G4') follows from the fact that all the exponents are bigger than two. When the coefficient of highest order term at infinity is positive (G2b) is satisfied.

3.1. Pure-powers. If $G(s) = -as^p$, then (G1) is satisfied because $G < 0$ and (G2b) holds if $p < 6$. Then the function $L(s) = a(p-2)(6-p)s^p$ is non-negative, while $V = 2as^{p-2}$ does not have local maxima, implying that (G5) is satisfied.

3.2. Combined pure-powers. Firstly, we consider the case $G(s) = -as^p + bs^q$ which clearly achieves negative values. Ω is the bounded interval $(0, s_V)$. If $p > 6$, then L is negative in a neighbourhood of the origin. Then (G3) is not satisfied. If $p = 6$, then $L > 0$ on $(0, +\infty)$ and (G3) holds. If $p < 6$, then L achieves negative values; the function V is bounded and has a single local maximum. Therefore, (G5) is satisfied and (G3) is satisfied if the (unique) zero of L occurs *after* the local maximum of V, which is the unique zero of V'. We will show that $V'(s_0) = 0$ implies $L(s_0) > 0$. In fact, $V'(s_0) = 0$ gives

$$2a(p-2)s_0^{p-3} - 2b(q-2)s_0^{q-3} = 0.$$

If we multiply it by s_0^3, and substitute $2b(q-2)s_0^q$ with $2a(p-2)s_0^p$ in L, we obtain $L(s_0) = a(p-2)(q-p)s_0^p > 0$ which implies (G3). In fact, the strict inequality $p < 6$ (which was required in [**11**, §5]) is not needed. If $G = -as^p - bs^q$ then $G < 0$ which implies (G1). For (G2b) to hold, we need $q < 6$. Then $L > 0$ on $(0, +\infty)$. (G5) is satisfied because $V' > 0$ on $(0, +\infty)$, so $A = \emptyset$. Finally, if $G(s) = as^p - bs^q$, $q < 6$. Since V' goes to $+\infty$, $\Omega = (0, +\infty)$. However, L achieves negative values. Therefore (G3) is not satisfied.

3.3. Three pure-power combinations. The cases with three negative coefficients are ruled out as in 3.2. Then all the assumptions are fulfilled. (G1) can be easily checked except for the case $(\varepsilon_a, \varepsilon_b, \varepsilon_c) = (1, -1, 1)$ in §3.3.4. The following remark will be useful in §3.3.1, §3.3.4 and §3.3.6: given the function

$$k(s) := A - Bs^{q-p} + Cs^{r-p}$$

with $p < q < r$ and A, B, C positive real numbers there holds

$$\inf(k) \geq 0 \iff A \geq B^{\frac{r-p}{r-q}} C^{\frac{p-q}{r-q}} d_* \tag{3.1}$$

where

$$d_* := \left[\left(\frac{q-p}{r-p} \right)^{\frac{q-p}{r-p}} - \left(\frac{q-p}{r-p} \right)^{\frac{r-p}{r-q}} \right] > 0. \tag{3.2}$$

It is obtained by evaluating k on its unique minimum.

3.3.1. $G(s) = -as^p - bs^q + cs^r$. Clearly (G5) holds, because the set A is a singleton. If $p \geq 6$, then $L < 0$ in a neighborhood of the origin. Therefore (G3) does not hold because Ω contains small neighborhoods of the origin, as shown in FIGURE 2. For the case $p < 6$ it is convenient to divide V' and L by the leading terms (which are positive) and use the substitution $t = s^{q-p}$. As in §3.2, we need to know the behavior of L at the unique zero of V'. We set

$$g(t) := 1 + \frac{b(q-2)(6-q)}{a(p-2)(6-p)} t - \frac{c(r-2)(6-r)}{a(p-2)(6-p)} t^{\frac{r-p}{q-p}}$$

$$h(t) := 1 + \frac{b(q-2)}{a(p-2)} t - \frac{c(r-2)}{a(p-2)} t^{\frac{r-p}{q-p}}.$$

If $q \leq 6$, then g has at most one zero and h has exactly one zero. Let t_0 be the unique zero of h. From $h(t_0) = 0$ we obtain

$$\frac{c(r-2)}{a(p-2)} t_0^{\frac{r-p}{q-p}} = \frac{b(q-2)}{a(p-2)} t_0 + 1.$$

Then

$$\begin{aligned} g(t_0) &= 1 + \frac{b(q-2)(6-q)}{a(p-2)(6-p)} t_0 - \frac{6-r}{6-p} \left(\frac{b(q-2)}{a(p-2)} t_0 + 1 \right) \\ &= \frac{b(q-2)(r-q)}{a(p-2)(6-p)} t_0 + 1 - \frac{6-r}{6-p} > 0. \end{aligned}$$

Then (G3) holds. The behavior of g and h is represented in FIGURE 3. When $q > 6$, the argument above does not work, because g could have two zeroes. Then, suppose that $g(t_1) = 0$, that is

$$\frac{c(r-2)(r-6)}{a(p-2)(6-p)} t_1^{\frac{r-p}{q-p}} = \frac{b(q-2)(q-6)}{a(p-2)(6-p)} t_1 - 1.$$

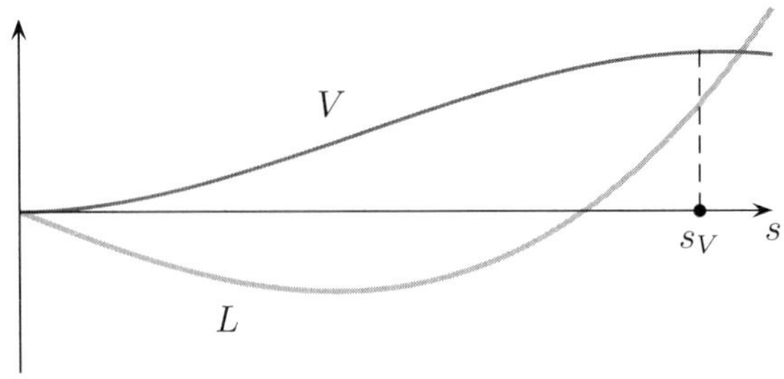

FIGURE 2. $L < 0$ on a subset of $\Omega = (0, s_V)$

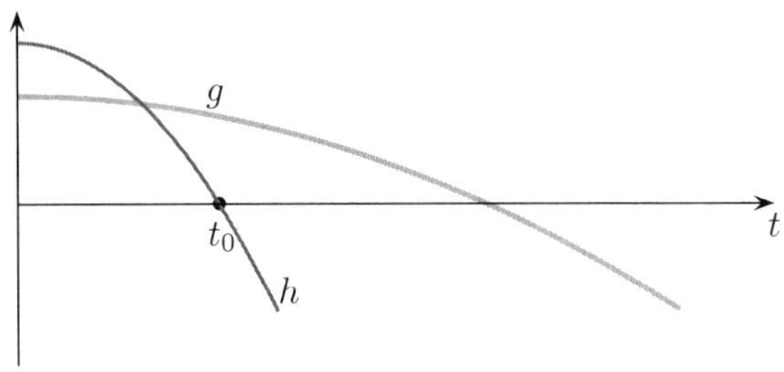

FIGURE 3. The zero of L occurs after the first zero of V'

$$\begin{aligned} h(t_1) &= 1 + \frac{b(q-2)}{a(p-2)}t_1 - \frac{b(q-2)(q-6)}{a(p-2)(r-6)}t_1 + \frac{6-p}{r-6} \\ &= \frac{r-p}{r-6} + \frac{b(q-2)}{a(p-2)} \cdot \frac{r-q}{r-6}t_1 > 0. \end{aligned}$$

Therefore, (G3) holds if and only if $\inf(g) \geq 0$. According to (3.1), that is equivalent to (3.8).

3.3.2. $G(s) = -as^p + bs^q + cs^r$. (G5) always holds as V has a single local maximum. If $p = 6$, then $L > 0$ everywhere. For $p \neq 6$ we can define the functions g and h in the same fashion as in §3.3.1

$$\begin{aligned} g(t) &:= 1 - \frac{b(q-2)(6-q)}{a(p-2)(6-p)}t - \frac{c(r-2)(6-r)}{a(p-2)(6-p)}t^{\frac{r-p}{q-p}} \\ h(t) &:= 1 - \frac{b(q-2)}{a(p-2)}t - \frac{c(r-2)}{a(p-2)}t^{\frac{r-p}{q-p}}. \end{aligned}$$

If $p < 6$ then $g \geq h$ because each coefficient of g is larger than the corresponding coefficient of h. Therefore, the first zero of L occurs after the first zero of V', and (G3) holds. If $p > 6$, then L is negative in a neighborhood of the origin, therefore (G3) does not hold, Figure 2.

3.3.3. $G(s) = as^p - bs^q - cs^r$. For (G2b) to hold, $r < 6$ must be satisfied. $\Omega = (0, +\infty)$, while $\inf(L) < 0$. Then (G5) holds, but (G3) does not.

3.3.4. $G(s) = as^p - bs^q + cs^r$. For (G1) to hold we need $\inf(G) < 0$. If we set $k := [as^p]^{-1}G$, the equivalence (3.1) gives

$$a < b^{\frac{r-p}{r-q}} c^{\frac{p-q}{r-q}} d_*. \tag{3.3}$$

If $p \leq 6$, then (G3) does not hold, because L is negative in a neighborhood of the origin, as in Figure 2. Before looking at the case $p > 6$ it is useful to observe that from (G1) we have $\sup(V') > 0$. On the contrary, $\inf(-V') \geq 0$. We apply (3.1) to $k := [2s^{p-3}]^{-1}V'$ and obtain

$$a(p-2) \geq [b(q-2)]^{\frac{r-p}{r-q}} [c(r-2)]^{\frac{p-q}{r-q}} d_*. \tag{3.4}$$

Dividing term-wise (3.4) by (3.3), we obtain

$$p-2 > (q-2)^{\frac{r-p}{r-q}} (r-2)^{\frac{p-q}{r-q}}. \tag{3.5}$$

By exponentiating both terms to $r-q$, dividing by $(p-2)^{r-q}$ and applying the variable changes $z = p-2$, $y = q-2$ and $x = r-2$, (3.5) reads $M(x,y,z) > 1$, with $0 < z < y < x$, which contradicts Lemma 3.2. Then $\sup(V') > 0$ and (G5) holds too.

When $p > 6$ we need to compare L and V'. Since $\sup(V') > 0$, it has two distinct zeroes. We will show that L is negative in the first zero of V', as in Figure 4. We set

$$\begin{aligned} g(t) &:= 1 - \frac{b(q-2)(6-q)}{a(p-2)(6-p)} t + \frac{c(r-2)(6-r)}{a(p-2)(6-p)} t^{\frac{r-p}{q-p}} \\ h(t) &:= 1 - \frac{b(q-2)}{a(p-2)} t + \frac{c(r-2)}{a(p-2)} t^{\frac{r-p}{q-p}}. \end{aligned}$$

We call t_1 the first zero of h. Since $hV' < 0$ on $(0,+\infty)$, $h(t_1) = 0$ and $h'(t_1) < 0$. From $h(t_1) = 0$, we obtain

$$\frac{c(r-2)}{a(p-2)} t_1^{\frac{r-p}{q-p}} = \frac{b(q-2)}{a(p-2)} t_1 - 1 \tag{3.6}$$

which we can substitute into the inequality $t_1 h'(t_1) < 0$ and obtain

$$-\frac{b(q-2)}{a(p-2)} t_1 + \frac{r-p}{q-p} \left(\frac{b(q-2)}{a(p-2)} t_1 - 1 \right) < 0$$

which gives

$$t_1 < \frac{a(p-2)(r-p)}{b(q-2)(r-q)}. \tag{3.7}$$

Therefore, from (3.6) and (3.7)

$$\begin{aligned} g(t_1) &= 1 - \frac{b(q-2)(6-q)}{a(p-2)(6-p)} t_1 + \frac{6-r}{6-p} \left(\frac{b(q-2)}{a(p-2)} t_1 - 1 \right) \\ &= \frac{b(q-2)(r-q)}{a(p-2)(p-6)} t_1 - \frac{r-p}{p-6} \\ &< \frac{b(q-2)(r-q)}{a(p-2)(p-6)} \cdot \frac{a(p-2)(r-p)}{b(q-2)(r-q)} - \frac{r-p}{p-6} = 0. \end{aligned}$$

Then, regardless of the values of the exponents, (G3) is never met, while (G5) holds.

3.3.5. $G(s) = as^p + bs^q - cs^r$. (G2b) implies $r < 6$. Therefore, $\inf(L) < 0$ on $(0, \varepsilon)$ for ε suitably small and (G3) does not hold.

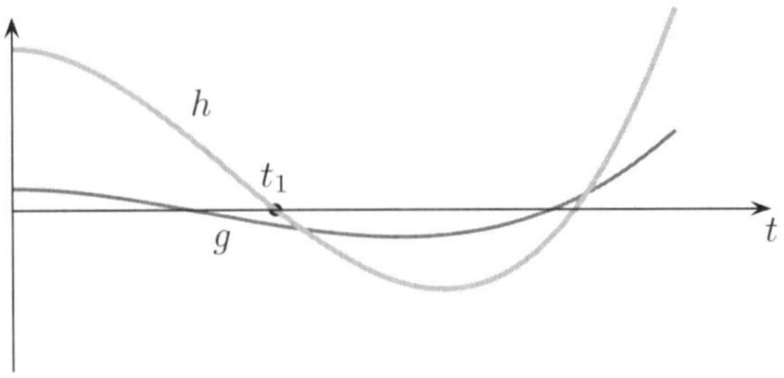

FIGURE 4. The zero of L occurs before the first zero of V', from §3.3.4.

3.3.6. $G(s) = -as^p + bs^q - cs^r$. (G2b) forces $r < 6$. Since V is not bounded, $\Omega = (0, +\infty)$. If we apply (3.1) with $k := [a(p-2)(6-p)s^p]^{-1}L$, we need

$$a(p-2)(6-p) \geq b^{\frac{r-p}{r-q}}(q-2)^{\frac{r-p}{r-q}}(6-q)^{\frac{r-p}{r-q}} \cdot c^{\frac{p-q}{r-q}}(r-2)^{\frac{p-q}{r-q}}(6-r)^{\frac{p-q}{r-q}}d_* \tag{3.8}$$

where d_* has been defined in (3.2). (3.8) can be true or false depending on whether a is large or small, respectively. Therefore, in this section we just show that whenever (G3) is satisfied, (G5) is satisfied too. In fact, if (G5) does not hold, then $\inf(V') < 0$. By applying (3.1) with $k := [a(p-2)s^{p-3}]V'$, we obtain

$$a(p-2) < b^{\frac{r-p}{r-q}}(q-2)^{\frac{r-p}{r-q}}c^{\frac{p-q}{r-q}}(r-2)^{\frac{p-q}{r-q}}d_*. \tag{3.9}$$

We divide term-wise (3.9) by (3.8) and obtain

$$(6-p) > (6-q)^{\frac{r-p}{r-q}}(6-r)^{\frac{p-q}{r-q}}. \tag{3.10}$$

By exponentiating both terms to $r-q$, dividing by $(6-p)^{r-q}$ and applying the variable changes $x = 6-p$, $y = 6-q$ and $z = 6-r$, (3.10) reads $M(x,y,z) > 1$ with $0 < z < y < x$ which contradicts Lemma 3.2. The set A is empty or a singleton depending on whether $\inf(V') \geq 0$ or not. We give a proof of the lemma we referred to in §3.3.4 and §3.3.6.

LEMMA 3.2. *Let M and D be the function and domain defined as*

$$M(x,y,z) = y^{z-x}z^{x-y}x^{y-z}, \quad D := \{0 < z \leq y \leq x\}.$$

Then $\sup_D(M) = 1$ and $M < 1$ in the interior of D. Moreover, for every (x,y,z) in D, $M(x,y,z) = 1$ if and only if $x = y$ or $y = z$.

PROOF. We have

$$M(x,y,z) = \frac{z^{x-y}x^{y-z}}{y^{x-z}} = \left(\frac{z}{y}\right)^{x-y}\left(\frac{x}{y}\right)^{y-z} = \left[\left(\frac{z}{y}\right)^{\frac{x}{y}-1}\left(\frac{x}{y}\right)^{1-\frac{z}{y}}\right]^y.$$

In order to show that $M < 1$ it is enough to prove that $M^{1/y} < 1$. We substitute $\frac{z}{y}$ with a and $\frac{x}{y}$ with b. Then $M^{1/y} = a^{b-1}b^{1-a}$, while $0 < a \leq 1$ and $b \geq 1$. We define $H(a,b) = a^{b-1}b^{1-a}$. If $b = 1$, then $H = 1$. If $b > 1$, then $H(1,b) = 1$, while the limit of $H(\cdot, b)$ is zero as $a \to 0$. Also,

$$\begin{aligned}\partial_a H(a,b) &= (b-1)a^{b-2}b^{1-a} - \ln(b)a^{b-1}b^{1-a} \\ &= a^{b-2}b^{1-a}(b-1-a\ln(b)) \geq a^{b-2}b^{1-a}(1-a)\ln(b)\end{aligned}$$

which is positive because $b > 1$. Then, $H(a,b) < 1$ if $b > 1$ and $a < 1$, while $H(a,1) = H(1,b) = 1$. Therefore, $M < 1$ unless $x = y$ or $z = y$, in which case $M = 1$. □

ε_a	ε_b	ε_c	$6-p$	$6-q$	$6-r$	$\#A$	Ω	Assumptions	Section
$-$	0	0	$+$			0	$(0,+\infty)$	(G3) $\wedge$ (G5)	3.1
$-$	$+$	0	0 or $+$			1	$(0,s_V)$	(G3) $\wedge$ (G5)	3.2
$+$	$-$	0	$+$	$+$		0	$(0,+\infty)$	$\neg$(G3) $\wedge$ (G5)	3.2
$-$	$-$	0	$+$	$+$		0	$(0,+\infty)$	(G3) $\wedge$ (G5)	3.2
$+$	$+$	$-$	$+$	$+$	$+$	0	$(0,+\infty)$	$\neg$(G3) $\wedge$ (G5)	3.3.5
$+$	$-$	$+$				1	$(0,s_V)$	$\neg$(G3) $\wedge$ (G5)	3.3.4
$+$	$-$	$-$	$+$	$+$	$+$	0	$(0,+\infty)$	$\neg$(G3) $\wedge$ (G5)	3.3.3
$-$	$+$	$+$	0 or $+$			1	$(0,s_V)$	(G3) $\wedge$ (G5)	3.3.2
$-$	$+$	$+$	$-$			1	$(0,s_V)$	$\neg$(G3) $\wedge$ (G5)	3.3.2
$-$	$+$	$-$	$+$	$+$	$+$	0 or 1	$(0,+\infty)$	(G3) $\Rightarrow$ (G5)	3.3.6
$-$	$-$	$+$	$-$ or 0			1	$(0,s_V)$	$\neg$(G3) $\wedge$ (G5)	3.3.1
$-$	$-$	$+$	$+$	$-$		1	$(0,s_V)$	(G3) $\Rightarrow$ (G5)	3.3.1
$-$	$-$	$+$	$+$	0 or $+$		1	$(0,s_V)$	(G3) $\wedge$ (G5)	3.3.1

THEOREM 3.3. *For every $\lambda > 0$ if the set $\mathcal{G}_\lambda \cap H^1_{r,+}$ is non-empty then it is a singleton, provided G*

(i) *is a pure-power with $\varepsilon_a < 0$*
(ii) *is a combined pure-power with $\varepsilon_a < 0$, or $\varepsilon_a > 0$ in sub-critical regime*
(iii) *is a three pure-power combination with $(\varepsilon_a, \varepsilon_b, \varepsilon_c) = (-1,1,1)$ and $p \leq 6$ or*
(iv) *$(\varepsilon_a, \varepsilon_b, \varepsilon_c) = (-1,-1,1)$ with $q \leq 6$ or $p < 6 < q$ provided* (3.8) *holds or*
(v) *$(\varepsilon_a, \varepsilon_b, \varepsilon_c) = (-1,1,-1)$ provided $r < 6$ and inequality* (3.8) *holds.*

PROOF. It follows from [**11**, Theorem 1.4]. □

Since the mentioned non-linearities (with $p < 6$) satisfy (G1) and (G2b), there exists $\lambda_* \geq 0$ such that the set $\mathcal{G}_\lambda \cap H^1_{r,+}$ is non-empty for every $\lambda > \lambda_*$, from [**4**, Theorem 2] or [**11**, Theorem 1.1].

References

[1] J. Albert and J. Angulo Pava, *Existence and stability of ground-state solutions of a Schrödinger-KdV system*, Proc. Roy. Soc. Edinburgh Sect. A **133** (2003), no. 5, 987–1029, DOI 10.1017/S030821050000278X. MR2018323

[2] A. Ambrosetti and G. Prodi, *A primer of nonlinear analysis*, Cambridge Studies in Advanced Mathematics, vol. 34, Cambridge University Press, Cambridge, 1993. MR1225101

[3] J. Bellazzini, V. Benci, C. Bonanno, and A. M. Micheletti, *Solitons for the nonlinear Klein-Gordon equation*, Adv. Nonlinear Stud. **10** (2010), no. 2, 481–499, DOI 10.1515/ans-2010-0211. MR2656691

[4] J. Bellazzini, V. Benci, M. Ghimenti, and A. M. Micheletti, *On the existence of the fundamental eigenvalue of an elliptic problem in $\mathbb{R}^N$*, Adv. Nonlinear Stud. **7** (2007), no. 3, 439–458, DOI 10.1515/ans-2007-0306. MR2340279

[5] H. Berestycki and P.-L. Lions, *Nonlinear scalar field equations. I. Existence of a ground state*, Arch. Rational Mech. Anal. **82** (1983), no. 4, 313–345, DOI 10.1007/BF00250555. MR695535

[6] J. Byeon, *Effect of symmetry to the structure of positive solutions in nonlinear elliptic problems*, J. Differential Equations **163** (2000), no. 2, 429–474, DOI 10.1006/jdeq.1999.3737. MR1758705

[7] J. Byeon, L. Jeanjean, and M. Mariş, *Symmetry and monotonicity of least energy solutions*, Calc. Var. Partial Differential Equations **36** (2009), no. 4, 481–492, DOI 10.1007/s00526-009-0238-1. MR2558325

[8] T. Cazenave and P.-L. Lions, *Orbital stability of standing waves for some nonlinear Schrödinger equations*, Comm. Math. Phys. **85** (1982), no. 4, 549–561. MR677997

[9] J. Dávila, M. del Pino, and I. Guerra, *Non-uniqueness of positive ground states of non-linear Schrödinger equations*, Proc. Lond. Math. Soc. (3) **106** (2013), no. 2, 318–344, DOI 10.1112/plms/pds038. MR3021464

[10] D. Garrisi, *On the orbital stability of standing-wave solutions to a coupled non-linear Klein-Gordon equation*, Adv. Nonlinear Stud. **12** (2012), no. 3, 639–658, DOI 10.1515/ans-2012-0311. MR2976057

[11] D. Garrisi and V. Georgiev, *Orbital stability and uniqueness of the ground state for the non-linear Schrödinger equation in dimension one*, Discrete Contin. Dyn. Syst. **37** (2017), no. 8, 4309–4328, DOI 10.3934/dcds.2017184. MR3642266

[12] T. Gou and L. Jeanjean, *Existence and orbital stability of standing waves for nonlinear Schrödinger systems*, Nonlinear Anal. **144** (2016), 10–22, DOI 10.1016/j.na.2016.05.016. MR3534090

[13] N. Ikoma, *Compactness of minimizing sequences in nonlinear Schrödinger systems under multiconstraint conditions*, Adv. Nonlinear Stud. **14** (2014), no. 1, 115–136, DOI 10.1515/ans-2014-0104. MR3158981

[14] P.-L. Lions, *The concentration-compactness principle in the calculus of variations. The locally compact case. I* (English, with French summary), Ann. Inst. H. Poincaré Anal. Non Linéaire **1** (1984), no. 2, 109–145. MR778970

[15] P.-L. Lions, *The concentration-compactness principle in the calculus of variations. The locally compact case. II* (English, with French summary), Ann. Inst. H. Poincaré Anal. Non Linéaire **1** (1984), no. 4, 223–283. MR778974

[16] C. Liu, N. V. Nguyen, and Z.-Q. Wang, *Existence and stability of solitary waves of an m-coupled nonlinear Schrödinger system*, J. Math. Study **49** (2016), no. 2, 132–148, DOI 10.4208/jms.v49n2.16.03. MR3518232

[17] K. McLeod and J. Serrin, *Uniqueness of positive radial solutions of* $\Delta u + f(u) = 0$ *in* $\mathbf{R}^n$, Arch. Rational Mech. Anal. **99** (1987), no. 2, 115–145, DOI 10.1007/BF00275874. MR886933

[18] N. V. Nguyen and Z.-Q. Wang, *Orbital stability of solitary waves for a nonlinear Schrödinger system*, Adv. Differential Equations **16** (2011), no. 9-10, 977–1000. MR2850761

[19] M. Ohta, *Stability of solitary waves for coupled nonlinear Schrödinger equations*, Nonlinear Anal. **26** (1996), no. 5, 933–939, DOI 10.1016/0362-546X(94)00340-8. MR1362765

[20] M. Shibata, *A new rearrangement inequality and its application for* L^2*-constraint minimizing problems*, Math. Z. **287** (2017), no. 1-2, 341–359, DOI 10.1007/s00209-016-1828-1. MR3694679

[21] M. Shibata, *Stable standing waves of nonlinear Schrödinger equations with a general nonlinear term*, Manuscripta Math. **143** (2014), no. 1-2, 221–237, DOI 10.1007/s00229-013-0627-9. MR3147450

[22] N. G. Vakhitov and A. A. Kolokolov, *Stationary solutions of the wave equation in a medium with nonlinearity saturation*, Radiophysics and Quantum Electronics **16** (1973), no. 7, 783–789.

[23] M. I. Weinstein, *Modulational stability of ground states of nonlinear Schrödinger equations*, SIAM J. Math. Anal. **16** (1985), no. 3, 472–491, DOI 10.1137/0516034. MR783974

School of Mathematics, University of Leeds, Room 11.01, Maths/Earth and Environment Building, LS2 9JT, Leeds, West Yorkshire, United Kingdom; and Inha University, Room 5S167, Building 5, Namgu Inharo 100, Incheon 22212 South Korea

Email address: d.garrisi@leeds.ac.uk

Department of Mathematics, University of Pisa, Largo Bruno Pontecorvo 5 I - 56127 Pisa, Italy; Faculty of Science and Engineering, Waseda University, 3-4-1, Okubo, Shinjuku-ku, Tokyo 169-8555 Japan; and Institute of Mathematics and Informatics, Bulgarian Academy of Sciences, Acad. Georgi Bonchev Str., Block 8, 1113 Sofia, Bulgaria

Email address: georgiev@dm.unipi.it

Contemporary Mathematics
Volume **725**, 2019
https://doi.org/10.1090/conm/725/14551

Below-threshold solutions of a focusing energy-critical heat equation in $\mathbb{R}^4$

Stephen Gustafson and Dimitrios Roxanas

ABSTRACT. We review recent results on the global well-posedness and decay versus blow-up and non-decay dichotomy, below the stationary solution, for a focusing energy-critical heat equation in $\mathbb{R}^4$. Our methods are inspired by recent progress towards the *soliton resolution conjecture* for dispersive and wave equations, providing an alternative way of studying critical parabolic problems.

CONTENTS

1. Introduction
2. Analytical preliminaries
3. Asymptotic decay of global solutions
4. Minimal blow-up solution
5. Rigidity
References

1. Introduction

We consider the Cauchy problem for the focusing, energy-critical nonlinear heat equation in four space dimensions:

$$\begin{cases} u_t = \Delta u + |u|^2 u \\ u(0,x) = u_0(x) \in \dot{H}^1(\mathbb{R}^4) \end{cases} \tag{1.1}$$

for $u(x,t) \in \mathbb{R}$ (or $\mathbb{C}$) with initial data in the energy space

$$\dot{H}^1(\mathbb{R}^4) = \{u \in L^4(\mathbb{R}^4;\mathbb{C}) \mid \|u\|_{\dot{H}^1}^2 = \int_{\mathbb{R}^4} |\nabla u(x)|^2\, dx < \infty\}.$$

2010 *Mathematics Subject Classification.* Primary 35K05, 35B40, 35B65.
Key words and phrases. Nonlinear heat equation, concentration compactness, regularity.
The first author's research is supported through the NSERC Discovery Grant 22124-12. The second author is supported by the European Research Council (grant no. 637995 "ProbDynDispEq"). The results reported in this article were obtained as part of the second author's PhD thesis at the University of British Columbia. The authors would like to thank the anonymous referee for comments that improved the presentation of the paper.

The associated *energy*, is defined for $u \in \dot{H}^1$ as

$$E(u) = \int_{\mathbb{R}^4} \left(\frac{1}{2}|\nabla u|^2 - \frac{1}{4}|u|^4 \right) dx,$$

and is (formally) dissipated along solutions of (1.1):

$$\frac{d}{dt}E(u(t)) = -\int_{\mathbb{R}^4} |u_t|^2 \, dx \leq 0. \tag{1.2}$$

We refer to the gradient term in E as the *kinetic energy*, and the second term as the *potential energy*. The fact that the potential energy is negative reflects the focusing nature of the nonlinearity. Problem (1.1) is *energy-critical* in the sense that the scaling

$$u_\lambda(t,x) = \lambda u(\lambda^2 t, \lambda x), \quad \lambda > 0 \tag{1.3}$$

leaves invariant the equation, the potential energy, and in particular the kinetic energy, which is the square of the *energy norm* $\|\cdot\|_{\dot{H}^1}$.

Static solutions of (1.1), which solve the elliptic equation

$$\Delta W + |W|^2 W = 0 \tag{1.4}$$

will be shown to play a key role in our analysis. The function

$$W = W(x) = \frac{1}{(1 + \frac{|x|^2}{8})} \in \dot{H}^1(\mathbb{R}^4), \quad \notin L^2(\mathbb{R}^4)$$

is a well-known solution. Its scalings by (1.3), and spatial translations of these are again static solutions, and multiples of these are well-known to be the unique extremizers of the Sobolev inequality

$$\forall u \in \dot{H}^1, \; \|u\|_{L^4} \leq C\|\nabla u\|_{L^2}, \quad C = \frac{\|W\|_{L^4}}{\|\nabla W\|_{L^2}} = \|\nabla W\|_{L^2}^{-\frac{1}{2}} \text{ the best constant.} \tag{1.5}$$

For time-dependent solutions a suitable local existence theory affirms the existence of a unique smooth solution $u \in C(I; \dot{H}^1(\mathbb{R}^4))$ on a maximal time interval $I = [0, T_{max}(u_0))$.

The main result of our work states that initial data lying "below" W gives rise to global smooth solutions of (1.1) which decay to zero:

THEOREM 1.1. *(**[10]**) Let $u_0 \in \dot{H}^1(\mathbb{R}^4)$ satisfy*

$$E(u_0) \leq E(W), \qquad \|\nabla u_0\|_{L^2} < \|\nabla W\|_{L^2}. \tag{1.6}$$

Then the solution u of (1.1) is global ($T_{max}(u_0) = \infty$) and satisfies

$$\lim_{t \to \infty} \|u(t)\|_{\dot{H}^1} = 0. \tag{1.7}$$

The conditions (1.6) define a non-empty set, since it includes all initial data of sufficiently small kinetic energy. Moreover, conditions (1.6) are sharp for global existence and decay in several senses. Firstly, if the kinetic energy inequality is replaced by equality, W itself provides a non-decaying (though still global) solution. Secondly, if the kinetic energy inequality is reversed, and under the additional technical assumption $u_0 \in L^2(\mathbb{R}^4)$, by a slight variant of a classical argument [**15**], we find that the solution blows up in finite time:

THEOREM 1.2. *([**10**]) Let $u_0 \in H^1(\mathbb{R}^4)$ with*

$$E(u_0) < E(W), \qquad \|\nabla u_0\|_{L^2} \geq \|\nabla W\|_{L^2}.$$

Then the solution u of (1.1) has finite maximal lifespan: $T_{max}(u_0) < \infty$.

Thirdly, for any $a^* > 0$, [**20**] constructed finite-time blow-up solutions with initial data $u_0 \in H^1(\mathbb{R}^4)$ satisfying $E(W) < E(u_0) < E(W) + a^*$. See also [**6**] for formal constructions of blow-up solutions close to W.

It follows from classical variational bounds – see Lemma 2.3 – and energy dissipation (1.2), that any solution u on a time interval $I = [0, T)$ whose initial data satisfies (1.6), necessarily satisfies

$$\sup_{t \in I} \|\nabla u\|_{L^2} < \|\nabla W\|_{L^2}. \tag{1.8}$$

So it will suffice to show that the conclusions of Theorem 1.1 hold for any solution satisfying (1.8). Indeed, we will prove:

(1) If $I = [0, \infty)$ and (1.8) holds, then $\lim_{t \to \infty} \|\nabla u(t)\|_{L^2} = 0$. This is given as Theorem 3.1.
(2) For any solution satisfying (1.8), $T_{max}(u(0)) = \infty$. This is given as Corollary 1.

That static solutions provide the natural threshold for global existence and decay, as in (1.8), is a classical phenomenon for critical equations, particularly well-studied in the setting of parabolic problems, mostly on compact domains (eg.,[**21**]), via "blow-up"-type arguments: first, failure of a solution to extend smoothly is shown, by a local regularity estimate, to imply (kinetic) energy concentration; then, near a point of concentration, rescaled subsequences are shown to converge locally to a non-trivial static solution; finally, elliptic/variational considerations prohibit non-trivial static solutions below the threshold.

The main goals of our work were, first, to establish the global-regularity-below-threshold result Theorem 1.1 on the full space $\mathbb{R}^4$; second, to do so avoiding the classical strategy sketched above, and instead following Kenig-Merle's [**12, 13**] "concentration-compactness plus rigidity" approach to critical *dispersive* equations, similar to Kenig-Koch's [**11**] implementation for the Navier-Stokes equations.

An outline of this review article and our proof strategy follows. In Section 2 we briefly discuss the local well-posedness theory and also give some technical results we will appeal to later. In Section 3, we sketch the energy-norm decay of *global* solutions which satisfy (1.8); this is the content of Theorem 3.1. The strategy is that employed for the Navier-Stokes equations in [**7**]: reduce the problem to establishing the decay of *small* solutions (which is a refinement of the local theory) by exploiting the L^2–dissipation relation, using a solution-splitting argument to overcome the fact that the solution fails to lie in L^2. Notice that unlike the case of Navier-Stokes, our theorem is only true for below threshold global solutions, W itself being a global non-decaying solution.

The rest of the paper, Sections 4 and 5, concerns the implementation of the Kenig-Merle roadmap, appropriately adapted to the necessities of our parabolic framework. The proof is by contradiction: we assume Theorem 1.1 is false.

Acting on this belief allows one to construct a very special solution. This is what we sketch in Section 4: we prove the existence and compactness (modulo symmetries) of a "critical" element - a counterexample to global existence and decay- which is *minimal* with respect to $\sup_t \|\nabla u(t)\|_{L^2}$. For this part we follow closely the work [**14**] concerning the energy-critical NLS. For a precise statement see Theorem 4.1. The technical tools are a profile decomposition compatible with the heat equation (described in Proposition 2) and a perturbation result for the nonlinear heat equation, based on the local theory (Proposition 1).

Finally, in Section 5, we exclude the possibility of a compact solution with finite maximal existence time in Theorem 5.1. In fact we reach a much stronger conclusion than required for the proof of Theorem 1.1, since it excludes *compact* finite-time blowup at *any* kinetic energy level – that is, it does not require (1.8). This part is based on more classical parabolic tools. We first show that the centre of compactness remains bounded, by exploiting energy dissipation. Then a local small-energy regularity criterion, together with the backwards uniqueness and unique continuation theorems of [**5**], as in [**11**], imply the *triviality* of the critical element. This contradicts results implied by our original assumption and proves the global existence and decay of below threshold solutions.

REMARK 1.3. We expect Theorem 1.1 to extend to the energy-critical problem for the nonlinear heat equation in general dimension $d \geq 3$:

$$\begin{aligned} u_t &= \Delta u + |u|^{\frac{4}{d-2}} u \\ u(t_0, x) &= u_0(x) \in \dot{H}^1(\mathbb{R}^d). \end{aligned} \tag{1.9}$$

For simplicity of presentation, in [**10**] we gave the proof only for the case $d = 4$. The result can be easily transferred to solutions of (1.9) for $d = 3$. The proof should also carry over to $d \geq 5$ with some extra work to estimate the low-power nonlinearity as in [**22**].

REMARK 1.4. Our proof makes no use of any parabolic comparison principles, and so applies to complex-valued solutions.

There is a vast literature on the semilinear heat equation $u_t = \Delta u + |u|^{p-1}u$. We content ourselves here with a very brief mention of some results in the critical setting and refer the reader to [**19**] for a more comprehensive review of the literature. We have already mentioned the finite-time blow-up constructions [**6, 20**], and we point to recent constructions of infinite-time blowup (bubbling) on $\mathbb{R}^3$ [**18**]. We finally mention the recent result [**3**], where a complete classification of solutions sufficiently close to the stationary solution W is provided for $d \geq 7$: such solutions either exhibit Type-I blow-up, dissipate to zero, or converge to (a slightly rescaled, translated) W. In particular, Type II blow-up is ruled out in $d \geq 7$ near W. For another related work see [**9**] for a critical case of the m-corotational Harmonic Map Heat Flow.

2. Analytical preliminaries

2.1. Local Theory. We first make precise what we mean by a *solution* in the energy space:

DEFINITION 2.1. A function $u : I \times \mathbb{R}^4 \to \mathbb{C}$ on a time interval $I = [0, T)$ $(0 < T \leq \infty)$ is a solution of (1.1) if $u \in (C_t \dot{H}_x^1 \cap L_{t,x}^6)([0, t] \times \mathbb{R}^4)$; $\nabla u \in L_{x,t}^3([0, t] \times \mathbb{R}^4)$;

$D^2u,\ u_t \in L^2_t L^2_x([0,t]\times\mathbb{R}^4)$ for all $t\in I$; and the Duhamel formula

$$u(t) = e^{t\Delta}u_0 + \int_0^t e^{(t-s)\Delta}F(u(s))ds, \tag{2.1}$$

is satisfied for all $t \in I$, where $F(u) = |u|^2 u$. We refer to the interval I as the lifespan of u. We say that u is a maximal-lifespan solution if the solution cannot be extended to any strictly larger interval. We say that u is a global solution if $I = \mathbb{R}^+ := [0,+\infty)$.

We will often measure the space-time size of solutions on a time interval I in $L^6_{x,t}$, denoting

$$S_I(u) := \int_I\int_{\mathbb{R}^4} |u(t,x)|^6 dxdt, \quad \|u\|_{S(I)} := S_I(u)^{\frac{1}{6}} = \left(\int_I\int_{\mathbb{R}^4}|u(t,x)|^6 dxdt\right)^{\frac{1}{6}}.$$

A local well-posedness theory in the energy space $\dot{H}^1(\mathbb{R}^4)$, analogous to that for the corresponding critical nonlinear Schrödinger equation (see e.g.,[**4**]), is easily constructed, based on the Sobolev inequality and space-time estimates for the heat equation on $\mathbb{R}^4$ (we refer to [**19**] for more details),

$$\begin{aligned}
&\|e^{t\Delta}\phi\|_{L^p_x(\mathbb{R}^4)} \lesssim t^{-2(1/a-1/p)}\|\phi\|_{L^a}, \qquad 1\le a\le p\le\infty\\
&\|e^{t\Delta}\phi\|_{L^q_tL^p_x(\mathbb{R}^+\times\mathbb{R}^4)} \lesssim \|\phi\|_{L^a}, \qquad \frac{1}{q}+\frac{2}{p}=\frac{2}{a}, \quad 1<a\le p\\
&\|\int_0^t e^{(t-s)\Delta}f(s)ds\|_{L^q_tL^p_x(\mathbb{R}^+\times\mathbb{R}^4)} \lesssim \|f\|_{L^{\tilde{q}'}_tL^{\tilde{p}'}_x(\mathbb{R}^+\times\mathbb{R}^4)},\\
&\qquad \frac{1}{q}+\frac{2}{p}=\frac{1}{\tilde{q}}+\frac{2}{\tilde{p}}=1, \qquad \frac{1}{q}+\frac{1}{q'}=\frac{1}{p}+\frac{1}{p'}=1,
\end{aligned} \tag{2.2}$$

and $(\tilde{q}',\tilde{p}')$ the dual to any admissible pair $(\tilde{q},\tilde{p})$.

We also refer the reader to [**2, 23**] for a treatment of the Cauchy problem in the critical Lebesque space $L^{\frac{2d}{d-2}}$; the arguments directly adapt to show well-posedness in $\dot{H}^1$. One can use a fixed-point argument to construct local-in-time solutions for arbitrary initial data in $\dot{H}^1(\mathbb{R}^4)$; however, as usual when working in scaling-critical spaces, the time of existence depends on the profile of the initial data, not merely on its $\dot{H}^1$-norm. We summarize:

THEOREM 2.2. *([**10**]) (Local well-posedness) Assume $u_0\in\dot{H}^1(\mathbb{R}^4)$.*

(1) *(Local existence) There exists a unique, maximal-lifespan solution to the Cauchy Problem (1.1) in $I\times\mathbb{R}^4, I=[0,T_{max}(u_0))$.*
(2) *(Continuous dependence) The solution depends continuously on the initial data (in both the $\dot{H}^1$ and the S_I-induced topologies). Furthermore, T_{max} is a lower-semicontinuous function of the initial data.*
(3) *(Blow-up criterion) If $T_{max}(u_0)<+\infty$, then $\|u\|_{S([0,T_{max}(u_0)))}=+\infty$.*
(4) *(Energy dissipation) The energy $E(u(t))$ is a non-increasing function in time. More precisely, for $0<t<T_{max}$,*

$$E(u(t)) + \int_0^t\int_{\mathbb{R}^4} |u_t|^2\, dx\, dt = E(u_0). \tag{2.3}$$

(5) *(Small data global existence) There is* $\epsilon_0 > 0$ *such that if* $\|e^{t\Delta}u_0\|_{S(\mathbb{R}^+)} \leq \epsilon_0$*, the solution* u *is global,* $T_{max}(u_0) = \infty$*, and moreover*

$$\|u\|_{S(\mathbb{R}^+)} + \|\nabla u\|_{(L_t^\infty L_x^2 \cap L_{x,t}^3)(\mathbb{R}^+\times\mathbb{R}^4)} + \|D^2 u\|_{L_{x,t}^2(\mathbb{R}^+\times\mathbb{R}^4)} \lesssim \epsilon_0. \tag{2.4}$$

This occurs in particular when $\|u_0\|_{\dot{H}^1(\mathbb{R}^4)}$ *is sufficiently small.*

An extension of the proof of the local existence theorem implies the following stability result (see, e.g., [**14**]):

PROPOSITION 1. *(Perturbation result) For every* $E, L > 0$ *and* $\epsilon > 0$ *there exists* $\delta > 0$ *with the following property: assume* $\tilde{u} : I \times \mathbb{R}^4 \to \mathbb{R}$, $I = [0, T)$*, is an approximate solution to (1.1) in the sense that*

$$\|\nabla e\|_{L_{t,x}^{\frac{3}{2}}(I\times\mathbb{R}^4)} \leq \delta, \quad e := \tilde{u}_t - \Delta\tilde{u} - |\tilde{u}|^2\tilde{u},$$

and also

$$\|\tilde{u}\|_{L_t^\infty \dot{H}_x^1(I\times\mathbb{R}^4)} \leq E \quad \text{and} \quad \|\tilde{u}\|_{S(I)} \leq L,$$

then if $u_0 \in \dot{H}_x^1(\mathbb{R}^4)$ *is such that*

$$\|u_0 - \tilde{u}(0)\|_{\dot{H}_x^1(\mathbb{R}^4)} \leq \delta,$$

there exists a solution $u : I \times \mathbb{R}^4 \to \mathbb{R}$ *of (1.1) with* $u(0) = u_0$*, and such that*

$$\|u - \tilde{u}\|_{L_t^\infty \dot{H}_x^1(I\times\mathbb{R}^4)} + \|u - \tilde{u}\|_{S(I)} \leq \epsilon.$$

2.2. Variational estimates. Here we record some elementary -yet crucial for our work- variational inequalities:

LEMMA 2.3. *(Variational Estimates)*

(1) *If*

$$\|\nabla u_0\|_{L^2}^2 \leq \|\nabla W\|_{L^2}^2, \quad E(u_0) \leq (1-\delta_0)E(W), \quad \delta_0 > 0,$$

then there exists $\bar{\delta} = \bar{\delta}(\delta_0) > 0$ *such that for all* $t \in [0, T_{max}(u_0))$*, the solution of (1.1) satisfies*

$$\int |\nabla u(t)|^2 \leq (1-\bar{\delta}) \int |\nabla W|^2. \tag{2.5}$$

(2) *If (2.5) holds, then*

$$\int (|\nabla u(t)|^2 - |u(t)|^4)dx \geq \bar{\delta} \int |\nabla u(t)|^2 \tag{2.6}$$

and moreover $E(u(t)) \geq 0$.

See, e.g., Lemma 3.4/Theorem 3.9 in [**12**].

REMARK 2.4. The bound (1.8) also holds when $E(u_0) = E(W), \|\nabla u_0\|_{L^2} < \|\nabla W\|_{L^2}$. Note that although the equality case for the energy is not covered by the above Lemma, energy dissipation and continuity of the flow reduce matters to the previous situation.

2.3. Profile decomposition. The following proposition is the main tool (along with the Perturbation Proposition 1) used to establish the existence of a critical element (as sketched in Section 4). The idea is to characterize the loss of compactness in some critical embedding. It can be traced back to ideas in [**16**] and their modern "evolution" counterparts [**1**], [**17**], [**12**] and [**13**]. The result we use in our work is the following:

PROPOSITION 2. *(Profile Decomposition) Let $\{u_n\}_n$ be a bounded sequence of functions in $\dot{H}^1(\mathbb{R}^4)$. Then, after possibly passing to a subsequence (but we still refer to it as u_n), there exists a family of functions $\{\phi^j\}_{j=1}^{\infty} \subset \dot{H}^1$, scales $\lambda_n^j > 0$ and centers $x_n^j \in \mathbb{R}^4$ such that:*

$$u_n(x) = \sum_{j=1}^{J} \frac{1}{\lambda_n^j} \phi^j\Big(\frac{x - x_n^j}{\lambda_n^j}\Big) + w_n^J(x),$$

$w_n^J \in \dot{H}^1(\mathbb{R}^4)$ is such that:

$$\lim_{J\to\infty} \limsup_{n} \|e^{t\Delta} w_n^J\|_{L^6_{t,x}(\mathbb{R}^+\times\mathbb{R}^4)} = 0, \tag{2.7}$$

$$\lambda_n^j w_n^J(\lambda_n^j x + x_n^j) \rightharpoonup 0, \quad in\ \dot{H}^1(\mathbb{R}^4), \quad \forall j \le J. \tag{2.8}$$

Moreover, the scales are asymptotically orthogonal, in the sense that

$$\frac{\lambda_n^j}{\lambda_n^i} + \frac{\lambda_n^i}{\lambda_n^j} + \frac{|x_n^i - x_n^j|^2}{\lambda_n^j \lambda_n^i} \to +\infty, \quad \forall i \neq j. \tag{2.9}$$

Furthermore, for all $J \ge 1$ we have the following decoupling properties:

$$\|u_n\|_{\dot{H}^1}^2 = \sum_{j=1}^{J} \|\phi^j\|_{\dot{H}^1}^2 + \|w_n^J\|_{\dot{H}^1}^2 + o_n(1) \tag{2.10}$$

and

$$E(u_n) = \sum_{j=1}^{J} E(\phi^j) + E(w_n^J) + o_n(1). \tag{2.11}$$

For more details about the proof see [**19**].

3. Asymptotic decay of global solutions

In this section we summarize the steps of the proof of the following theorem:

THEOREM 3.1. *([**10**]) If $u \in C([0,\infty); \dot{H}^1(\mathbb{R}^4))$ is a solution to equation (1.1) which moreover satisfies*

$$\sup_{t\ge 0} \|\nabla u(t)\|_{L^2} < \|\nabla W\|_{L^2}, \tag{3.1}$$

then

$$S_{\mathbb{R}_+}(u) < \infty \quad and \quad \lim_{t\to\infty} \|u(t)\|_{\dot{H}^1} = 0.$$

The general strategy, drawn from the techniques of [**7**] for the Navier-Stokes equations, is as follows.

Step 1: for initial data $\|u_0\|_{\dot{H}^1} \ll 1$, we show that for the corresponding solution

$$\|u(t)\|_{\dot{H}^1} \xrightarrow{t\to+\infty} 0.$$

This is a refinement of the small data theory. The proof requires a careful splitting of the Duhamel term, and an approximation by well-localized functions to exploit the fast decay of the heat kernel.

Step 2: if we now further assume $u_0 \in H^1$:

$$\frac{d}{dt}\frac{1}{2}\|u(t)\|_{L^2}^2 = \int_{\mathbb{R}^4} [u^4 - |\nabla u|^2] dx \leq -\left[1 - \left(\frac{\|\nabla u(t)\|_{L^2}}{\|\nabla W\|_{L^2}}\right)^2\right] \|\nabla u\|_{L^2}^2,$$

and because of the *threshold assumption*, from (2.6), there is a constant $C > 0$ such that

$$\|u(t)\|_{L^2}^2 + C\|\nabla u\|_{L^2_{t,x}(\mathbb{R}_+ \times \mathbb{R}^4)}^2 \leq \|u_0\|_{L^2}^2.$$

But the solution is assumed to be *global* so there must be some $t_1 > 0$, large enough at which $\|\nabla u(t_1)\|_{L^2}$ is as small as we like. We then call upon the first step.

Step 3: to remove the H^1-assumption on the data we split (in frequency space, separating high from low frequencies) $u_0 = w_0 + v_0$, where $\|w_0\|_{\dot{H}^1}$ is small and $\|v_0\|_{H^1}$ is large. A perturbation argument coupled again with L^2-dissipation concludes the proof.

4. Minimal blow-up solution

For any $0 \leq E_0 \leq \|\nabla W\|_2^2$, we define

$$L(E_0) := \sup\{S_I(u) \mid u \text{ a solution of (1.1) on } I \text{ with } \sup_{t\in I} \|\nabla u(t)\|_2^2 \leq E_0\},$$

where $I = [0, T)$ denotes the existence interval of the solution in question. $L : [0, \|\nabla W\|_2^2] \to [0, \infty]$ is a continuous, non-decreasing function with $L(\|\nabla W\|_2^2) = \infty$. Moreover, from the small-data theory (2.4),

$$L(E_0) \lesssim E_0^3 \text{ for } E_0 \leq \epsilon_0.$$

Thus, there exists a unique critical kinetic energy $E_c \in (0, \|\nabla W\|_2^2]$ such that

$$L(E_0) < \infty \text{ for } E_0 < E_c, \quad L(E_0) = \infty \text{ for } E_0 \geq E_c.$$

In particular, if $u : I \times \mathbb{R}^4 \to \mathbb{R}$ is a maximal-lifespan solution, then

$$\sup_{t\in I} \|\nabla u(t)\|_2^2 < E_c \implies u \text{ is global, and } \|u\|_{S(\mathbb{R}^+)} \leq L(\sup_{t\in I} \|\nabla u(t)\|_2^2) < \infty.$$

This section describes the implementation of the first part of the Kenig-Merle scheme, which is usually referred to as the "concentration-compactness step". The goal is to extract a "minimal" counterexample to global existence/decay: this is a maximal-lifespan solution that "lives" on the putative threshold (see the theorem below for a precise statement) and has infinite $S-$norm over its lifespan. This solution is special in that it enjoys exceptional compactness properties.

In particular, we prove the following theorem:

THEOREM 4.1. *([**10**]) There is a maximal-lifespan solution $u_c : I \times \mathbb{R}^4 \to \mathbb{R}$ to (1.1) such that $\sup_{t\in I} \|\nabla u_c(t)\|_{L^2}^2 = E_c$, $\|u_c\|_{S(I)} = +\infty$. Moreover, there are $x(t) \in \mathbb{R}^4, \lambda(t) \in \mathbb{R}^+$, such that*

$$K = \left\{\frac{1}{\lambda(t)} u_c\left(t, \frac{x - x(t)}{\lambda(t)}\right) \mid t \in I\right\} \tag{4.1}$$

is precompact in $\dot{H}^1$.

For the proof of this theorem we closely follow the arguments in [**14**]. The extraction of this minimal blow-up solution (and its compactness up to scaling and translation) will be a consequence of the following proposition:

PROPOSITION 3. *Let $u_n : I_n \times \mathbb{R}^4$ be a sequence of solutions to (1.1) such that*

$$\limsup_n \sup_{t\in I_n} \|\nabla u_n\|_2^2 = E_c \text{ and } \lim_{n\to\infty} \|u_n\|_{S(I_n)} = +\infty. \tag{4.2}$$

where I_n are of the form $[0, T_n)$. Denote the initial data by $u_n(x, 0) = u_{n,0}(x)$. Then the sequence $\{u_{n,0}\}_n$ converges, modulo scaling and translations, in $\dot{H}^1$ (up to an extraction of a subsequence).

The proof of this proposition requires the combined use of two ingredients: the profile decomposition Proposition 2, and the perturbation Proposition 1. Roughly speaking, one wants to evolve the static profile decomposition in time to be able to decompose nonlinear solutions into simpler building blocks.

One performs a profile decomposition on a sequence satisfying the assumptions of Proposition 3 with the goal of showing that only *one profile* is responsible for the behaviour we require. There are two main steps:

Step 1: show there is at least one "bad" profile. To show this, one uses Proposition 1 to construct appropriate approximate solutions. The smallness of the error of such an approximation is a consequence of the asymptotic orthogonality of the profiles and statement (2.7).

We then face the same problem as in the case of the focusing energy-critical NLS: the kinetic energy is not conserved, so it is conceivable that the $S-$norm of several profiles can be large over short times, while their kinetic energy does not achieve the critical value until later. To finish the proof of the proposition we have to take one more step:

Step 2: prove that only one profile is responsible for the blow-up. The proof requires a combinatorial argument, which in turn implies that, for later times, the kinetic energy of a sum of profiles (asymptotically) decomposes as the sum of the kinetic energies of individual profiles. Recall that the analogous statement for the *initial data* is already encoded in our static profile decomposition Proposition 2.

We now show how one uses Proposition 3 to prove Theorem 4.1.

PROOF. By the definition of E_c we can find a sequence of solutions $u_n : I_n \times \mathbb{R}^4 \to \mathbb{R}$, with I_n compact, so that

$$\sup_n \sup_{t\in I_n} \|\nabla u_n(t)\|_{L^2}^2 = E_c \text{ and } \lim_n \|u_n\|_{S(I_n)} = +\infty.$$

An application of Proposition 3 shows that the corresponding sequence of initial data converges strongly, modulo symmetries, to some $\phi \in \dot{H}^1$. By rescaling and translating u_n, we may in fact assume $u_{n,0} := u_n(0, \cdot) \xrightarrow{\dot{H}^1} \phi$.

Let $u_c : I \times \mathbb{R}^4 \to \mathbb{R}$ be the maximal-lifespan solution with initial data ϕ. Since $u_{n,0} \xrightarrow{\dot{H}^1} \phi$, employing the stability Proposition 1, $I \subset \liminf I_n$, and $\|u_n - u_c\|_{L_t^\infty \dot{H}_x^1(K\times\mathbb{R}^4)} \xrightarrow{n\to\infty} 0$, for all compact $K \subset I$. Thus, by (4.2):

$$\sup_{t\in I} \|\nabla u_c(t)\|_{L^2}^2 \leq E_c. \tag{4.3}$$

Applying the stability Proposition 1 once again we can also see that $\|u_c\|_{S(I)} = \infty$. Hence, by the definition of the critical kinetic energy level, E_c,

$$\sup_{t\in I} \|\nabla u_c(t)\|_{L^2}^2 \geq E_c. \tag{4.4}$$

In conclusion,

$$\sup_{t\in I} \|\nabla u_c(t)\|_{L^2}^2 = E_c \tag{4.5}$$

and

$$\|u_c\|_{S(I)} = +\infty. \tag{4.6}$$

Finally, the compactness modulo symmetries (4.1) follows from another application of Proposition 3. □

5. Rigidity

To finally reach a contradiction, we need to implement the second part of the Kenig-Merle approach, which is usually called the "rigidity" part. One wants to show that the solution we found in the previous section has properties that are incompatible with properties of solutions to the nonlinear heat equation and thus show it cannot exist. This is the equation-specific part of the "concentration-compactness plus rigidity" argument, while the first part is very general and widely applicable.

In this section we mostly focus on ruling out finite-time blowup of compact (modulo symmetries) solutions, which is the content of the theorem below. Note this is a considerably stronger statement than we require, since it is not limited to solutions with below-threshold kinetic energy:

THEOREM 5.1. *([**10**]) If u is a solution to (1.1) on maximal existence interval $I = [0, T_{max})$, such that $K := \left\{ \frac{1}{\lambda(t)} u(t, \frac{x - x(t)}{\lambda(t)}) \mid t \in I \right\}$ is precompact in $\dot{H}^1$ for some $x(t) \in \mathbb{R}^4$, $\lambda(t) \in \mathbb{R}^+$, then $T_{max} = +\infty$.*

As a corollary, we can complete the proof of the main result Theorem 1.1 by showing:

COROLLARY 1. *For any solution satisfying (1.8), $T_{max}(u(0)) = \infty$.*

PROOF. By Theorem 5.1, the solution u_c produced by Theorem 4.1 must be global: $T_{max}(u_c(0)) = \infty$. But since $\|u_c\|_{S(\mathbb{R}^+)} = \infty$, Theorem 3.1 shows $E_c = \|\nabla W\|_2^2$, and the Corollary follows. □

In view of this the rest of the section is devoted to the discussion of the proof of Theorem 5.1. Due to the lack of L^2–conservation and the irreversibility of the flow we build a more involved "solving backwards" procedure than in [**12**] for the NLS. Our proof is inspired by the work of Kenig and Koch [**11**] for the Navier-Stokes system, and it is based on more classical parabolic tools – local smallness regularity, backwards uniqueness, and unique continuation – though implemented in a somewhat different way. In particular, we will make use of the following two results, see [**5**] and the references therein:

THEOREM 5.2. *(Backwards Uniqueness) Fix any* R, δ, M, *and* $c_0 > 0$. *Let* $Q_{R,\delta} := (\mathbb{R}^4 \backslash B_R(0)) \times (-\delta, 0)$, *and suppose a vector-valued function* v *and its distributional derivatives satisfy* $v, \nabla v, \nabla^2 v \in L^2(\Omega)$ *for any bounded subset* $\Omega \subset Q_{R,\delta}$, $|v(x,t)| \leq e^{M|x|^2}$ *for all* $(x,t) \in Q_{R,\delta}$, $|v_t - \Delta v| \leq c_0(|\nabla v| + |v|)$ *on* $Q_{R,\delta}$, *and* $v(x,0) = 0$ *for all* $x \in \mathbb{R}^4 \setminus B_R(0)$. *Then* $v \equiv 0$ *in* $Q_{R,\delta}$.

THEOREM 5.3. *(Unique Continuation) Let* $Q_{r,\delta} := B_r(0) \times (-\delta, 0)$, *for some* $r, \delta > 0$, *and suppose a vector-valued function* v *and its distributional derivatives satisfy* $v, \nabla v, \nabla^2 v \in L^2(Q_{r,\delta})$ *and there exist* $c_0, C_k > 0, (k \in \mathbb{N})$ *such that* $|v_t - \Delta v| \leq c_0(|\nabla v| + |v|)$ *a.e. on* $Q_{r,\delta}$ *and* $|v(x,t)| \leq C_k(|x| + \sqrt{-t})^k$ *for all* $(x,t) \in Q_{r,\delta}$. *Then* $v(x,0) \equiv 0$ *for all* $x \in B_r(0)$.

As well, we establish the following:

LEMMA 5.4. *(Local Smallness Regularity Criterion) For any* $k \in \mathbb{N}$, *there are* $\epsilon_0 > 0$ *and* C *such that: if* u *is a solution of equation (*1.1*) on* Q_1, *where* $Q_r := B_r(0) \times (-r^2, 0)$ *for* $r > 0$, *and satisfies*

$$\epsilon := \|u\|_{L_t^\infty(\dot{H}_x^1 \cap L_x^4)(Q_1)} < \epsilon_0,$$

then u *is smooth on* $\overline{Q_{\frac{1}{2}}}$ *with bounds*

$$\max_{\overline{Q_{\frac{1}{2}}}} |D^k u| \leq C\epsilon.$$

We proceed now with the proof of Theorem 5.1. Let us assume that the conclusion is false, i.e., $T_{\max} < +\infty$.

Step 1 (Compactness): By compactness in $\dot{H}^1$, and the continuous embedding $\dot{H}^1 \hookrightarrow L^4$, for every $\epsilon > 0$, there is an $R_\epsilon > 0$ such that for all $t \in I := [0, T_{\max})$:

$$\int_{|x - x(t)| \geq \frac{R_\epsilon}{\lambda(t)}} \left(|\nabla u_c(t,x)|^2 + |u_c(t,x)|^4\right) dx < \epsilon. \tag{5.1}$$

Fix any $\{t_n\} \subset [0, T_{\max}), t_n \nearrow T_{\max}$, and let $\lambda_n = \lambda(t_n) \to \infty$ (the other possibilities are easily excluded) and $\{x_n\} = \{x(t_n)\} \subset \mathbb{R}^4$, so that, up to subsequence,

$$v_n(x) = \frac{1}{\lambda_n} u_c\left(\frac{x - x_n}{\lambda_n}, t_n\right) \xrightarrow{\dot{H}^1} \bar{v}, \text{ for some } \bar{v} \in \dot{H}^1,$$

and also in L^4 by Sobolev embedding.

Step 2 (local asymptotic vanishing of the L^2−norm): as in [**11**] we prove

LEMMA 5.5. *For any* $R > 0$,

$$\lim_{n \to \infty} \int_{|x| \leq R} |u_c(x, t_n)|^2 dx = 0.$$

Now, unlike the Navier-Stokes case ([**11**]), in our proof the smallness required for the regularity result will come from the compactness (uniform smallness of the tails). If one were to impose more symmetry, e.g., working with radial solutions, energy concentration can only occur at the spatial origin; however, since we are not willing to simplify in this way, we have to account for the possibility that the solution concentrates its energy at different points as time increases, and in particular to deal with the scenario where the centre of concentration moves off to infinity. As we will see below, this possibility would be particularly bad since we could not ensure the required smallness in a uniform way: for our argument to work we need to excise a large -but fixed- ball around the spatial origin and look for smallness (and regularity) on its exterior. Fortunately, this bad scenario cannot happen as shown in

Step 3 (controlling the motion of the centre of concentration): we prove that the centre of compactness $x(t)$ is bounded:

PROPOSITION 4. $\sup_{0\leq t<T_{max}} |x(t)| < \infty.$

We will define the following quantity

$$\underline{E} := \inf_{t\in[0,T_{\max})} E(u_c(t)). \tag{5.2}$$

For the proof we first assumed that $\underline{E} > 0$, and later showed that this is indeed the case for compact blowing-up solutions, *without any size restriction.* Note that under the assumptions of our Theorem 1.1, i.e., in the below threshold case, we certainly have that $\underline{E} > 0$. This can be easily deduced by the variational estimates in Lemma 2.3 and the small data theory.

The proof is technical and will not be discussed here, for more details we refer to [**10**]. Let us only mention that the energy dissipation relation

$$E(u(t_2)) + \int_{t_1}^{t_2} \|u_t\|_{L^2}^2\, ds = E(u(t_1)) \leq E(u(0)) \tag{5.3}$$

for $t_2 > t_1 > 0$, plays a crucial role, especially for the case where $\underline{E} > 0$.

Ruling out the complementary possibilities $\underline{E} \leq 0$ requires a different argument which bears some similarities with ideas in the rigidity argument of Kenig and Merle for NLS ([**12**]).

Step 4 (smallness on a cylinder far from the spatial origin): Since $|x(t)|$ remains bounded while $\lambda(t) \xrightarrow{t\to T_{\max}} \infty$, by the compactness we can find an $R_0 > 0$ large enough such that for all $x, |x| \geq R_0$:

$$\|u_c\|_{L_t^\infty \dot{H}_x^1 \cap L_t^\infty L_x^4(\Omega_{T_{\max}})} < \epsilon_0,$$

where $\Omega_{T_{\max}} := (0, T_{\max}) \times B_{\sqrt{T_{\max}}}(x_0)$.

Step 5 (conclusion of the proof): by an appropriate scaling and shifting argument, the Regularity Lemma 5.4 shows that u_c is smooth on $\Omega := (\mathbb{R}^4 \setminus B_{R_0}(0)) \times [\frac{3}{4}T_{\max}, T_{\max}]$, with uniform bounds on derivatives. Since u is continuous up to $T_{\max}$ outside B_{R_0}, Lemma 5.5 implies that $u_c(x, T_{\max}) \equiv 0$, in the exterior of this ball. Since u_c is bounded and smooth in Ω, an application of the Backwards Uniqueness Theorem 5.2 implies that $u_c \equiv 0$ in Ω. Define $\tilde{\Omega} := \mathbb{R}^4 \times (\frac{3}{4}T_{\max}, \frac{7}{8}T_{\max}]$. Applying the Unique Continuation Theorem 5.3 on a cylinder of sufficiently large spatial radius, centered at a point of Ω, implies $u_c \equiv 0$ in $\tilde{\Omega}$. By the uniqueness guaranteed by the local well-posedness theory we get that $u_c \equiv 0$, which contradicts (4.6), thus finishing the proof.

References

[1] H. Bahouri and P. Gérard, *High frequency approximation of solutions to critical nonlinear wave equations*, Amer. J. Math. **121** (1999), no. 1, 131–175. MR1705001

[2] H. Brezis and T. Cazenave, *A nonlinear heat equation with singular initial data*, J. Anal. Math. **68** (1996), 277–304, DOI 10.1007/BF02790212. MR1403259

[3] C. Collot, F. Merle, and P. Raphaël, *Dynamics near the ground state for the energy critical nonlinear heat equation in large dimensions*, Comm. Math. Phys. **352** (2017), no. 1, 215–285, DOI 10.1007/s00220-016-2795-4. MR3623259

[4] T. Cazenave and F. B. Weissler, *The Cauchy problem for the critical nonlinear Schrödinger equation in H^s*, Nonlinear Anal. **14** (1990), no. 10, 807–836, DOI 10.1016/0362-546X(90)90023-A. MR1055532

[5] L. Iskauriaza, G. A. Serëgin, and V. Shverak, *$L_{3,\infty}$-solutions of Navier-Stokes equations and backward uniqueness* (Russian, with Russian summary), Uspekhi Mat. Nauk **58** (2003), no. 2(350), 3–44, DOI 10.1070/RM2003v058n02ABEH000609; English transl., Russian Math. Surveys **58** (2003), no. 2, 211–250. MR1992563

[6] S. Filippas, M. A. Herrero, and J. J. L. Velázquez, *Fast blow-up mechanisms for sign-changing solutions of a semilinear parabolic equation with critical nonlinearity*, R. Soc. Lond. Proc. Ser. A Math. Phys. Eng. Sci. **456** (2000), no. 2004, 2957–2982, DOI 10.1098/rspa.2000.0648. MR1843848

[7] I. Gallagher, D. Iftimie, and F. Planchon, *Non-explosion en temps grand et stabilité de solutions globales des équations de Navier-Stokes* (French, with English and French summaries), C. R. Math. Acad. Sci. Paris **334** (2002), no. 4, 289–292, DOI 10.1016/S1631-073X(02)02255-0. MR1891005

[8] P. Gérard, *Description du défaut de compacité de l'injection de Sobolev* (French, with French summary), ESAIM Control Optim. Calc. Var. **3** (1998), 213–233, DOI 10.1051/cocv:1998107. MR1632171

[9] S. Gustafson and D. Roxanas, Global regularity and asymptotic convergence for the higher-degree 2d corotational harmonic map heat flow to $\mathbb{S}^2$, preprint, arXiv:1711.06476.

[10] S. Gustafson and D. Roxanas, *Global, decaying solutions of a focusing energy-critical heat equation in $\mathbb{R}^4$*, J. Differential Equations **264** (2018), no. 9, 5894–5927, DOI 10.1016/j.jde.2018.01.023. MR3765769

[11] C. E. Kenig and G. S. Koch, *An alternative approach to regularity for the Navier-Stokes equations in critical spaces* (English, with English and French summaries), Ann. Inst. H. Poincaré Anal. Non Linéaire **28** (2011), no. 2, 159–187, DOI 10.1016/j.anihpc.2010.10.004. MR2784068

[12] C. E. Kenig and F. Merle, *Global well-posedness, scattering and blow-up for the energy-critical, focusing, non-linear Schrödinger equation in the radial case*, Invent. Math. **166** (2006), no. 3, 645–675, DOI 10.1007/s00222-006-0011-4. MR2257393

[13] C. E. Kenig and F. Merle, *Global well-posedness, scattering and blow-up for the energy-critical focusing non-linear wave equation*, Acta Math. **201** (2008), no. 2, 147–212, DOI 10.1007/s11511-008-0031-6. MR2461508

[14] R. Killip and M. Visan, *The focusing energy-critical nonlinear Schrödinger equation in dimensions five and higher*, Amer. J. Math. **132** (2010), no. 2, 361–424, DOI 10.1353/ajm.0.0107. MR2654778

[15] H. A. Levine, *Some nonexistence and instability theorems for solutions of formally parabolic equations of the form $Pu_t = -Au + \mathcal{F}(u)$*, Arch. Rational Mech. Anal. **51** (1973), 371–386, DOI 10.1007/BF00263041. MR0348216

[16] P.-L. Lions, *The concentration-compactness principle in the calculus of variations. The limit case. I*, Rev. Mat. Iberoamericana **1** (1985), no. 1, 145–201, DOI 10.4171/RMI/6. MR834360

[17] F. Merle and L. Vega, *Compactness at blow-up time for L^2 solutions of the critical nonlinear Schrödinger equation in 2D*, Internat. Math. Res. Notices **8** (1998), 399–425, DOI 10.1155/S1073792898000270. MR1628235

[18] M. del Pino, M. Musso, and J. Wei, Infinite time blow-up for the 3-dimensional energy critical heat equation, preprint, arXiv:1705.01672.

[19] D. Roxanas, Long-time dynamics for the energy-critical Harmonic Map Heat Flow and Nonlinear Heat Equation, (2017), PhD Thesis, University of British Columbia.

[20] R. Schweyer, *Type II blow-up for the four dimensional energy critical semi linear heat equation*, J. Funct. Anal. **263** (2012), no. 12, 3922–3983, DOI 10.1016/j.jfa.2012.09.015. MR2990063

[21] M. Struwe, *On the evolution of harmonic mappings of Riemannian surfaces*, Comment. Math. Helv. **60** (1985), no. 4, 558–581, DOI 10.1007/BF02567432. MR826871

[22] T. Tao and M. Visan, *Stability of energy-critical nonlinear Schrödinger equations in high dimensions*, Electron. J. Differential Equations (2005), No. 118, 28. MR2174550

[23] F. B. Weissler, *Local existence and nonexistence for semilinear parabolic equations in L^p*, Indiana Univ. Math. J. **29** (1980), no. 1, 79–102, DOI 10.1512/iumj.1980.29.29007. MR554819

Department of Mathematics, University of British Columbia, 1984 Mathematics Road, Vancouver, British Columbia, V6T 1Z2, Canada

Email address: gustaf@math.ubc.ca

School of Mathematics, The University of Edinburgh, James Clerk Maxwell Building, Rm 5605, The King's Buildings, Peter Guthrie Tait Road, Edinburgh, EH9 3FD, United Kingdom

Email address: dimitrios.roxanas@ed.ac.uk

Contemporary Mathematics
Volume **725**, 2019
https://doi.org/10.1090/conm/725/14557

A regularity upgrade of pressure

Dong Li and Xiaoyi Zhang

ABSTRACT. For the incompressible Euler equations the pressure formally scales as a quadratic function of velocity. We provide several optimal regularity estimates on the pressure by using regularity of velocity in various Sobolev, Besov and Hardy spaces. Our proof exploits the incompressibility condition in an essential way and is deeply connected with the classic Div-Curl lemma which we also generalise as a fractional Leibniz rule in Hardy spaces. To showcase the sharpness of results, we construct a class of counterexamples at several end-points.

1. Introduction

The n-dimensional ($n \geq 2$) incompressible Euler equation takes the form

$$\begin{cases} \partial_t u + (u \cdot \nabla) u = -\nabla p, \quad (t,x) \in (0,\infty) \times \mathbb{R}^n, \\ \nabla \cdot u = 0, \\ u\Big|_{t=0} = u_0, \end{cases} \tag{1.1}$$

where $u: [0,\infty) \times \mathbb{R}^n \to \mathbb{R}^n$, $p: [0,\infty) \times \mathbb{R}^n \to \mathbb{R}$ represent velocity and pressure of the underlying fluid respectively. In this work we shall not consider the Cauchy problem or wellposedness issues at all. Instead we regard u as a *given* solution to (1.1) in appropriate function spaces. Our main objective is to study the regularity properties of the pressure p in terms of the known velocity field u. By taking the divergence on both sides of the first equation in (1.1), we get

$$-\Delta p = \nabla \cdot \big((u \cdot \nabla) u \big). \tag{1.2}$$

Equation (1.2) will be our main object of study. To simplify the discussion we shall completely ignore the explicit time dependence and only focus on spatial regularity. Regarding (1.2) as a Poisson problem for the pressure p, it is well known that p is determined up to a harmonic part. This degree of freedom can be eliminated by supplying some decay conditions at spatial infinity. Alternatively to simplify matters, in the following discussion, we shall take the convention that p is identified with the expression

$$(-\Delta)^{-1} \nabla\nabla \big(u \otimes u \big) = \sum_{l,k} (-\Delta)^{-1} \partial_l \partial_k (u_l u_k)$$

which is well-defined by putting suitable assumptions on the velocity vector u (e.g. u is Schwartz). Our main focus is the quantitative estimates of p in terms of u in various functional spaces.

Note that since u is divergence-free, one can rewrite (1.2) as

$$-\Delta p = \sum_{k,l} \partial_k \partial_l (u_k u_l). \tag{1.3}$$

Alternatively, we can rewrite (1.3) as

$$-\Delta p = \sum_{k,l} (\partial_k u_l)(\partial_l u_k). \tag{1.4}$$

There are some subtle differences between the expressions (1.2)–(1.4). Assume that $u \in W^{1,q}(\mathbb{R}^n)$ for some $n < q < \infty$. If we only make use of (1.2), then

$$\nabla p = (-\Delta)^{-1} \nabla \nabla \cdot (u \cdot \nabla u),$$

which gives $\nabla p \in L^q(\mathbb{R}^n)$. By using (1.3) it is easy to check that $p \in L^q(\mathbb{R}^n)$. Thus we obtain $p \in W^{1,q}(\mathbb{R}^n)$. On the other hand, if we use (1.4), then clearly $p \in W^{2,\frac{q}{2}}(\mathbb{R}^n)$. This is apparently a better estimate in view of the embedding $W^{2,\frac{q}{2}}(\mathbb{R}^n) \hookrightarrow W^{1,q}(\mathbb{R}^n)$. As it turns out, this "upgrade of regularity" phenomenon is quite generic. For example it can be generalised to Sobolev spaces $W^{s,q}$ with $0 < s \leq 1$, $2 < q < \infty$ and even Besov spaces. We have the following theorem.

THEOREM 1.1 (Sobolev and Besov). *Suppose $u \in \mathcal{S}(\mathbb{R}^n)$. Then for $0 \leq s \leq 1$ and $1 < q < \infty$, we have*

$$\|p\|_{W^{2s,q}(\mathbb{R}^n)} \lesssim \|u\|^2_{W^{s,2q}(\mathbb{R}^n)}. \tag{1.5}$$

If $1 \leq q, r \leq \infty$ and $0 < s < 1$, then

$$\|p\|_{\dot{B}^{2s}_{q,r}(\mathbb{R}^n)} \lesssim \|u\|^2_{\dot{B}^{s}_{2q,2r}(\mathbb{R}^n)}. \tag{1.6}$$

In particular for Hölder spaces we have for $0 < s < 1$,

$$\|p\|_{\dot{B}^{2s}_{\infty,\infty}} \lesssim \|u\|^2_{\dot{B}^{s}_{\infty,\infty}}. \tag{1.7}$$

On the other hand, for $s = 1$ the corresponding Besov estimate does not hold and can be replaced by

$$\|p\|_{\dot{B}^{2}_{\infty,\infty}} \lesssim \|\nabla u\|^2_{\infty}. \tag{1.8}$$

REMARK. (1.8) is a simple consequence of (1.4) and thus we omit the proof.

REMARK. For Schwartz functions $f\colon \mathbb{R}^n \to \mathbb{R}^n$, $g\colon \mathbb{R}^n \to \mathbb{R}^n$ with the property $\nabla \cdot f = \nabla \cdot g = 0$, one can consider the bilinear operator

$$B(f,g) = \sum_{l,k=1}^{n} \partial_l \partial_k \Delta^{-1}(f_l g_k).$$

Same proof as in Theorem 1.1 yields that for $0 \leq s \leq 1$ and $1 < q < \infty$,

$$\|B(f,g)\|_{W^{2s,q}} \lesssim \|f\|_{W^{s,2q}} \|g\|_{W^{s,2q}};$$

and for $0 < s < 1$, $1 \leq q, r \leq \infty$,

$$\|B(f,g)\|_{\dot{B}^{2s}_{q,r}} \lesssim \|f\|_{\dot{B}^{s}_{2q,2r}} \|g\|_{\dot{B}^{s}_{2q,2r}}.$$

REMARK. The estimate (1.7) shows that for $0 < s < \frac{1}{2}$,

$$\|p\|_{\dot{C}^{2s}} \lesssim \|u\|^2_{\dot{C}^s};$$

and for $\frac{1}{2} < s < 1$,

$$\|\nabla p\|_{\dot{C}^{2s-1}} \lesssim \|u\|^2_{\dot{C}^s}.$$

One should recall that $\|f\|_{\dot{C}^s} \sim \|f\|_{\dot{B}^s_{\infty,\infty}}$ for $0 < s < 1$. Thus the Besov formulation connects these two estimates in a most natural way.

Note from (1.3) it is evident that $p = \mathcal{R}(O(u^2))$ where $\mathcal{R}$ is a Riesz-type operator. The usual product rule in Hölder spaces says that if f, $g \in C^s$, then $fg \in C^s$. Thus by (1.3) one should only expect $p \in C^s$ if $u \in C^s$. However here by using Theorem 1.1, one can prove that $p \in \dot{B}^{2s}_{\infty,\infty}$ as long as $0 < s < 1$. Roughly speaking, we are asserting that the $2s$-derivative on p can fall "evenly" into each composing velocity:

$$|\nabla|^{2s} p \sim \mathcal{R}(|\nabla|^s u |\nabla|^s u),$$

and there do not appear terms such as $O(|\nabla|^{s-} u |\nabla|^{s+} u)$.

There exists an analogue of Theorem 1.1 in Hardy space. The following theorem can be regarded as the case $q = 2$ in Theorem 1.1.

THEOREM 1.2 (Hardy space). *Let $0 < s \leq 1$. Then for $u \in \mathcal{S}(\mathbb{R}^n)$,*

$$\||\nabla|^{2s} p\|_{\mathcal{H}^1(\mathbb{R}^n)} \lesssim \|u\|^2_{W^{s,2}(\mathbb{R}^n)}. \tag{1.9}$$

In the case $s = 1$, the operator can be replaced by a general second order derivative $\partial^2 = \partial_i \partial_j$ for any $i, j \in \{1, \cdots, n\}$.

REMARK. For $0 < s < 1$ one has the stronger estimate $p \in \dot{B}^{2s}_{1,1}$ thanks to theorem 1.1.

In the proof of Theorem 1.2 we need to exploit the incompressibility of velocity which provides cancelation of some high frequency interaction terms in the nonlinearity. As it turns out, in the 3D case, this is deeply linked to the "Div-Curl" lemma in Coifman-Lions-Meyer-Semmes [**1**]. In its simplest formulation, the Div-Curl lemma asserts that if $\nabla \cdot f = \nabla \times g = 0$, then

$$\|f \cdot g\|_{\mathcal{H}^1} \lesssim \|f\|_p \|g\|_{p'},$$

where $1 < p < \infty$ and $p' = p/(p-1)$. In light of the proof in Theorem 1.2, we can obtain the following generalisation which can also be regarded as a fractional Leibniz rule in Hardy space.

THEOREM 1.3 (Generalised Div-Curl lemma). *Let $1 < p_1, p_2 < \infty$ and $-1 < s < \infty$. Then for any Schwartz f, g: $\mathbb{R}^3 \to \mathbb{R}^3$ with*

$$\nabla \cdot f = 0, \quad \nabla \times g = 0,$$

we have

$$\||\nabla|^s (f \cdot g)\|_{\mathcal{H}^1} \lesssim \||\nabla|^s f\|_{p_1} \|g\|_{p_1'} + \|f\|_{p_2'} \||\nabla|^s g\|_{p_2}$$

whenever the RHS is finite. When $-1 < s \leq 0$, we have

$$\||\nabla|^s (f \cdot g)\|_{\mathcal{H}^1} \lesssim \min\{\||\nabla|^s f\|_{p_1} \|g\|_{p_1'}, \ \|f\|_{p_2'} \||\nabla|^s g\|_{p_2}\}.$$

REMARK. For $s > 0$ one does not need the "Div-Curl" condition to derive the estimate. The only nontrivial case is $-1 < s \leq 0$.

In Theorem 1.1, for the Besov case we only considered the regime $0 < s < 1$ and left out the cases $s = 0$ and $s = 1$. It is instructive to investigate these end-point cases. As it turns out, the corresponding regularity estimate for pressure fails in general. To clarify this point we construct counterexamples.

THEOREM 1.4. *The estimate* (1.7) *fails in the case* $s = 0, 1$. *More specifically, for any* $\varepsilon > 0$, *there exists divergence free* $u \in \mathcal{S}(\mathbb{R}^n)$ *for with*

$$\|u\|_{\dot{B}^1_{\infty,\infty}} + \|u\|_{L^2} \le 1,$$

but

$$\|p\|_{\dot{B}^2_{\infty,\infty}} > \frac{1}{\varepsilon}.$$

A similar statement holds for $s = 0$. *Here* $\mathcal{S}(\mathbb{R}^n)$ *is the class of Schwartz functions.*

REMARK. Our construction shows that the inflation occurs at high frequencies.

In the estimate (1.7), when $s = \frac{1}{2}$, $p \in \dot{B}^1_{\infty,\infty}$. The norm of $\dot{B}^1_{\infty,\infty}$ is much weaker than the norm of C^1. Of course it is not difficult to construct a function which lies in $\dot{B}^1_{\infty,\infty}$ but not in C^1. However, it is not obvious to construct a $p \notin C^1$ starting from a divergence free $u \in C^{\frac{1}{2}}$. The following theorem gives such a result.

THEOREM 1.5. *There exists* $u \in C^{\frac{1}{2}} \cap L^2$ *for which* $p \notin C^1$.

The last result says the estimate (1.9) fails at the endpoint case $s = 0$.

PROPOSITION 1.6. *There exists divergence-free* $u \in \mathcal{S}(\mathbb{R}^n)$ *such that* $p \notin L^1(\mathbb{R}^n)$.

We should remark that the case of domains with appropriate boundary conditions can be explored further and we plan to address it elsewhere.

The rest of this paper is organised as follows. In Section 2 we introduce some basic notation and collect some preliminary estimates. In Section 3 we give the regularity estimates of pressure in aforementioned function spaces. In Section 4 we give the construction of counterexamples at various end-point cases.

Acknowledgements. D. Li was supported in part by a start-up grant from HKUST and HK RGC grant 16307317. X. Zhang was supported by Simons Collaboration grant.

2. Preliminaries

For any real number $a \in \mathbb{R}$, we denote by $a+$ the quantity $a + \epsilon$ for sufficiently small $\epsilon > 0$. The numerical value of ϵ is unimportant and the needed smallness of ϵ is usually clear from the context. The notation $a-$ is similarly defined.

For any two quantities X and Y, we denote $X \lesssim Y$ if $X \le CY$ for some constant $C > 0$. Similarly $X \gtrsim Y$ if $X \ge CY$ for some $C > 0$. We denote $X \sim Y$ if $X \lesssim Y$ and $Y \lesssim X$. The dependence of the constant C on other parameters or constants are usually clear from the context and we will often suppress this dependence. We shall denote $X \lesssim_{Z_1, Z_2, \cdots, Z_k} Y$ if $X \le CY$ and the constant C depends on the quantities $Z_1, \cdots, Z_k$.

We denote by $\mathcal{S}(\mathbb{R}^n)$ the space of Schwartz functions and $\mathcal{S}'(\mathbb{R}^n)$ the space of tempered distributions. For any function $f : \mathbb{R}^n \to \mathbb{R}$, we use $\|f\|_{L^q(\mathbb{R}^n)}$, $\|f\|_{L^q}$ or sometimes $\|f\|_q$ to denote the usual Lebesgue L^q norm for $0<q\le\infty$. For $s>0$, $1<q<\infty$, we recall the Sobolev norms

$$\|f\|_{\dot W^{s,q}(\mathbb{R}^n)} = \||\nabla|^s f\|_q, \quad \|f\|_{W^{s,q}(\mathbb{R}^n)} = \|f\|_q + \||\nabla|^s f\|_q.$$

For a sequence of real numbers $(a_j)_{j=-\infty}^{\infty}$, we denote

$$(a_j)_{l_j^q} = \|(a_j)_{j\in\mathbb{Z}}\|_{l^q} = \begin{cases} (\sum_{j\in\mathbb{Z}} |a_j|^q)^{\frac{1}{q}}, & \text{if } 0<q<\infty, \\ \sup_j |a_j|, & \text{if } q=\infty. \end{cases}$$

We shall often use mixed-norm notation. For example, for a sequence of functions $f_j:\ \mathbb{R}^n\to\mathbb{R}$, we will denote (below $0<r<\infty$)

$$\|(f_j)_{l_j^r}\|_q = \|(\sum_j |f_j(x)|^r)^{\frac{1}{r}}\|_{L_x^q(\mathbb{R}^n)},$$

with obvious modification for $q=\infty$.

We use the following convention for the Fourier transform:

$$(\mathcal{F}f)(\xi) = \hat f(\xi) = \int_{\mathbb{R}^n} e^{-ix\cdot\xi} f(x)dx.$$

$\mathcal{F}^{-1}$ is the inverse Fourier transform:

$$\mathcal{F}^{-1}g(x) = \frac{1}{(2\pi)^n}\int_{\mathbb{R}^n} e^{ix\cdot\xi} g(\xi)d\xi.$$

For $s\in\mathbb{R}$, the fractional Laplacian $|\nabla|^s = (-\Delta)^{s/2}$ corresponds to the multiplier $|\xi|^s$ on the Fourier side (whenever it is well-defined). Sometimes we also denote $|\nabla|^s$ as D^s.

We first introduce the Littlewood-Paley operators.

Let $\tilde\phi\in C_c^\infty(\mathbb{R}^n)$ be such that

$$\tilde\phi(\xi) = \begin{cases} 1, & |\xi|\le 1 \\ 0, & |\xi|>\frac{7}{6}. \end{cases}$$

Let $\phi_c(\xi) = \tilde\phi(\xi) - \tilde\phi(2\xi)$ which is supported on $\frac12\le|\xi|\le\frac76$. For any $f\in\mathcal{S}(\mathbb{R}^n)$, $j\in\mathbb{Z}$, define

$$\mathcal{F}(P_{\le j}f)(\xi) = \widehat{P_{\le j}f}(\xi) = \tilde\phi(2^{-j}\xi)\hat f(\xi),$$
$$\widehat{P_j f}(\xi) = \phi_c(2^{-j}\xi)\hat f(\xi), \qquad \xi\in\mathbb{R}^n.$$

We will denote $P_{>j} = I - P_{\le j}$ (I is the identity operator). Sometimes for simplicity of notation (and when there is no obvious confusion) we will write $f_j = P_j f$, $f_{\le j} = P_{\le j} f$ and $f_{a\le\cdot\le b} = \sum_{a\le j\le b} f_j$. By using the support property of ϕ, we have $P_jP_{j'} = 0$ whenever $|j-j'|>1$. This property will be useful in product decompositions. For example the Bony paraproduct for a pair of functions f,g take the form

$$fg = \sum_{i\in\mathbb{Z}} f_i\tilde g_i + \sum_{i\in\mathbb{Z}} f_i g_{\le i-2} + \sum_{i\in\mathbb{Z}} g_i f_{\le i-2},$$

where $\tilde g_i = g_{i-1}+g_i+g_{i+1}$.

The fattened operators $\tilde P_j$ are defined by

$$\tilde P_j = \sum_{l=-n_1}^{n_2} P_{j+l},$$

where $n_1 \geq 0$, $n_2 \geq 0$ are some finite integers whose values play no role in the argument.

We will often use the following Bernstein inequalities without explicit mentioning.

PROPOSITION 2.1 (Bernstein inequality). *Let $1 \leq p \leq q \leq \infty$. For any $f \in L^p(\mathbb{R}^n)$, $j \in \mathbb{Z}$, we have*

$$\begin{aligned}
&\|P_{\leq j} f\|_q + \|P_j f\|_q \lesssim 2^{jn(\frac{1}{p}-\frac{1}{q})}\|f\|_p;\\
&\||\nabla|^s P_{\leq j} f\|_p \lesssim 2^{js}\|f\|_p, \quad \forall s \geq 0;\\
&\|P_j f\|_p \sim 2^{-js}\||\nabla|^s P_j f\|_p, \quad \forall s \in \mathbb{R};\\
&\|P_{>j} f\|_p \lesssim 2^{-js}\||\nabla|^s f\|_p, \quad \forall s \geq 0.
\end{aligned}$$

For $s \in \mathbb{R}$, $1 \leq p, q \leq \infty$, the homogeneous Besov $\dot{B}^s_{p,q}$ (semi)-norm is given by

$$\|f\|_{\dot{B}^s_{p,q}(\mathbb{R}^n)} = \|(2^{js}\|f_j\|_p)\|_{l^q_j} = \begin{cases} \left(\sum_{j\in\mathbb{Z}} 2^{jqs}\|P_j f\|^q_{L^p_x(\mathbb{R}^n)}\right)^{\frac{1}{q}}, & \text{if } 1 \leq q < \infty;\\ \sup_{j\in\mathbb{Z}} 2^{js}\|P_j f\|_{L^p_x(\mathbb{R}^n)}, & \text{if } q = \infty. \end{cases}$$

For $s \in \mathbb{R}$, $1 \leq p$, $q \leq \infty$,

$$\|f\|_{B^s_{p,q}(\mathbb{R}^n)} = \|P_{\leq 0} f\|_p + \|(2^{js}\|f_j\|_p)\|_{l^q_j(j\geq 1)}.$$

For $s > 0$ and $1 \leq p, q \leq \infty$, it is easy to check that

$$\|f\|_{B^s_{p,q}} \sim \|f\|_p + \|f\|_{\dot{B}^s_{p,q}}.$$

PROPOSITION 2.2 (Continuity in "s"). *Let $1 \leq p, q \leq \infty$. Suppose $\|f\|_{B^s_{p,q}} < \infty$, then*

$$\lim_{\substack{\tilde{s}\to s\\ \tilde{s}<s}} \|f\|_{B^{\tilde{s}}_{p,q}} = \|f\|_{B^s_{p,q}}.$$

On the other hand, if for some $\epsilon_0 > 0$,

$$\sup_{\tilde{s}\in(s-\epsilon_0,s)} \|f\|_{B^{\tilde{s}}_{p,q}} < \infty,$$

then $\|f\|_{B^s_{p,q}} < \infty$, and

$$\lim_{\substack{\tilde{s}\to s\\ s-\epsilon_0<\tilde{s}<s}} \|f\|_{B^{\tilde{s}}_{p,q}} = \|f\|_{B^s_{p,q}}.$$

If for some $\epsilon_1 > 0$, $\|f\|_{B^{s+\epsilon_1}_{p,q}} < \infty$, then

$$\lim_{\substack{\tilde{s}\to s\\ \tilde{s}>s}} \|f\|_{B^{\tilde{s}}_{p,q}} = \|f\|_{B^s_{p,q}}.$$

REMARK. This simple proposition explains the folklore fact that the limit $\alpha \to 1$ for Hölder C^α, $0 < \alpha < 1$, is $B^1_{\infty,\infty}$; and the limit $\alpha \to 0$ for Hölder $C^{1,\alpha}$, $0 < \alpha < 1$ is $B^1_{\infty,\infty}$.

PROOF. If $q < \infty$, one can just use Monotone Convergence to obtain

$$\lim_{\substack{\tilde{s}\to s\\ \tilde{s}<s}} (2^{j\tilde{s}}\|P_j f\|_p)_{l^q_j(j\geq 1)} = (2^{js}\|P_j f\|_p)_{l^q_j(j\geq 1)}.$$

Also for $q < \infty$, by using Dominated convergence, we have

$$\lim_{\substack{\tilde{s}\to s\\ s<\tilde{s}<s+\epsilon_1}} (2^{j\tilde{s}}\|P_j f\|_p)_{l^q_j(j\geq 1)} = (2^{js}\|P_j f\|_p)_{l^q_j(j\geq 1)}.$$

If $q = \infty$, one observes that

$$\sup_{\tilde{s}<s}\sup_{j\ge 1} 2^{j\tilde{s}}\|P_j f\|_p = \sup_{j\ge 1} 2^{js}\|P_j f\|_p;$$

$$\lim_{\substack{\tilde{s}\to s\\ s<\tilde{s}<s+\epsilon_1}} \sup_{j\ge 1} 2^{j\tilde{s}}\|P_j f\|_p = \sup_{j\ge 1} 2^{js}\|P_j f\|_p.$$

□

We will use the following characterisation for the $\dot{C}^\alpha$ space. Recall that for $0<\alpha<1$,

$$\|f\|_{\dot{C}^\alpha} := \sup_{x\neq y}\frac{|f(x)-f(y)|}{|x-y|^\alpha},$$

and

$$\|f\|_{C^\alpha} = \|f\|_\infty + \|f\|_{\dot{C}^\alpha}.$$

LEMMA 2.3. *Let $0<\alpha<1$. Then for any f with $\|f\|_{\dot{C}^\alpha(\mathbb{R}^n)}<\infty$, we have*

$$\|f\|_{\dot{C}^\alpha} \sim \|f\|_{\dot{B}^\alpha_{\infty,\infty}}.$$

REMARK. Similarly for $\alpha = 1$ and $f\in\mathcal{S}(\mathbb{R}^n)$, one has the equivalence

$$\sup_{h\neq 0, x\in\mathbb{R}^n}\frac{|f(x+h)+f(x-h)-2f(x)|}{|h|} \sim \|f\|_{\dot{B}^1_{\infty,\infty}}.$$

REMARK. It follows easily that for $0<\alpha<1$, and any $f\in\mathcal{S}(\mathbb{R}^n)$, we have

$$\|\nabla f\|_{\dot{C}^\alpha} \sim \|\nabla f\|_{\dot{B}^\alpha_{\infty,\infty}} \sim \|f\|_{\dot{B}^{1+\alpha}_{\infty,\infty}}.$$

PROOF. Easy to check that $\|f\|_{\dot{B}^\alpha_{\infty,\infty}} \lesssim \|f\|_{\dot{C}^\alpha}$. To show the reverse inequality, one can fix $x\neq y$ and note that with the assumption $\|f\|_{\dot{C}^\alpha}<\infty$ one has

$$\lim_{J\to\infty}|(P_{\le -J}f)(x) - (P_{\le -J}f)(y)| \lesssim \lim_{J\to\infty}\|\nabla P_{\le -J}f\|_\infty\cdot|x-y| = 0.$$

This implies the point-wise identity:

$$f(x)-f(y) = \sum_{j\in\mathbb{Z}}((P_j f)(x) - (P_j f)(y)).$$

Thus

$$\begin{aligned}|f(x)-f(y)| &\le \sum_{j\le J_0}\|\nabla P_j f\|_\infty\cdot|x-y| + \sum_{j>J_0} 2\cdot\|P_j f\|_\infty\\ &\lesssim 2^{J_0(1-\alpha)}|x-y|\|f\|_{\dot{B}^\alpha_{\infty,\infty}} + 2^{-J_0\alpha}\|f\|_{\dot{B}^\alpha_{\infty,\infty}}.\end{aligned}$$

Choosing $2^{J_0}\sim|x-y|^{-1}$ then yields the result. □

LEMMA 2.4. *Let $0<\alpha<1$. Suppose $f\in\mathcal{S}'(\mathbb{R}^n)$ satisfies*

$$\|f\|_{\dot{B}^\alpha_{\infty,\infty}} = \sup_j(\|P_j f\|_\infty\cdot 2^{j\alpha}) < \infty.$$

Then

$$f = \tilde{f} + p,$$

where p is a polynomial, and $\tilde{f}$ is a continuous function satisfying

$$\|\tilde{f}\|_{\dot{C}^\alpha} \sim \|f\|_{\dot{B}^\alpha_{\infty,\infty}}.$$

PROOF. Denote $f_J = \sum_{-J\le j\le J} P_j f$. Clearly $\|f_J\|_{\dot{C}^\alpha} \lesssim \|f\|_{\dot{B}^\alpha_{\infty,\infty}}$. Set

$$\tilde{f}_J(x) = f_J(x) - f_J(0).$$

Note that $|\tilde{f}_J(x)| \lesssim 1 + |x|^\alpha$. By Arzelà-Ascoli one can extract a subsequence $\tilde{f}_{J_k}$ converging locally uniformly to a limit $\tilde{f}$. Clearly $\|\tilde{f}\|_{\dot{C}^\alpha} \lesssim \|f\|_{\dot{B}^\alpha_{\infty,\infty}}$. By the previous lemma we have $\|\tilde{f}\|_{\dot{C}^\alpha} \sim \|\tilde{f}\|_{\dot{B}^\alpha_{\infty,\infty}}$. Fix any $j_0 \in \mathbb{Z}$. Clearly if J is large we have $P_{j_0}\tilde{f}_J = P_{j_0} f$. By Lebesgue Dominated Convergence and sending J_k to infinity we obtain $P_{j_0}\tilde{f} = P_{j_0} f$ for any $j_0 \in \mathbb{Z}$. Thus $\tilde{f}$ and f differ at most by a polynomial. □

The following corollary records the usual fact that for $0 < \alpha < 1$ and $f \in \mathcal{S}'(\mathbb{R}^n)$, one has the equivalence $\|f\|_{B^\alpha_{\infty,\infty}} \sim \|f\|_{C^\alpha}$.

COROLLARY 2.5. *Let $0 < \alpha < 1$. Suppose f is continuous and bounded on $\mathbb{R}^n$. Then*

$$\|f\|_{\dot{B}^\alpha_{\infty,\infty}} \sim \|f\|_{\dot{C}^\alpha}.$$

PROOF. If $\|f\|_{\dot{C}^\alpha} < \infty$, clearly then $\|f\|_{\dot{B}^\alpha_{\infty,\infty}} \lesssim \|f\|_{\dot{C}^\alpha}$. Conversely if $\|f\|_{\dot{B}^\alpha_{\infty,\infty}} < \infty$, then by the previous lemma, we have $f = \tilde{f} + p$, where p is a polynomial, and $\|\tilde{f}\|_{\dot{C}^\alpha} \sim \|f\|_{\dot{B}^\alpha_{\infty,\infty}}$. Since f is continuous and bounded, by considering the limit of the ratios $f(x)/|x|$ and $\tilde{f}(x)/|x|$ as $|x| \to \infty$, it follows easily that p must be a constant. Thus $f = \tilde{f} + \text{const}$ from which the desired result follows. □

REMARK. Alternatively one can use the condition $\|f\|_\infty < \infty$ to obtain that for any $x \ne y$,

$$\begin{aligned}\lim_{J\to\infty} |P_{\le -J} f(x) - P_{\le -J} f(y)| &\lesssim \lim_{J\to\infty} \|\nabla P_{\le -J} f\|_\infty |x-y| \\ &\lesssim \lim_{J\to\infty} 2^{-J} \|f\|_\infty |x-y| = 0.\end{aligned}$$

Repeating the proof in Lemma 2.3 then yields easily $\|f\|_{\dot{C}^\alpha} \lesssim \|f\|_{\dot{B}^\alpha_{\infty,\infty}}$.

We also need the following maximal function estimates.

LEMMA 2.6 (Fefferman-Stein inequality, [**2**]). *Let $f = (f_j)_{j=1}^\infty$ be a sequence of functions in $\mathbb{R}^n$. Let $1 < q, r < \infty$. There exists a positive constant $A_{r,q,n} > 0$ such that*

$$\Big\| \Big(\sum_{j=1}^\infty |\mathcal{M}_1 f_j|^r\Big)^{\frac{1}{r}} \Big\|_q \le A_{r,q,n} \Big\| \Big(\sum_{j=1}^\infty |f_j|^r\Big)^{\frac{1}{r}} \Big\|_q.$$

Here $\mathcal{M}_1 g$ is the maximal function defined by (here $|Q|$ denotes n-dimensional Lebesgue measure of Q): $(\mathcal{M}_1 g)(x) = \sup_Q \frac{1}{|Q|}\int_Q |g(y)|dy$, where the sup *is taken over all cubes Q centered at x.*

REMARK 2.7. Of course the maximal operator $\mathcal{M}_1$ can be replaced by the usual uncentered maximal operator $\mathcal{M}$ on balls:

$$(\mathcal{M}g)(x) = \sup_B \frac{1}{|B|}\int_B |g(y)|dy,$$

where the sup is taken over all balls B containing the point x.

As an immediate consequence of Lemma 2.6, we have the following estimate.

COROLLARY 2.8. *Let $\tilde{P}_j$ be a family of Littlewood-Paley type projection operators adapted to frequency $|\xi| \sim 2^j$, and obey the bound (often this is easy to check)*

$$\sup_{j\in\mathbb{Z}} |(\tilde{P}_j g)(x)| \le \text{const} \cdot (\mathcal{M}g)(x), \qquad \forall\, g.$$

Then we have

$$\Big\| \big(\sum_j |\tilde{P}_j f_j|^r\big)^{\frac{1}{r}} \Big\|_q \lesssim \Big\| \big(\sum_j |f_j|^r\big)^{\frac{1}{r}} \Big\|_q, \quad \forall\, 1 < q, r < \infty. \tag{2.1}$$

Several acceptable choices of $\tilde{P}_j$ include, for example, the fattened projection: $\tilde{P}_j = P_{j-1} + P_j + P_{j+1}$ and the projections involving derivatives: $\tilde{P}_j = 2^{-js}|\nabla|^s P_j$, $s > 0$. In later sections, we shall often use this inequality (mostly in the case $r = 2$) without explicit mentioning. Note that alternatively one can prove (2.1) by using the usual vector-valued CZ operator theory.

Note that in (2.1), $q \in (1, \infty)$. Next we shall discuss an analogue of this estimate when $q \le 1$. Such estimate will be used in dealing with estimates in Hardy spaces. And we start by a short review of Hardy spaces.[1]

We recall the definition of Hardy space $\mathcal{H}^1$:

$$\mathcal{H}^1(\mathbb{R}^n) = \Big\{ f \in L^1(\mathbb{R}^n) : \ \mathcal{R}_j f \in L^1, \quad \forall\, j = 1, \cdots, n \Big\},$$

where $\mathcal{R}_j$ are the Riesz transforms ($\hat{R}_j(\xi) = -i\xi_j/|\xi|$) on $\mathbb{R}^n$ (for $n = 1$ it is just the Hilbert transform) with the norm

$$\|f\|_{\mathcal{H}^1} = \|f\|_1 + \sum_{j=1}^n \|\mathcal{R}_j f\|_1.$$

We shall not need the atomic or maximal function characterization of $\mathcal{H}^1$. On the other hand, it is well-known that (see e.g. Chapter 6.4 on p37 of [**3**])

$$\|f\|_{\mathcal{H}^1} \sim \|\mathcal{S}f\|_1,$$

where $\mathcal{S}f$ is the usual Littlewood-Paley square function $\mathcal{S}f = (\sum_{j\in\mathbb{Z}} |P_j f|^2)^{\frac{1}{2}}$. In view of this equivalence, we shall conveniently use the definition

$$\|f\|_{\mathcal{H}^1} = \|\mathcal{S}f\|_1 \tag{2.2}$$

in this note without explicit mentioning.

LEMMA 2.9. *Suppose $u \in \mathcal{S}(\mathbb{R}^n)$ with $\operatorname{supp}(\hat{u}) \subset \{\xi : |\xi| < t\}$ for some $t > 0$. Then for any $0 < r < \infty$,*

$$\sup_{z\in\mathbb{R}^n} \frac{|u(x-z)|}{(1+t|z|)^{\frac{n}{r}}} \lesssim_{r,n} (\mathcal{M}(|u|^r)(x))^{\frac{1}{r}}.$$

REMARK. To understand the meaning of the lemma one can take $x = 0$ and $t = 1$. Thanks to frequency localization, for given z the value $u(z)$ is mainly controlled by the L^r norm of $|u|$ in a ball of size $O(|z|)$ centered at the origin. The factor $|z|^{\frac{n}{r}}$ then provides the needed normalization for the maximal function.

REMARK. See Lemma 2.3 of [**4**] for a proof under a much weaker condition that u is a tempered distribution (i.e. one can replace $\mathcal{S}(\mathbb{R}^n)$ by $\mathcal{S}'(\mathbb{R}^n)$).

[1] To keep it simple, we only discuss the estimates in Hardy space $\mathcal{H}^1$, the extension to general Hardy space $\mathcal{H}^p$ can be done without much effort.

PROOF. By scaling we may assume $t = 1$. By using translation we only need to consider the case $x = 0$. Assume first $1 \le r < \infty$. Since $u = P_{<2}u$, we have

$$u(z) = \int \psi(z-y)u(y)dy,$$

where ψ is the Schwartz function corresponding to $P_{<2}$. Clearly then

$$\begin{aligned}|u(z)| &\le \int |\psi(z-y)|^{\frac{1}{r}}|u(y)||\psi(z-y)|^{1-\frac{1}{r}}dy \\ &\lesssim \Big(\int |\psi(z-y)||u(y)|^r dy\Big)^{\frac{1}{r}} \\ &\lesssim (1+|z|)^{\frac{n}{r}}(\mathcal{M}(|u|^r)(0))^{\frac{1}{r}},\end{aligned}$$

where in the last step we used the simple inequality

$$\int |\psi(z-y)||f(y)|dy \lesssim (1+|z|)^n(\mathcal{M}f)(0).$$

Now consider $0 < r < 1$. Clearly

$$\begin{aligned}|u(z)| &\le \int |\psi(z-y)||u(y)|^r|u(y)|^{1-r}dy \\ &\le \int |\psi(z-y)||u(y)|^r(1+|y|)^{\frac{n}{r}(1-r)}dy \sup_x \frac{|u(x)|^{1-r}}{(1+|x|)^{\frac{n}{r}(1-r)}} \\ &\lesssim (1+|z|)^{\frac{n}{r}}\mathcal{M}(|u|^r)(0)\cdot \sup_x \frac{|u(x)|^{1-r}}{(1+|x|)^{\frac{n}{r}(1-r)}}.\end{aligned}$$

This then easily leads to the desired inequality. □

COROLLARY 2.10. *Let $0 < r < \infty$ and $c_1 > 0$. Then for any $j \in \mathbb{Z}$ and Schwartz f with $\mathrm{supp}(\hat{f}) \subset \{\xi : |\xi| < c_1 2^j\}$, we have*

$$|(P_{\le j}f)(x)| \lesssim_{r,c_1,n} (\mathcal{M}(|f|^r)(x))^{\frac{1}{r}}.$$

In the above estimate the operator $P_{\le j}$ can be replaced by P_j or any $\tilde{P}_{\le j}$, $\tilde{P}_j$.

PROOF. The case $1 \le r < \infty$ is trivial. In particular it follows from the estimate $|f(x)| \lesssim \mathcal{M}f(x) \lesssim_r ((\mathcal{M}|f|^r)(x))^{\frac{1}{r}}$ when $r > 1$. Now consider $0 < r < 1$ and without loss of generality assume $j = 1$. Clearly

$$\begin{aligned}|(P_{\le 1}f)(x)| &\le \int |\psi(y)||f(x-y)|dy \\ &\lesssim \int |\psi(y)|(1+|y|)^{\frac{n}{r}}\frac{|f(x-y)|}{(1+|y|)^{\frac{n}{r}}}dy \\ &\lesssim \sup_y \frac{|f(x-y)|}{(1+|y|)^{\frac{n}{r}}} \lesssim ((\mathcal{M}(|f|^r))(x))^{\frac{1}{r}},\end{aligned}$$

where in the last step we have used Lemma 2.9 (this is where the frequency localisation condition on f is needed). □

REMARK. In Corollary 2.10, the frequency localisation condition $\{|\xi| \lesssim 2^j\}$ is not needed for $1 \le r < \infty$. On the other hand, for $0 < r < 1$, this condition is indeed needed. For a simple counterexample, one can take (in 1D) $f(y) = \chi_{\delta<|y|<1}|y|^{-1}$ for which $\mathcal{M}(|f|^r)$ is finite. On the other hand, $(P_{\le 1}f)(0)$ can be made arbitrarily large by taking $\delta \to 0+$.

LEMMA 2.11. *Let $0 < q < \infty$, $0 < b < \infty$ and $c_1 > 0$. Let $\{f_j\}$, $j \in \mathbb{Z}$ be a sequence of functions such that* $\operatorname{supp}(\widehat{f_j}) \subset \{\xi : |\xi| < c_1 2^j\}$. *Then*

$$\Big\| \Big(\sum_{j\in\mathbb{Z}} |\tilde{P}_j f_j|^b \Big)^{\frac{1}{b}} \Big\|_{L^q} \lesssim_{q,b,c_1,n} \Big\| \Big(\sum_{j\in\mathbb{Z}} |f_j|^b \Big)^{\frac{1}{b}} \Big\|_{L^q}.$$

PROOF. Note that by Corollary 2.10,

$$|(\tilde{P}_j f_j)(x)| \lesssim (\mathcal{M}(|f_j|^r))(x))^{\frac{1}{r}}.$$

Then by taking $0 < r < \min\{q, b\}$ and using Lemma 2.6, we get

$$\begin{aligned} \|(\tilde{P}_j f_j)_{l_j^b}\|_{L_x^q} &\lesssim \|((\mathcal{M}(|f_j|^r))^{\frac{1}{r}})_{l_j^b}\|_{L_x^q} \\ &\lesssim \|(\mathcal{M}(|f_j|^r))_{l_j^{\frac{b}{r}}}\|_{L_x^{\frac{q}{r}}}^{\frac{1}{r}} \\ &\lesssim \|(|f_j|^r)_{l_j^{\frac{b}{r}}}\|_{L_x^{\frac{q}{r}}}^{\frac{1}{r}} \lesssim \|(f_j)_{l_j^b}\|_{L_x^q}. \end{aligned}$$

□

We will need to use the following useful estimates sometimes without explicit mentioning.

LEMMA 2.12. *Let $1 \le q, r \le \infty$. If $s > 0$, then*

$$\|(2^{js} P_{>j} f)_{l_j^r}\|_{L_x^q} \lesssim \|(2^{js} P_j f)_{l_j^r}\|_{L_x^q}; \tag{2.3}$$

$$\|(2^{js} \|P_{>j} f\|_{L_x^q})\|_{l_j^r} \lesssim \|(2^{js} \|P_j f\|_{L_x^q})\|_{l_j^r}. \tag{2.4}$$

If $0 < s < 1$ and $1 \le q < \infty$, then

$$\|(2^{j(s-1)} \nabla P_{\le j} f)_{l_j^r}\|_{L_x^q} \lesssim \|(2^{js} P_j f)_{l_j^r}\|_{L_x^q}. \tag{2.5}$$

If $0 < s < 1$ and $1 \le q \le \infty$, then

$$\|(2^{j(s-1)} \|\nabla P_{\le j} f\|_{L_x^q})\|_{l_j^r} \lesssim \|(2^{js} \|P_j f\|_{L_x^q})\|_{l_j^r}. \tag{2.6}$$

PROOF. The inequality (2.3) follows from Young's inequality in l^r space. For (2.4) one can first use triangle inequality in L_x^q and then use Young's inequality in l^r space to conclude.

For (2.5), just observe that by Young's inequality in l^r, we have

$$(2^{j(s-1)} \nabla P_{\le j} f)_{l_j^r} \lesssim (2^{js} \tilde{P}_j P_j f)_{l_j^r},$$

where $\tilde{P}_j$ has frequency localised to $\{\xi : |\xi| \sim 2^j\}$. The desired result then follows from Lemma 2.11 (here we need $q < \infty$).

The proof of (2.6) is simpler and therefore omitted. □

3. Regularity of pressure

PROOF OF THEOREM 1.1. Proof of (1.5). We first notice that the case $s = 0$ follows from (1.3), the case $s = 1$ follows from (1.4) and standard bounds for Riesz type operators. Therefore we may assume $0 < s < 1$ and estimate only the homogeneous $\dot{W}^{2s,q}$-norm of p.

Denote $R_{lk} = (-\Delta)^{-1}\partial_l\partial_k$ and make the decomposition

$$p = \sum_{l,k}\sum_j \Big(R_{lk}(P_{\le j-2}u_l P_j u_k) + R_{lk}(P_j u_l P_{\le j-2}u_k) + R_{lk}(P_j u_l \tilde{P}_j u_k) \Big)$$

$$=: (A) + (B) + (C). \tag{3.1}$$

where $\tilde{P}_j = P_{j-1} + P_j + P_{j+1}$.

For the first piece, by using frequency localisation and the fact that

$$\sum_k \partial_k (P_{\le j-2}u_l P_j u_k) = \sum_k (\partial_k P_{\le j-2}u_l) u_k,$$

we obtain

$$\begin{aligned}\||\nabla|^{2s}(A)\|_q &\lesssim \|(2^{j(2s-1)}|\nabla P_{\le j-2}u|\cdot|P_j u|)_{l_j^2}\|_q \\ &\lesssim \|(2^{j(s-1)}\nabla P_{\le j-2}u)_{l_j^2}\|_{2q}\cdot\|(2^{js}P_j u)_{l_j^2}\|_{2q} \lesssim \|u\|^2_{\dot{W}^{s,2q}},\end{aligned}$$

where in the last step we have used Lemma 2.12.

The estimate of (B) is similar to (A) and therefore omitted. For (C) we write u_l and u_k simply as u, and estimate

$$\begin{aligned}\||\nabla|^{2s}(C)\|_q &\lesssim \|(2^{2ms}|\sum_{j\ge m-10} P_j u \tilde{P}_j u|)_{l_m^2}\|_q \\ &\lesssim \|(2^{js}P_j u)_{l_j^2}\|_{2q}\|(2^{js}\tilde{P}_j u)_{l_j^2}\|_{2q} \lesssim \|u\|^2_{\dot{W}^{s,2q}}.\end{aligned}$$

Proof of (1.6). Note here $0 < s < 1$. We use the same decomposition of p as above. For the first piece, one has

$$\begin{aligned}\|(A)\|_{\dot{B}^{2s}_{q,r}} &\lesssim \|(2^{j(2s-1)}\|\nabla P_{\le j-2}u\|_{2q}\|P_j u\|_{2q})\|_{l_j^r} \\ &\lesssim \|(2^{j(s-1)}\|\nabla P_{\le j-2}u\|_{2q})\|_{l_j^{2r}}\cdot\|u\|_{\dot{B}^s_{2q,2r}} \lesssim \|u\|^2_{\dot{B}^s_{2q,2r}}.\end{aligned}$$

The estimate of (B) is identical and omitted. For (C) we have (here we only need to require $s > 0$)

$$\|(C)\|_{\dot{B}^{2s}_{q,r}} \lesssim \|(2^{2ms}\|\sum_{j\ge m-10} P_j u\tilde{P}_j u\|_q)\|_{l_m^r} \lesssim \|u\|^2_{\dot{B}^s_{2q,2r}}.$$

□

REMARK. The estimate (1.5) can be easily derived using the new fractional Leibniz rule introduced in [**4**]. Namely by using Corollary 1.4 in [**4**] and taking $s_1 = s_2 = s$, $A^{2s}_{lk} = |\nabla|^{2s}(-\Delta)^{-1}\partial_l\partial_k$, and noting that $0 < s < 1$, we obtain

$$\|A^{2s}_{lk}(u_l u_k) - u_k A^{2s}_{lk}u_l - u_l A^{2s}_{lk}u_k\|_{\frac{q}{2}} \lesssim \||\nabla|^s u_l\|_q\||\nabla|^s u_k\|_q \lesssim \||\nabla|^s u\|^2_q.$$

Summing in l and k (inside the norm) and using incompressibility then yields the result.

PROOF OF THEOREM 1.2. As was already mentioned before the case $0 < s < 1$ follows from (1.6) and the embedding of $\dot{B}^0_{1,1}$ into $\mathcal{H}^1$. Thus we only need to consider the case $s = 1$. By using (3.1) and the same manipulations therein, we have

$$\begin{aligned}\||\nabla|^2(A)\|_{\mathcal{H}^1} &\lesssim \|(2^j|\nabla P_{\le j-2}u|\cdot|P_j u|)_{l_j^2}\|_1 \lesssim \|(|\nabla P_{\le j-2}u|)_{l_j^\infty}\|_2\|(2^j P_j u)_{l_j^2}\|_2 \\ &\lesssim \|\nabla u\|_2^2.\end{aligned}$$

For the third piece, a close inspection of the Besov case in Theorem 1.1 shows that

$$\|(C)\|_{\dot{B}^2_{1,1}} \lesssim \|\nabla u\|_2^2.$$

Since $\||\nabla|^2(C)\|_{\mathcal{H}^1} \lesssim \||\nabla|^2(C)\|_{\dot{B}^0_{1,1}}$, the desired result follows easily. □

PROOF OF THEOREM 1.3. We first decompose the product $f \cdot g$ as

$$f \cdot g = \underbrace{\sum_{j\in\mathbb{Z}} f_{\le j-2} \cdot g_j}_{=:(a)} + \underbrace{\sum_{j\in\mathbb{Z}} f_j \cdot g_{\le j-2}}_{=:(b)} + \underbrace{\sum_{j\in\mathbb{Z}} \tilde{f}_j \cdot g_j}_{=:(c)},$$

where $\tilde{f}_j = P_{j-1}f + P_j f + P_{j+1} f$. We only need to estimate (a) and (c) since the estimate of (b) is similar to (a). For (a), note that by frequency localisation, we have

$$\begin{aligned}\||\nabla|^s(a)\|_{\mathcal{H}^1} &\lesssim \|(2^{js}|f_{\le j-2}|\cdot|g_j|)_{l_j^2}\|_1 \lesssim \|(2^{js}g_j)_{l_j^2}\|_{p_2} \cdot \|(f_{\le j-2})_{l_j^\infty}\|_{p_2'} \\ &\lesssim \||\nabla|^s g\|_{p_2} \cdot \|f\|_{p_2'}.\end{aligned}$$

One should note that for this estimate we do not have severe constraints on the exponent s. On the other hand if $-1 < s \le 0$, then

$$\||\nabla|^s(a)\|_{\mathcal{H}^1} \lesssim \|(2^{js}|f_{\le j-2}|)_{l_j^\infty}\|_{p_1}\|(g_j)_{l_j^2}\|_{p_1'} \lesssim \||\nabla|^s f\|_{p_1}\|g\|_{p_1'}.$$

Thus

$$\||\nabla|^s(a)\|_{\mathcal{H}^1} \lesssim \begin{cases} \||\nabla|^s g\|_{p_2} \cdot \|f\|_{p_2'}, & \text{if } s > 0; \\ \min\{\||\nabla|^s f\|_{p_1}\|g\|_{p_1'},\ \||\nabla|^s g\|_{p_2}\|f\|_{p_2'}\}, & \text{if } -1 < s \le 0. \end{cases}$$

Now for the diagonal piece (c), we first note that for $s > 0$ the "Div-Curl" condition is not needed since

$$\begin{aligned}\||\nabla|^s(c)\|_{\mathcal{H}^1} &\lesssim \||\nabla|^s(c)\|_{\dot{B}^0_{1,1}} \\ &\lesssim \sum_k 2^{ks} \sum_{j \ge k-10} \|\tilde{f}_j g_j\|_1 \\ &\lesssim \min\{\||\nabla|^s f\|_{p_1}\|g\|_{p_1'},\ \||\nabla|^s g\|_{p_2}\|f\|_{p_2'}\}.\end{aligned}$$

In the regime $-1 < s \le 0$ we need to exploit the "Div-Curl" condition. Since $\nabla \times g = 0$ we can write $g = \nabla\psi$ for a scalar potential ψ. Since $\nabla \cdot f = 0$, we have

$$(c) = \sum_j \tilde{f}_j \cdot g_j = \sum_j \nabla \cdot (\tilde{f}_j \psi_j).$$

Clearly by frequency localisation,

$$\begin{aligned}\||\nabla|^s(c)\|_{\mathcal{H}^1} &\lesssim \||\nabla|^s(c)\|_{\dot{B}^0_{1,1}} \lesssim \|(2^{k(s+1)} \sum_{j\ge k-10} |f_j|\cdot|\psi_j|)_{l_k^1}\|_1 \\ &\lesssim \|(2^{j(s+1)}|f_j|\cdot|\psi_j|)_{l_j^1}\|_1 \\ &\lesssim \min\{\||\nabla|^s f\|_{p_1}\|g\|_{p_1'},\ \||\nabla|^s g\|_{p_2}\|f\|_{p_2'}\}.\end{aligned}$$

□

REMARK. We record here a quite useful (albeit simple) estimate extracted from the proof above. For any $\varphi \in C_c^\infty(\mathbb{R}^n)$, $\psi \in C_c^\infty(\mathbb{R}^n)$ with support *localised to an annulus* (say $\{|\xi| \sim 1\}$), define

$$\widehat{P_j^\varphi f}(\xi) = \varphi(2^{-j}\xi)\hat{f}(\xi), \quad \widehat{P_j^\psi f}(\xi) = \psi(2^{-j}\xi)\hat{f}(\xi).$$

Then for any $s > 0$, any $s_1, s_2 \geq 0$ with $s_1 + s_2 = s$, and any $f \in \mathcal{S}(\mathbb{R}^n)$, $g \in \mathcal{S}(\mathbb{R}^n)$, we have

$$\begin{aligned}\||\nabla|^s(\sum_j P_j^\varphi f P_j^\psi g)\|_{\mathcal{H}^1} &\lesssim \|\sum_j P_j^\varphi f P_j^\psi g\|_{\dot{B}^s_{1,1}} \\ &\lesssim_{s,\varphi,\psi,n,r,s_1,s_2} \||\nabla|^{s_1} f\|_r \||\nabla|^{s_2} g\|_{r'},\end{aligned} \tag{3.2}$$

where $1 < r < \infty$, $r' = \frac{r}{r-1}$.

Similarly for any $s > 0$, any $s_1, s_2 \geq 0$ with $s_1 + s_2 = s$, any $1 < p < \infty$, any $1 < p_1, p_2 < \infty$ with $\frac{1}{p} = \frac{1}{p_1} + \frac{1}{p_2}$, we have

$$\||\nabla|^s(\sum_j P_j^\varphi f P_j^\psi g)\|_p \lesssim_{s,\varphi,\psi,n,p,p_1,p_2,s_1,s_2} \||\nabla|^{s_1} f\|_{p_1} \||\nabla|^{s_2} g\|_{p_2}. \tag{3.3}$$

Also (the following corresponds to $p_2 = \infty$)

$$\||\nabla|^s(\sum_j P_j^\varphi f P_j^\psi g)\|_p \lesssim_{s,\varphi,\psi,n,p,s_1,s_2} \||\nabla|^{s_1} f\|_p \||\nabla|^{s_2} g\|_{\dot{B}^0_{\infty,\infty}}. \tag{3.4}$$

Inspired by the "Div-Curl" theorem and the considerations above, we record below a simple fractional Leibniz rule in the frequency-localised context which is quite useful for practical purposes. For $s > 0$, denote by A^s a differential operator such that its symbol $\widehat{A^s}(\xi)$ is a homogeneous function of degree s and $\widehat{A^s}(\xi) \in C^\infty(\mathbb{S}^{n-1})$ (for example: $\widehat{A^s}(\xi) = i|\xi|^{s-1}\xi_1$ which corresponds to $A^s = |\nabla|^{s-1}\partial_1$, or $\widehat{A^s}(\xi) = |\xi|^s$ which corresponds to $A^s = |\nabla|^s = D^s$). We have the following proposition.

PROPOSITION 3.1. *Let $s > 0$. Then for any $1 < p_1, p_2 < \infty$ with $\frac{1}{p_1} + \frac{1}{p_2} = 1$, any $s_1, s_2 \geq 0$ with $s_1 + s_2 = s$, and any f, $g \in \mathcal{S}(\mathbb{R}^n)$, we have*

$$\begin{aligned}&\|A^s(fg) - \sum_{j\in\mathbb{Z}} A^s(f_{\leq j-2}g_j) - \sum_{j\in\mathbb{Z}} A^s(f_j g_{\leq j-2})\|_{\mathcal{H}^1} \\ &\quad \lesssim \|A^s(fg) - \sum_{j\in\mathbb{Z}} A^s(f_{\leq j-2}g_j) - \sum_{j\in\mathbb{Z}} A^s(f_j g_{\leq j-2})\|_{\dot{B}^0_{1,1}} \\ &\quad \lesssim_{A^s,s,s_1,s_2,p,p_1,p_2,n} \|D^{s_1} f\|_{p_1} \|D^{s_2} g\|_{p_2}.\end{aligned}$$

For any $1 < p < \infty$, any $1 < p_1, p_2 < \infty$ with $\frac{1}{p_1} + \frac{1}{p_2} = \frac{1}{p}$, any $s_1, s_2 \geq 0$ with $s_1 + s_2 = s$, and any f, $g \in \mathcal{S}(\mathbb{R}^n)$, we have

$$\begin{aligned}&\|A^s(fg) - \sum_{j\in\mathbb{Z}} A^s(f_{\leq j-2}g_j) - \sum_{j\in\mathbb{Z}} A^s(f_j g_{\leq j-2})\|_p \\ &\quad \lesssim_{A^s,s,s_1,s_2,p,p_1,p_2,n} \|D^{s_1} f\|_{p_1} \|D^{s_2} g\|_{p_2}.\end{aligned}$$

Also

$$\begin{aligned}&\|A^s(fg) - \sum_{j\in\mathbb{Z}} A^s(f_{\leq j-2}g_j) - \sum_{j\in\mathbb{Z}} A^s(f_j g_{\leq j-2})\|_p \\ &\quad \lesssim_{A^s,s,s_1,s_2,p,n} \|D^{s_1} f\|_p \|D^{s_2} g\|_{\dot{B}^0_{\infty,\infty}}.\end{aligned}$$

REMARK. Since $f,\ g \in \mathcal{S}(\mathbb{R}^n)$, one does not need to worry about the convergence issues. For example, it is easy to check that $\sum_j f_{\le j-2} g_j \in H^m(\mathbb{R}^n)$, $\sum_j \tilde{f}_j g_j \in W^{m,2}(\mathbb{R}^n)$, for all $m \ge 0$, and

$$A^s(\sum_j f_{\le j-2} g_j) = \sum_j A^s(f_{\le j-2} g_j),$$
$$A^s(\sum_j \tilde{f}_j g_j) = \sum_j A^s(\tilde{f}_j g_j),$$

where the above equality holds point-wisely for $x \in \mathbb{R}^n$ and in stronger norms.

PROOF OF PROPOSITION 3.1. This follows from the para-product decomposition, (3.2), (3.3) and (3.4). □

COROLLARY 3.2. *Let $s > 0$. For any $1 < r_1, r_2 < \infty$, we have*

$$\|A^s(fg)\|_{\mathcal{H}^1} \lesssim_{A^s, s, n, r_1, r_2} \|D^s f\|_{r_1} \|g\|_{\frac{r_1}{r_1-1}} + \|f\|_{r_2} \|D^s g\|_{\frac{r_2}{r_2-1}}.$$

For any $1 < p < \infty$, any $1 < p_1, p_3 < \infty$, $1 < p_2, p_4 \le \infty$ with $\frac{1}{p} = \frac{1}{p_1} + \frac{1}{p_2} = \frac{1}{p_3} + \frac{1}{p_4}$, we have

$$\|A^s(fg)\|_p \lesssim_{A^s, s, n, p, p_1, p_2, p_3, p_4} \|D^s f\|_{p_1} \|g\|_{p_2} + \|D^s g\|_{p_3} \|f\|_{p_4}.$$

PROOF. This follows easily from Proposition 3.1. For example, one observes that

$$\|\sum_j A^s(f_{\le j-2} g_j)\|_p \lesssim \|(2^{js} g_j)_{l_j^2}\|_{p_3} \|(f_{\le j-2})_{l_j^\infty}\|_{p_4} \lesssim \|D^s g\|_{p_3} \|f\|_{p_4}.$$

The other cases are similarly estimated. We omit the details. □

4. Counterexamples

PROOF OF THEOREM 1.4, CASE $s = 1$. Step 1: Some simplification. We shall take $u = (u_1, u_2, 0, 0, \cdots, 0)$, where

$$u_1 = -\partial_2 \phi, \quad u_2 = \partial_1 \phi,$$

and $\phi : \mathbb{R}^n \to \mathbb{R}$ will be chosen later. Note that by construction u is divergence-free. Now

$$\begin{aligned} -\Delta p &= \sum_{j,k} \partial_j u_k \partial_k u_j \\ &= (\partial_1 u_1)^2 + (\partial_2 u_2)^2 + 2 \partial_1 u_2 \partial_2 u_1 \\ &= 2\Big((\partial_{12}\phi)^2 - \partial_{11}\phi \partial_{22}\phi\Big). \end{aligned} \tag{4.1}$$

Clearly, the task now is to find a scalar function $\phi \in \mathcal{S}(\mathbb{R}^n)$ with $\|\phi\|_2 + \|\phi\|_{\dot{B}^2_{\infty,\infty}} \le 1$, such that

$$\|(\partial_{12}\phi)^2 - \partial_{11}\phi \partial_{22}\phi\|_{\dot{B}^0_{\infty,\infty}} \gg 1.$$

As a matter of fact, we shall show that for some $j_1 \gg 1$,

$$\|P_{j_1}\big((\partial_{12}\phi)^2 - \partial_{11}\phi \partial_{22}\phi\big)\|_\infty \gg 1,$$

i.e. the norm inflation occurs at a high frequency block.

Step 2: There exists $\phi_0 \in \mathcal{S}(\mathbb{R}^n)$ with $\mathrm{supp}(\widehat{\phi_0}) \subset \{\xi : \frac{2}{3} \le |\xi| \le \frac{5}{6}\}$ and

$$\Big((\partial_{12}\phi_0)^2 - \partial_{11}\phi_0\partial_{22}\phi_0\Big)(0) \neq 0.$$

Indeed one can take $\widehat{\phi_0}(\xi)$ as a radial real-valued function in ξ with

$$\int \xi_1\xi_2\widehat{\phi_0}(\xi)d\xi = 0,$$

$$\int \xi_1^2\widehat{\phi_0}(\xi)d\xi \neq 0,$$

$$\int \xi_2^2\widehat{\phi_0}(\xi)d\xi \neq 0.$$

These yield $(\partial_{12}\phi_0)(0) = 0$, $(\partial_{11}\phi_0)(0) \cdot (\partial_{22}\phi_0)(0) \neq 0$.

Step 3: Now take $J \gg 1$, J being an even integer and

$$\phi(x) = \sum_{j=100}^{J} 2^{-2j^2}\phi_0(2^{j^2}x).$$

Easy to check that (note that $\mathrm{supp}(\hat{\phi}_0) \subset \{\frac{2}{3} \le |\xi| \le \frac{5}{6}\} \subset \{\frac{1}{2} < |\xi| < \frac{7}{6}\}$)

$$\|\phi\|_2 + \|\phi\|_\infty + \|\phi\|_{\dot{B}^2_{\infty,\infty}} \lesssim 1. \tag{4.2}$$

Then clearly

$$\begin{aligned}
&(\partial_{12}\phi(x))^2 \\
&= \sum_{j=100}^{J} (\partial_{12}\phi_0)(2^{j^2}x)(\partial_{12}\phi_0)(2^{j^2}x) \\
&\quad + 2\sum_{j_1=100}^{J} (\partial_{12}\phi_0)(2^{j_1^2}x)\Big(\sum_{100\le j_2<j_1} (\partial_{12}\phi_0)(2^{j_2^2}x)\Big).
\end{aligned} \tag{4.3}$$

Similarly

$$\begin{aligned}
&\partial_{11}\phi(x)\partial_{22}\phi(x) \\
&= \sum_{j=100}^{J} (\partial_{11}\phi_0)(2^{j^2}x)(\partial_{22}\phi_0)(2^{j^2}x) \\
&\quad + \sum_{j_1=100}^{J} (\partial_{11}\phi_0)(2^{j_1^2}x)\Big(\sum_{100\le j_2<j_1} (\partial_{22}\phi_0)(2^{j_2^2}x)\Big) \\
&\quad + \sum_{j_1=100}^{J} (\partial_{22}\phi_0)(2^{j_1^2}x)\Big(\sum_{100\le j_2<j_1} (\partial_{11}\phi_0)(2^{j_2^2}x)\Big).
\end{aligned} \tag{4.4}$$

Thus

$$\begin{aligned}
&(\partial_{12}\phi(x))^2 - \partial_{11}\phi(x)\partial_{22}\phi(x) \\
&= H_1(x) + H_2(x),
\end{aligned} \tag{4.5}$$

where

$$H_1(x) = \sum_{j=100}^{J} \Big((\partial_{12}\phi_0)^2 - \partial_{11}\phi_0\partial_{22}\phi_0 \Big)(2^{j^2}x), \tag{4.6}$$

is the "diagonal piece", and $H_2(x)$ is the off-diagonal piece.

We shall show

$$\|P_{\le 1}H_1\|_\infty + \|H_1\|_{\dot{B}^0_{\infty,\infty}} \lesssim 1, \quad \|P_{j_1}H_2\|_\infty \gg 1, \quad \text{for some } j_1 \gg 1.$$

Step 4: the diagonal piece. One can rewrite H_1 as

$$\begin{aligned} 2H_1(x) &= \partial_1\Big(\sum_{j=100}^{J} 2^{-j^2}(\partial_2\phi_0\partial_{12}\phi_0 - \partial_1\phi_0\partial_{22}\phi_0)(2^{j^2}x)\Big) \\ &\quad + \partial_2\Big(\sum_{j=100}^{J} 2^{-j^2}(-\partial_2\phi_0\partial_{11}\phi_0 + \partial_1\phi_0\partial_{12}\phi_0)(2^{j^2}x)\Big). \end{aligned}$$

Thanks to the above expression, it is easy to check that

$$\|P_{\le 1}H_1\|_\infty \lesssim \sum_{j\ge 100} 2^{-j^2} \lesssim 1.$$

Note that on the Fourier side, we have roughly speaking

$$\widehat{H_1}(\xi) \sim \xi \sum_{j=100}^{J} c_j 1_{|\xi|\le \frac{5}{3}2^{j^2}}.$$

It is then easy to check that for each $l \ge 2$ (note that P_l has frequency localised to $\frac{1}{2}2^l \le |\xi| \le \frac{7}{6}2^l$),

$$\|P_l H_1\|_\infty \lesssim 2^l \sum_{j\ge\sqrt{l-2}} 2^{-j^2} \lesssim 1.$$

Step 5: The off-diagonal piece. Take $j_0 = J/2$ and $j_1 = j_0^2$. It is not difficult to check that

$$\begin{aligned} P_{j_1}H_2 &= 2(\partial_{12}\phi_0)(2^{j_0^2}x) \sum_{100\le j_2<j_0} (\partial_{12}\phi_0)(2^{j_2^2}x) \\ &\quad - (\partial_{11}\phi_0)(2^{j_0^2}x) \sum_{100\le j_2<j_0} (\partial_{22}\phi_0)(2^{j_2^2}x) \\ &\quad - (\partial_{22}\phi_0)(2^{j_0^2}x) \sum_{100\le j_2<j_0} (\partial_{11}\phi_0)(2^{j_2^2}x). \end{aligned}$$

Thus

$$|(P_{j_1}H_2)(0)| \gtrsim j_0|(\partial_{12}\phi_0)(0)^2 - (\partial_{11}\phi_0)(0)(\partial_{22}\phi_0)(0)| \gg 1.$$

This yields the desired result. □

PROOF OF THEOREM 1.4, CASE $s = 0$. We shall take $u=(-\partial_2\phi, \partial_1\phi, 0, \cdots, 0)$ where $\phi : \mathbb{R}^n \to \mathbb{R}$ will be specified momentarily.

Step 1. First we choose Schwartz $g : \mathbb{R}^n \to \mathbb{R}$ with $\mathrm{supp}(\hat{g}) \subset \{\xi : \frac{1}{2} \le |\xi| \le 2\}$ such that

$$\Big(P_{j_0}\Delta^{-1}\partial_{11}(g^2)\Big)(0) \ne 0, \qquad \text{for any } j_0 \in \mathbb{Z}.$$

This can be easily achieved since $\Delta^{-1}\partial_{11}$ corresponds to the symbol $\xi_1^2/|\xi|^2$ and one can just take $\hat{g}$ to be non-negative.

Step 2. Let $J \gg 1$ and define

$$\phi(x) = \sum_{j=1}^{J} \frac{1}{k_j} \cos(k_j x_2) g(x),$$

where k_j are dyadic numbers which are well separated and will be chosen sufficiently large. Note that $\|\phi\|_{\dot{B}^1_{\infty,\infty}} + \|\phi\|_2 \lesssim 1$.

Step 3. Non-diagonal term does not matter. Note that $\|u\|_{\dot{B}^0_{\infty,\infty}} \lesssim 1$. Recall

$$-p = \sum_{l,k} \Delta^{-1}\partial_l\partial_k(u_l u_k).$$

Clearly

$$\begin{aligned}
&\|\sum_{l,k} \Delta^{-1}\partial_l\partial_k(\sum_j P_{\le j-2}u_l P_j u_k)\|_{\dot{B}^0_{\infty,\infty}} \\
&\lesssim \sup_j \sum_{l,k} \|\Delta^{-1}\partial_l(\partial_k P_{\le j-2}u_l P_j u_k)\|_\infty \lesssim \|u\|^2_{\dot{B}^0_{\infty,\infty}} \lesssim 1.
\end{aligned}$$

Step 4. Inflation through the diagonal terms. In terms of ϕ, we have

$$-p = \Delta^{-1}\partial_{11}((\partial_2\phi)^2) + \Delta^{-1}\partial_{22}((\partial_1\phi)^2) - 2\Delta^{-1}\partial_{12}(\partial_1\phi\partial_2\phi).$$

Clearly

$$\partial_2\phi = \sum_{j=1}^{J}\big(-\sin(k_j x_2)g(x) + \frac{1}{k_j}\cos(k_j x_2)\partial_2 g\big).$$

Then

$$\begin{aligned}
(\partial_2\phi)^2 &= \text{“Non-diagonal terms”} \\
&\quad + \sum_{j=1}^{J}\big(\sin^2(k_j x_2)g(x)^2 + \frac{1}{k_j^2}\cos^2(k_j x_2)(\partial_2 g)^2\big) \\
&= \text{“Non-diagonal terms”} \\
&\quad + \frac{1}{2}Jg(x)^2 + \frac{1}{2}\big(\sum_{j=1}^{J}\frac{1}{k_j^2}\big)(\partial_2 g)^2 \\
&\quad + \frac{1}{2}\sum_{j=1}^{J}\big(-\cos(2k_j x_2)g(x)^2 + \frac{1}{k_j^2}\cos(2k_j x_2)(\partial_2 g)^2\big).
\end{aligned}$$

It follows that

$$\|-p - \frac{J}{2}\Delta^{-1}\partial_{11}(g^2)\|_{\dot{B}^0_{\infty,\infty}} \lesssim 1.$$

Thus for any $j_0 \in \mathbb{Z}$, by taking J large, we obtain

$$|(P_{j_0}p)(0)| \gtrsim \text{const} \cdot J \gg 1.$$

This clearly implies the desired inflation of $\|p\|_{\dot{B}^0_{\infty,\infty}}$ in high frequency. $\square$

We now consider the case $u \in \dot{C}^{\frac{1}{2}} = \dot{B}^{\frac{1}{2}}_{\infty,\infty}$.

PROPOSITION 4.1. *For any $\epsilon > 0$, there exists divergence free $u \in \mathcal{S}(\mathbb{R}^n)$ with $\|u\|_{\dot{B}^{\frac{1}{2}}_{\infty,\infty}} + \|u\|_2 \le 1$ such that*

$$\|\nabla p\|_\infty > \frac{1}{\epsilon}.$$

PROOF OF PROPOSITION 4.1.

Step 1: Take a scalar function $\phi_0 \in \mathcal{S}(\mathbb{R}^n)$ with $\operatorname{supp}(\widehat{\phi_0}) \subset \{\xi : \frac{2}{3} \le |\xi| \le \frac{5}{6}\}$. Let $J \gg 1$ and

$$\phi(x) = \sum_{j=1000}^{J} N_j^{-\frac{3}{2}} \phi_0(N_j x),$$

where $N_j = 2^{a_j}$, $a_j = 3^{j^2}$. Note that by construction we have

$$N_{j-1} < N_j^{\frac{1}{2}}, \qquad \forall j. \tag{4.7}$$

Denote $\nabla^\perp = (-\partial_2, \partial_1, 0, \cdots, 0)$. Then define

$$u = \nabla^\perp \phi = \sum_{j=1000}^{J} \underbrace{N_j^{-\frac{1}{2}} (\nabla^\perp \phi_0)(N_j x)}_{=:v_j}.$$

Easy to check that

$$\|u\|_{\dot{B}^{\frac{1}{2}}_{\infty,\infty}} + \|u\|_2 \lesssim 1.$$

On the other hand,

$$\begin{aligned}
\nabla p &= (-\Delta)^{-1} \nabla\nabla \cdot ((u \cdot \nabla) u) \\
&= \sum_j (-\Delta)^{-1} \nabla\nabla \cdot ((v_j \cdot \nabla) v_j) && (4.8) \\
&\quad + \sum_j (-\Delta)^{-1} \nabla\nabla \cdot ((v_j \cdot \nabla) v_{\le j-1}) && (4.9) \\
&\quad + \sum_j (-\Delta)^{-1} \nabla\nabla \cdot ((v_{\le j-1} \cdot \nabla) v_j), && (4.10)
\end{aligned}$$

where we have denoted $v_{\le j-1} = \sum_{i \le j-1} v_i$.

By (4.7), frequency localization and Bernstein, we have

$$\|(4.9)\|_\infty \lesssim \sum_j N_j^{-\frac{1}{2}} N_{j-1}^{\frac{1}{2}} \lesssim 1.$$

For (4.10), we shall transform it to a similar form as in (4.9). Note that for any two vector functions f, g with $\nabla \cdot f = \nabla \cdot g = 0$, we have

$$\begin{aligned}
\nabla \cdot ((f \cdot \nabla) g) &= \sum_{k,l} \partial_k f_l \partial_l g_k \\
&= \sum_l \partial_l \Big(\sum_k (\partial_k f_l) g_k\Big).
\end{aligned}$$

Thanks to the above transformation, it is clear that we have

$$\nabla \cdot ((v_{\le j-1} \cdot \nabla) v_j) = O(\partial(\partial v_{\le j-1} v_j)).$$

Thus it can be estimated in the same way as in (4.9), and we have

$$\|(4.10)\|_\infty \lesssim_j N_j^{-\frac{1}{2}} N_{j-1}^{\frac{1}{2}} \lesssim 1.$$

Step 2: From step 1, it is clear that the dominant contribution to $\|\nabla p\|_\infty$ comes from the diagonal piece (4.8). Denote (4.8) as $H(x)$. Then by using scaling, we have

$$\begin{aligned} H(0) &= \sum_j \Big((-\Delta)^{-1} \nabla\nabla \cdot ((v_j \cdot \nabla) v_j) \Big)(0) \\ &= (J - 999) \Big((-\Delta)^{-1} \nabla\nabla \cdot ((\nabla^\perp \phi_0) \cdot \nabla)(\nabla^\perp \phi_0) \Big)(0). \end{aligned}$$

Thus we only need to show the existence of ϕ_0 such that

$$|\Big((-\Delta)^{-1} \nabla\nabla \cdot ((\nabla^\perp \phi_0) \cdot \nabla)(\nabla^\perp \phi_0) \Big)(0)| \neq 0.$$

This is not difficult to do and we sketch the detail below.

First by using the proof of Theorem 1.4 (see Step 2 therein), one can find $\tilde{\phi}_0 \in \mathcal{S}(\mathbb{R}^n)$ with $\operatorname{supp}(\widehat{\tilde{\phi}_0}) \subset \{ \xi : \frac{2}{3} \le |\xi| \le \frac{5}{6} \}$ such that

$$\begin{aligned} &\Big(\nabla \cdot ((\nabla^\perp \tilde{\phi}_0 \cdot \nabla)(\nabla^\perp \tilde{\phi}_0)) \Big)(0) \\ &= 2 \Big((\partial_{12} \tilde{\phi}_0)^2 - \partial_{11} \tilde{\phi}_0 \partial_{22} \tilde{\phi}_0 \Big)(0) \neq 0. \end{aligned}$$

Take a smooth bump function g such that

$$\int_{\mathbb{R}^n} \Big(\nabla \cdot ((\nabla^\perp \tilde{\phi}_0 \cdot \nabla)(\nabla^\perp \tilde{\phi}_0)) \Big)(x) g(x) dx \neq 0.$$

Clearly then

$$\int_{\mathbb{R}^n} \Big((-\Delta)^{-1} \nabla\nabla \cdot ((\nabla^\perp \tilde{\phi}_0 \cdot \nabla)(\nabla^\perp \tilde{\phi}_0)) \Big)(x) \cdot \nabla g(x) dx \neq 0.$$

From this, one sees that there exists x_* such that

$$\Big((-\Delta)^{-1} \nabla\nabla \cdot ((\nabla^\perp \tilde{\phi}_0 \cdot \nabla)(\nabla^\perp \tilde{\phi}_0)) \Big)(x_*) \neq 0.$$

Define $\phi_0(x) = \tilde{\phi}_0(x + x_*)$. Since translation in physical space is equivalent to modulation in frequency space (which does not change frequency localisation), it is not difficult to check that ϕ_0 satisfies all the required properties. □

The next proposition requires a slightly different construction, in the sense that the frequency supports are no longer disjoint. We only need to work in the physical space.

Proposition 4.2. *For any $\epsilon > 0$, there exists divergence free $u \in \mathcal{S}(\mathbb{R}^n)$ with $\|\nabla u\|_\infty + \|u\|_2 \le 1$ such that*

$$\|\nabla p\|_{C^1} > \frac{1}{\epsilon}.$$

Proof of Proposition 4.2. Take a smooth bump function $\phi_0 \in C_c^\infty(\mathbb{R}^n)$ with

$$\operatorname{supp}(\phi_0) \subset \{ x : c_1 < |x| < c_2 \}$$

for some constants $0 < c_1 < c_2 < \infty$. Define

$$\phi(x) = \sum_{j=1}^{J} k_j^{-2} \phi_0(k_j x),$$

where $J \gg 1$, and k_j are dyadic numbers sufficiently large such that each summand has disjoint support. Denote $\nabla^\perp = (-\partial_2, \partial_1, 0, \cdots, 0)$ and

$$u = \nabla^\perp \phi = \sum_{j=1}^{J} \underbrace{k_j^{-1} (\nabla^\perp \phi_0)(k_j x)}_{=:v_j}.$$

Obviously $\|u\|_{C^1} + \|u\|_2 \lesssim 1$ (since the rescaled copies of $\nabla^\perp \phi_0$ have disjoint supports). On the other hand, by using scaling,

$$\begin{aligned}
(\partial_{11} p)(0) &= \Big((-\Delta)^{-1} \partial_{11} \nabla \cdot ((u \cdot \nabla) u) \Big)(0) \\
&= \sum_{j=1}^{J} \Big((-\Delta)^{-1} \partial_{11} \nabla \cdot (v_j \cdot \nabla v_j) \Big)(0) \\
&= J \Big((-\Delta)^{-1} \partial_{11} \nabla \cdot ((\nabla^\perp \phi_0 \cdot \nabla) \nabla^\perp \phi_0) \Big)(0).
\end{aligned}$$

Thus we only need to choose ϕ_0 compactly supported on an annulus such that

$$|\Big((-\Delta)^{-1} \partial_{11} \nabla \cdot ((\nabla^\perp \phi_0 \cdot \nabla) \nabla^\perp \phi_0) \Big)(0)| \neq 0.$$

This is not difficult to achieve and we sketch the details below.

Step 1. Simplification. We shall choose ϕ_0 to have support away from the origin. This way one does not need to worry about the singular kernels in the integration by part argument below. Denote $v = \nabla^\perp \phi_0$. Then

$$\begin{aligned}
&(\Delta^{-1} \partial_{11} \nabla \cdot (v \cdot \nabla v))(0) \\
&= \sum_{1 \le j,k \le 2} \int v_j v_k \partial_j \partial_k (\underbrace{\partial_{11} \Delta^{-1} \delta_0}_{=:a(x)}) dx \\
&= \int (v_1^2 \partial_{11} a + v_2^2 \partial_{22} a - 2 v_1 v_2 \partial_{12} a) dx.
\end{aligned}$$

Step 2. Case dimension $n = 2$. Note that

$$a(x) = C_1 \cdot ((x_1^2 + x_2^2)^{-1} - 2(x_1^2 + x_2^2)^{-2} x_1^2),$$

where $C_1 > 0$ is an absolute constant. Denote $\tilde{a}(x) = a(x)/C_1$. Easy to verify that

$$(\partial_{11} \tilde{a})(1,0) = -6, \quad (\partial_{22} \tilde{a})(1,0) = 6, \quad (\partial_{12} \tilde{a})(1,0) = 0.$$

Now define $x_* = (1,0)$ and let

$$\phi_0(x) = \cos(k x_2) c(\frac{x - x_*}{\eta}),$$

where c is a bump function localised to the unit ball, $\eta > 0$ will be taken sufficiently small (so that $|(\partial_{11} \tilde{a})(x)| \sim 1$ for $|x - x_*| < \eta$) and k will be taken sufficiently large.

It is then easy to check (recall $v_1 = -\partial_2\phi_0$)

$$\begin{aligned}&\Big|\int (v_1^2\partial_{11}\tilde{a} + v_2^2\partial_{22}\tilde{a} - 2v_1v_2\partial_{12}\tilde{a})dx\Big|\\ &\geq \text{const}\cdot k^2 - O(k) \gtrsim 1,\end{aligned}$$

if k is sufficiently large. This settles the 2D case.

Step 3. Case dimension $n \geq 3$. Denote $x_* = (1, 0, \cdots, 0)$. Easy to check that

$$\partial_{11}a(x)\Big|_{x=x_*} = -\,\text{const}\cdot\partial_{x_1}^4(x_1^{-(n-2)})\Big|_{x_1=1} \neq 0.$$

Again let

$$\phi_0(x) = \cos(kx_2)c(\frac{x - x_*}{\eta})$$

will yield

$$\begin{aligned}&\Big|\int (v_1^2\partial_{11}\tilde{a} + v_2^2\partial_{22}\tilde{a} - 2v_1v_2\partial_{12}\tilde{a})dx\Big|\\ &\geq \text{const}\cdot k^2 - O(k) \gtrsim 1.\end{aligned}$$

This settles the case for $n \geq 3$. □

PROOF OF PROPOSITION 1.6. Define $\nabla^{\perp} = (-\partial_2, \partial_1, 0, \cdots, 0)$. Let $\phi \in C_c^\infty(\mathbb{R}^n)$ be such that

$$\begin{aligned}&\int_{\mathbb{R}^n} (\partial_1\phi)^2 dx \neq \int_{\mathbb{R}^n} (\partial_2\phi)^2 dx,\\ &\int_{\mathbb{R}^n} \partial_1\phi\partial_2\phi dx = 0.\end{aligned}$$

For example, one can take $\phi(x) = b(x)\cos(kx_1)$, where $b \in C_c^\infty(\mathbb{R}^n)$ is radial, and k is sufficiently large.

Define $u = \nabla^{\perp}\phi$. By using a computation similar to that in (4.1), we have

$$p = 2(-\Delta)^{-1}\Big(\underbrace{(\partial_{12}\phi)^2 - \partial_{11}\phi\partial_{22}\phi}_{:=g(x)}\Big).$$

Easy to check that $\hat{g}(0) = 0$ and $(\partial_\xi\hat{g})(0) = 0$. The latter is due to the fact that

$$\int_{\mathbb{R}^n} ((\partial_{12}\phi)^2 - \partial_{11}\phi\partial_{22}\phi)x_j dx = 0, \quad \forall 1 \leq j \leq n.$$

To derive this, one can use the simple identity (below $\beta = \beta(x)$ denotes a smooth weight function)

$$\begin{aligned}\int \partial_{11}\phi\partial_{22}\phi\beta(x)dx = &\int (\partial_{12}\phi)^2\beta\\ &+ \frac{1}{2}\int (-(\partial_2\phi)^2\partial_{11}\beta - (\partial_1\phi)^2\partial_{22}\beta + 2\partial_1\phi\partial_2\phi\partial_{12}\beta)dx.\end{aligned}$$

On the other hand, by using the identity above, it is not difficult to check that

$$\int_{\mathbb{R}^n} g(x)x_1^2dx = \int_{\mathbb{R}^n} (\partial_2\phi)^2dx,$$

$$\int_{\mathbb{R}^n} g(x)x_2^2dx = \int_{\mathbb{R}^n} (\partial_1\phi)^2dx,$$

$$\int_{\mathbb{R}^n} g(x)x_1x_2dx = -\int_{\mathbb{R}^n} \partial_1\phi\partial_2\phi dx = 0.$$

Thus near $\xi = 0$,

$$\hat{g}(\xi) = c_1\xi_1^2 + c_2\xi_2^2 + O(|\xi|^4),$$

where $c_1 \neq c_2$. Clearly then $\hat{g}(\xi)/|\xi|^2$ is not continuous at $\xi = 0$. This immediately implies that $p \notin L^1(\mathbb{R}^n)$. □

References

[1] R. Coifman, P.-L. Lions, Y. Meyer, and S. Semmes, *Compensated compactness and Hardy spaces* (English, with English and French summaries), J. Math. Pures Appl. (9) **72** (1993), no. 3, 247–286. MR1225511

[2] C. Fefferman and E. M. Stein, *Some maximal inequalities*, Amer. J. Math. **93** (1971), 107–115, DOI 10.2307/2373450. MR0284802

[3] L. Grafakos, *Modern Fourier analysis*, 2nd ed., Graduate Texts in Mathematics, vol. 250, Springer, New York, 2009. MR2463316

[4] D. Li. *On Kato-ponce and fractional Leibniz.* To appear in Revista Matemática Iberoamericana.

Department of Mathematics, The Hong Kong University of Science and Technology, Clear Water Bay, Kowloon, Hong Kong

Email address: madli@ust.hk

Department of Mathematics, University of Iowa, 14 MacLean Hall, Iowa City, Iowa 52242

Email address: xiaoyi-zhang@uiowa.edu

Contemporary Mathematics
Volume **725**, 2019
https://doi.org/10.1090/conm/725/14559

On large future-global-in-time solutions to energy-supercritical nonlinear wave equation

Shuang Miao

ABSTRACT. In a recent work we constructed future-global-in-time large solution to the Cauchy problem for semi-linear wave equation on $\mathbb{R}^{3+1}$ with nonlinear terms satisfying the null conditions. In this article we show that the method also applies to energy-supercritical nonlinear wave equation with power type nonlinearity. In addition to constructing a large global-in-time solution, we also show that the solution constructed decays to zero as $t \to \infty$.

CONTENTS

1. Introduction
1.1. Historical results
1.2. The short pulse data and main results
1.3. Notations
1.4. Comments on the proof
2. Short pulse region
2.1. Main a priori estimates
2.2. Higher order estimates
2.3. Improved estimates on C_δ
2.4. Existence of short pulse Cauchy data
3. Small data region
3.1. Klainerman-Sobolev inequality revisited
3.2. A priori energy estimates
Acknowledgments
References

1. Introduction

We consider the Cauchy problem of the following nonlinear wave equation on $\mathbb{R}^{3+1}$:

$$\Box\phi = \pm|\phi|^{p-1}\phi, \quad p > 5. \tag{1.1}$$

Key words and phrases. Large data, global solution, wave equations.

Here, $\Box = -\partial_t^2 + \Delta$ is the standard wave operator. The function ϕ is real-valued. The equation (1.1) admits a conserved energy

$$E(\phi(t)) = \frac{1}{2}\int_{\mathbb{R}^3} |\partial_t\phi(t,x)|^2 + |\nabla_x\phi(t,x)|^2 \pm \frac{2}{p+1}|\phi(t,x)|^{p+1}dx \tag{1.2}$$

(1.1) is called *defocusing*, if there is a plus sign in front of the nonlinearity, otherwise it is called *focusing*. In view of the Sobolev embedding $H^1(\mathbb{R}^3) \hookrightarrow L^6(\mathbb{R}^3)$, one refers to the range $p < 5$ as the energy-subcritical regime, to $p = 5$ as the energy-critical regime and to $p > 5$ as the energy-supercritical regime. In this article we consider the energy-supercritical cases, regardless of whether it is focusing or defocusing.

The study of long time behavior of the solution to (1.1) for large initial data has been studied extensively in the literature. The purpose of the current paper is to exhibit a class of large Cauchy data for (1.1) which also leads to global classical solutions.

1.1. Historical results.

1.1.1. *Previous works on nonlinear wave equation with power type nonlinearity.* The study of the Cauchy problem for (1) has a long history. For the defocusing case, Rauch [**13**] proved global existence for arbitrary smooth data for subcritical equations and for small energy smooth data in the critical case. Struwe [**15**] obtained global existence for large but radially symmetric data in the critical case, then Grillakis [**6**] removed the radially symmetric condition on the data. After that Shatah and Struwe [**14**] studied the energy-class solutions. See also [**1**], [**2**] and [**16**] for further results. The study of focusing case is initiated by Krieger and Schlag [**10**] as well as Kenig and Merle [**7**]. The study of energy-supercritical nonlinear wave equation with large global solution has become active in recent years. See [**8**], [**17**], [**11**], [**3**], [**4**] and [**20**] as well as references therein. In [**3, 4**] the authors also consider large global solutions propagating along outgoing null cones. They use Strichartz estimates by looking at the parametrix of the linear wave operator. While in the present work, we use pure energy estimates without referring to the parametrix of the linear operator at all.

1.1.2. *Method of using short pulse data.* Our method in constructing global-in-time large solutions is different from the above mentioned works. Instead, we use the *short pulse* data which dates back to Christodoulou's work [**5**]. In this work, the dynamical formation of black hole in vacuum spacetime is proved. See also an alternative approach and generalization in [**9**]. Later on, this method is used in constructing large global solutions to nonlinear wave equation with nonlinearity satisfying null condition. Earlier works include [**18**] and [**19**], in which the construction of the large data is indirect. The authors essentially impose the characteristic data on the past null infinity and solve the inverse scattering problem all the way up to a finite time to construct the initial Cauchy data for backward evolution. This indirectness issue is resolved in [**12**] in which the authors solve the Cauchy problem forward in time directly. For a more detailed discussion of the short pulse method, see [**12**]. In this paper, we show that the method in [**12**] also applies to constructing large global solutions to energy-supercritical nonlinear wave equation (1.1).

1.2. The short pulse data and main results. We use (x_0, x_1, x_2, x_3) to denote the standard Cartesian coordinates (t, x, y, z) on $\mathbb{R}^{3+1}$. In particular, ∂_0

stands for ∂_t. We study the Cauchy problem for the following nonlinear wave equation

$$\Box\phi = \pm|\phi|^{p-1}\phi, \quad (\phi, \partial_t\phi)\big|_{t=1} = (\phi_0, \phi_1), \quad p \geq 5. \tag{1.3}$$

We remark that in order to simplify some of the expressions appearing in the proof of the main theorem, we give the initial data at $\{t = 1\}$ rather than $\{t = 0\}$. Because of the invariance of the equation under time translations, this is equivalent as giving data on $\{t = 0\}$.

Before describing the short pulse data, we introduce some notations: r and θ are used to denote the usual radial and angular coordinates on $\mathbb{R}^3$. Let δ be a small positive constant which will be determined later. We identify the $\{t = 1\}$ hypersurface with $\mathbb{R}^3$ and divide it into three parts:

$$\{t = 1\} = B_{1-2\delta} \cup \big(B_1 - B_{1-2\delta}\big) \cup \big(\mathbb{R}^3 - B_1\big),$$

where B_r is the ball centered at the origin with radius r.

In the following, $f \lesssim g$ always means there exists a constant C such that $f \leq Cg$ holds. We consider the initial data (ϕ_0, ϕ_1) on $\{t = 1\}$ of (1.3) satisfying the following conditions (for some $\delta < 1$):

- On $B_{1-2\delta}$, we set $(\phi_0, \phi_1) \equiv (0, 0)$.
- On $B_1 - B_{1-2\delta}$,

(1.4)
$$\|\partial_r^k \nabla\!\!\!/^l \phi_0\|_{L^\infty(B_1 - B_{1-2\delta})} + \|\partial_r^{k-1} \nabla\!\!\!/^l \phi_1\|_{L^\infty(B_1 - B_{1-2\delta})} \leq C\delta^{\frac{1}{2}-k}, \quad 0 \leq k + l \leq 100,$$

  with $\nabla\!\!\!/$ denoting the covariant derivative on the spheres centered at origin, and ∂_r being the radial derivative. We also have

(1.5)
$$\|\left(\partial_t + \partial_r\right)^k \nabla\!\!\!/^l \partial_r^m \phi\|_{L^\infty(B_1 - B_{1-2\delta})} \leq C\delta^{\frac{1}{2}-m}, \quad k > 1, \quad 0 \leq k + l + m \leq 100.$$

  Here C in (1.4) and (1.5) is an absolute constant. In (1.5) the derivative $\partial_t^k\phi$ for $k > 1$ is given by the equation (1.3) and $\partial_t\phi = \phi_1$.
- On $\mathbb{R}^3 - B_1$, $(\phi_0, \phi_1) \equiv (0, 0)$.

REMARK 1.1. Let us now assume the existence of such a Cauchy data. After we derive the energy estimates in the so-called "short pulse region" (defined in Section 1.3), the existence of the Cauchy data follows as a straightforward corollary. We will give a more detailed argument at the end of Section 2.

For a fixed value of θ, the graph of ϕ_0 versus the variable r is as follows:

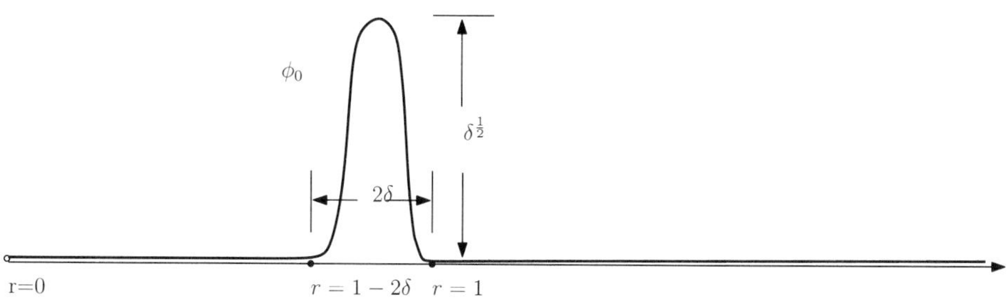

The pulse-like shape of the graph explains the name "short pulse" used for this data. The width of the pulse is 2δ and its amplitude is $\delta^{\frac{1}{2}}$, which is very large relative to the width if δ is small.

The requirements (1.4)-(1.5) look obscure. In fact, we have a natural geometric explanation of this choice, which also serves as heuristics to understand why one expects a future-global-in-time solution.

REMARK 1.2 (Geometric / Physical interpretation). In terms of the solution ϕ, it is easy to observe from our initial data satisfying (1.4)-(1.5) that

$$|(\partial_t + \partial_r)\phi|_{t=1}| \lesssim \delta^{\frac{1}{2}}, \quad |\nabla\!\!\!/ \phi|_{t=1}| \lesssim \delta^{\frac{1}{2}}. \tag{1.6}$$

We will prove that if δ is small enough, the smallness indicated by (1.6) can be propagated by showing

$$\|(\partial_t + \partial_r)\phi\|_{L^\infty(\Sigma_t)} \lesssim \delta^{\frac{1}{2}} t^{-\alpha_0}, \quad \|\nabla\!\!\!/ \phi\|_{L^\infty(\Sigma_t)} \lesssim \delta^{\frac{1}{4}} t^{-\alpha_0} \tag{1.7}$$

for some $\alpha_0 > 1$. Here we use Σ_t to denote the constant time slice. Recall that $L = \partial_t + \partial_r$ is the normal (with respect to the Minkowski metric) of the outgoing light cones $\{t - r = \text{constant}\}$ in $\mathbb{R}^{3+1}$. If we integrate $|(\partial_t + \partial_r)\phi|^2 + |\nabla\!\!\!/ \phi|^2$ on such an outgoing light cone C, the quantity

$$\int_C |(\partial_t + \partial_r)\phi|^2 + |\nabla\!\!\!/ \phi|^2 d\mu_C$$

measures the incoming energy through this light cone. Therefore, since we will choose δ to be very small, the conditions on the initial data is to keep the incoming energy as small as possible. Intuitively, we expect most of the energy will be emanated in the outgoing direction.

Now let us explain in which sense the data we described is large. The main equation (1.1) is invariant under the scaling transformation

$$\phi(t,x) \mapsto \lambda^{\frac{2}{p-1}} \phi(\lambda t, \lambda x), \quad \lambda > 0. \tag{1.8}$$

In other words, if $\phi(t,x)$ is a solution to (1.1), then $\phi_\lambda(t,x) := \lambda^{\frac{2}{p-1}} \phi(\lambda t, \lambda x)$ is also a solution to (1.1). For most $s \geq 0$, one can choose $\lambda > 0$ large or small such that the $\dot{H}^s$-norm of ϕ_λ becomes large or small (or the other way round). However, if $s = s_c := \frac{3}{2} - \frac{2}{p-1}$, the $\dot{H}^{s_c}$-norm of ϕ_λ does *not* depend on λ. In our case when $p > 5$, $s_c > 1$. By construction of our short pulse data, for any $s \geq 1$, the $\dot{H}^s$-norm of ϕ is large if $\delta > 0$ is sufficiently small. Therefore our data is large in the sense that it is independent of how large or how small the scaling parameter λ is, or equivalently, the scaling-invariant $\dot{H}^{s_c}$ is large.

We are now ready to state the main theorem of the paper:

MAIN THEOREM. *For any given pair of short pulse data* (ϕ_0, ϕ_1) *as described between* (1.3) *and Remark* 1.1, *there exists an absolute positive number* $\delta_0 < 1$, *such that for all* $\delta < \delta_0$, *the Cauchy problem*

$$\Box\phi = \pm|\phi|^{p-1}\phi, \quad p > 5,$$
$$(\phi, \partial_t\phi)\big|_{t=1} = (\phi_0, \phi_1)$$

admits a unique smooth solution ϕ *with lifespan* $[1, +\infty)$.

1.3. Notations. We review the basic geometry of Minkowski space $\mathbb{R}^{3+1}$. In particular, we discuss the standard double null (cone) foliations on $\mathbb{R}^{3+1}$ which plays a crucial role for the energy estimates.

Let $r = \sqrt{x_1^2 + x_2^2 + x_3^2}$. We define two optical functions u and $\underline{u}$ as follows

$$u = \frac{1}{2}(t - r), \quad \underline{u} = \frac{1}{2}(t + r).$$

For a given constant c, we use C_c to denote the level surface $u = c$ with an extra constraint that $t \geq 1$ (since we will construct a future-global-in-time solution starting from the initial hypersurface $\{t = 1\}$). These are called outgoing light cones. Thus, $\{C_u \big| u \in \mathbb{R}\}$ defines a foliation of $\mathbb{R}^{3+1}_{t\geq 1} \setminus \{r = 0\}$. We refer this as a null foliation because each leaf C_u is a null hypersurface with respect to the Minkowski metric.

Similarly, using the level sets of the optical function $\underline{u}$, we define another null foliation of $\mathbb{R}^{3+1}_{t\geq 1} \setminus \{r = 0\}$, denoted by $\{\underline{C}_{\underline{u}} \big| \underline{u} \in \mathbb{R}\}$. Each $\underline{C}_{\underline{u}}$ is a truncated incoming light cone. If $\underline{u} - u > 0$, the intersection $C_u \cap \underline{C}_{\underline{u}}$ is a round 2-sphere with radius $\underline{u} - u$, denoted by $S_{\underline{u},u}$. We say that the two foliations $\{\underline{C}_{\underline{u}} \big| \underline{u} \in \mathbb{R}\}$ and $\{C_u \big| u \in \mathbb{R}\}$ form a *double null foliation* of $\mathbb{R}^{3+1}_{t\geq 1} \setminus \{r = 0\}$.

We recall that the following two null vector fields

$$L = \partial_t + \partial_r, \quad \text{and} \quad \underline{L} = \partial_t - \partial_r$$

defined on $\mathbb{R}^{3+1}_{t\geq 1} \setminus \{r = 0\}$ are the normals of (also tangent to) C_u and $\underline{C}_{\underline{u}}$ respectively. In the following, null pair always refers to the pair of two null vector fields $(L, \underline{L})$.

The following picture depicts the outgoing null foliation C_u of $\mathbb{R}^{3+1}_{t\geq 1}$:

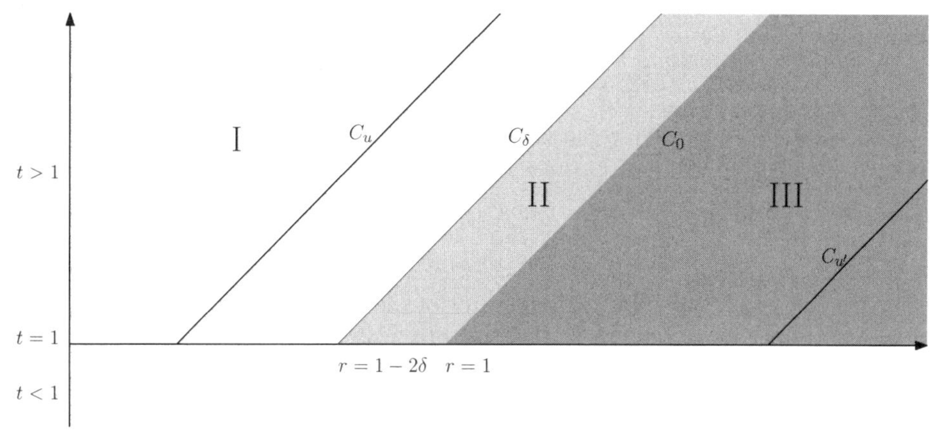

Since the foliation is spherically symmetric, we only draw the t and r components in the schematic diagram. The other pictures in the paper should also be understood in this way. In the above picture, a 45° line denotes an outgoing cone C_u. Two outgoing cones C_0 and C_δ divide $\mathbb{R}^{3+1}$ into three regions: the small data region, i.e. region I in the picture, the short pulse region, i.e. the region with light grey color, region II in the picture, and the region III in the picture, i.e. the region with dark grey color.

REMARK 1.3 (Vanishing Property on C_0). Recall that the short pulse data prescribed on $\{t = 1\}$ in the last subsection are identically zero on $\{r \geq 1\}$.

Therefore, according to the weak Huygens' principle, the solution of the main equation (1.3) vanishes identically in the region III (dark grey). In particular, the solution ϕ (if it exists) and its derivatives vanish on C_0.

We now pay more attention to the short pulse region (region II with light grey color). We use $D_{\underline{u},u}$ to denote the interior of the spacetime region enclosed by the hypersurfaces $\{t=1\}$, C_0, C_u and $\underline{C}_{\underline{u}}$, where $u \in [0, \delta]$ and $\underline{u} \geq 1-u$.

The following picture shows how $D_{\underline{u},u}$ looks like when $\underline{u} < 1$. A dashed 45° segment denotes an incoming cone $\underline{C}_{\underline{u}}$. A thickened black point denotes a 2-sphere $S_{\underline{u},u}$. An orthogonal pair of arrow denotes the null vector pair $(L, \underline{L})$.

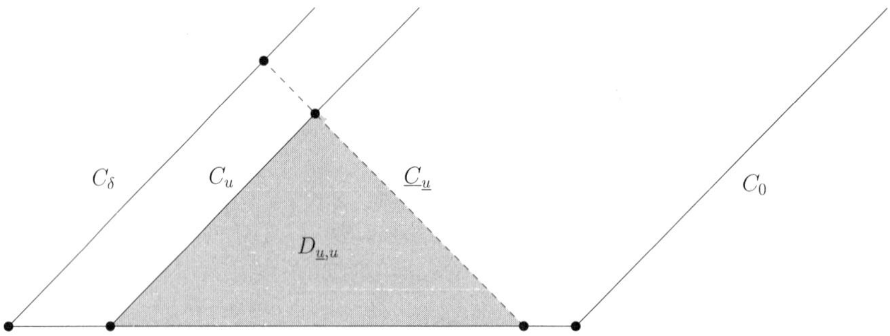

We remark that both $\{C_{u'} | 0 \leq u' \leq u\}$ and $\{\underline{C}_{\underline{u}'} | 1-u \leq \underline{u}' \leq \underline{u}\}$ foliate $D_{\underline{u},u}$.

In view of Remark 1.2, we also remark that the choice of the short pulse data is also adapted to the double null foliation in the short pulse region: the data are chosen in a way that very little energy propagates in the incoming direction through C_u's. We expect most of the energy will radiate through the $\underline{C}_{\underline{u}}$'s to the future null infinity.

For a given 2-sphere $S_{\underline{u},u}$, we use $\not{g}$ to denote the induced metric from the Minkowski metric on $S_{\underline{u},u}$. The intrinsic covariant derivative on $S_{\underline{u},u}$ is denoted by $\not\nabla$. This covariant derivative is closely related to the rotational symmetry of $\mathbb{R}^{3+1}$. Recall that the infinitesimal rotations are represented by the following three vector fields:

$$\Omega_{ij} = x_i \partial_j - x_j \partial_i, \quad \text{for } 1 \leq i < j \leq 3.$$

We use Ω as a short hand notation for an arbitrary choice from the above vector fields. We also use Ω^2 to denote an operator of the form $\Omega_{i'j'}\Omega_{ij}$; similarly for Ω^n. For a given function ϕ, we use $|\Omega\phi|$ to denote $\sum |\Omega_{ij}\phi|$ and use $|\Omega^2\phi|$ to denote $\sum |\Omega_{ij}\Omega_{i'j'}\phi|$, and so on. Therefore, by a direct computation, we obtain

$$|\Omega\phi| \sim r|\not\nabla\phi|,$$

where the size $\not\nabla\phi$ is measured with respect to $\not{g}$. Moreover, for all n, we have

$$|\Omega^n\phi| \sim_n r^n |\not\nabla^n \phi|.$$

In the rest of the paper, the number of derivatives that we impose on the solution is a fixed number which does not exceed, say, 30. Therefore the dependence on n in the above inequality is universal.

REMARK 1.4. In the short pulse region II, if δ is sufficiently small, then $|\underline{u}| \sim r$. Therefore, for all n, we have

$$|\Omega^n\phi| \sim |\underline{u}|^n |\not\nabla^n \phi|.$$

In particular, for all $p \geq 1$ and n, we have

$$\|\Omega^n\phi\|_{L^p(S_{\underline{u},u})} \sim |\underline{u}|^n \|\nabla\!\!\!/^n\phi\|_{L^p(S_{\underline{u},u})}.$$

1.4. Comments on the proof. We construct a solution ϕ in three steps:

- Step 1 We construct ϕ in the short pulse region II.
  The initial data for this region is given on initial hypersurface Σ_1. We expect to see the largeness of the data in the proof. In particular, the $\underline{L}$ derivative of the solution causes a loss of δ^{-1}. This makes the proof challenging and also different from the classical small data problem. The decay of derivatives of ϕ is another challenge which will be explained in detail in Section 1.4.2.
- Step 2 Smallness of ϕ on C_δ.
  Although ϕ constructed in Step 1 has large derivatives, we show that the derivatives of ϕ are indeed small on the inner boundary C_δ. This is a key intermediate step: since in the next step, ϕ restricted on C_δ gives initial characteristic data, this step allows one to reduce the problem to a small data problem in region I.
- Step 3 We construct ϕ in the small data region I.
  In region I, the problem is reduced to a small data problem. We can then use the classical approach to construct ϕ.

1.4.1. *Vector field method.* We will derive energy estimates for the main equation (1.3). Our approach is based on the classical vector field method and we briefly recall the main structure of the method as follows.

Let ϕ be a (scalar) solution for a non-homogenous wave equation $\Box\phi = \Phi$ on $\mathbb{R}^{3+1}$. The energy-momentum tensor associated to ϕ is $\mathbb{T}_{\alpha\beta}[\phi] = \nabla_\alpha\phi\nabla_\beta\phi - \frac{1}{2}g_{\alpha\beta}\nabla^\mu\phi\nabla_\mu\phi$ where $g_{\alpha\beta}$ is the Minkowski metric. In particular, in terms of the null pair $(L, \underline{L})$, we have $\mathbb{T}[\phi](L, L) = (L\phi)^2$, $\mathbb{T}[\phi](\underline{L}, \underline{L}) = (\underline{L}\phi)^2$ and $\mathbb{T}[\phi](L, \underline{L}) = |\nabla\!\!\!/\phi|^2$. Given a vector field X, we use ${}^{(X)}\pi_{\mu\nu} = \frac{1}{2}\mathcal{L}_X g_{\mu\nu}$ to denote its deformation tensor. The energy currents associated to ϕ are defined by $J^X_\alpha[\phi] = \mathbb{T}_{\alpha\mu}[\phi]X^\mu$ and $K^X[\phi] = \mathbb{T}^{\mu\nu}[\phi]\,{}^{(X)}\pi_{\mu\nu}$. The following divergence identity is the key to the energy estimates:

$$\nabla^\alpha J^X_\alpha[\phi] = K^X[\phi] + \Phi \cdot X\phi. \tag{1.9}$$

In applications, we integrate this identity on the spacetime region. This is equivalent to multiplying $\Box\phi = \Phi$ by $X\phi$ and then integrating by parts. This is the reason that we call X a *multiplier* vector field.

In the short pulse region II, we integrate (1.9) on $D_{\underline{u},u}$. Since ϕ and its derivatives vanish on C_0, this yields

$$\begin{aligned}&\int_{C_u} \mathbb{T}[\phi](X, L)d\underline{u}d\sigma + \int_{\underline{C}_{\underline{u}}} \mathbb{T}[\phi](X, \underline{L})dud\sigma \\ =&\frac{1}{2}\int_{\Sigma_1} \mathbb{T}[\phi](X, L+\underline{L})drd\sigma - \iint_{D_{\underline{u},u}} \left(K^X[\phi] - \Phi \cdot X\phi\right) dud\underline{u}d\sigma,\end{aligned} \tag{1.10}$$

where Σ_1 is the initial Cauchy hypersurface $\{t = 1\}$ and $d\sigma$ is the standard measure on $S_{\underline{u},u}$, a round sphere with radius $\underline{u} - u$.

In the short pulse region II, we will use two multiplier vector fields: $X = \underline{L}$ and $X = \underline{u}^\alpha L$, where the power $\alpha = 1 - \varepsilon_0$ and $\varepsilon_0 \in (0, \frac{1}{2})$ is a given constant.

The first plays a similar role to the time vector field ∂_t; the second plays a similar role to the vector field $S = t\partial_t + r\partial_r$ as a multiplier vectorfield.

For $X = \underline{L}$ and $X = \underline{u}^\alpha L$, the corresponding deformation tensors and energy currents are

- For $X = \underline{L}$, $\pi_{AB} = -\frac{1}{r}\slashed{g}_{AB}$ and $K = -\frac{1}{r}L\phi \cdot \underline{L}\phi$.
- For $X = \underline{u}^\alpha L$, $\pi_{L\underline{L}} = -\alpha\underline{u}^{\alpha-1}$, $\pi_{AB} = \frac{1}{r}\underline{u}^\alpha \slashed{g}_{AB}$ and $K = -\frac{\alpha}{2}\underline{u}^{\alpha-1}|\slashed{\nabla}\phi|^2 + \frac{1}{r}\underline{u}^\alpha L\phi \cdot \underline{L}\phi$.

respectively. We remark that indices A and B are used to denote a frame on $S_{\underline{u},u}$ and we only listed the nonzero components of the deformation tensors.

We will also need estimates for higher order derivatives for ϕ. To achieve this, we will commute the main equation (1.3) with certain vector fields, i.e. the *commutator vector fields.* These vector fields are essentially the Lie algebra of the conformal isometries of $\mathbb{R}^{3+1}$. We list all of them as follows:

$$\mathcal{Z} = \{\Omega_{ij}, \Omega_{0i}, \partial_t, \partial_i, S \big| i, j = 1, 2, 3, i \neq j\},$$

where $\Omega_{0i} = x_i\partial_t + t\partial_i$ and $S = t\partial_t + r\partial_r = \underline{u}L + u\underline{L}$. We also define the *good* and *bad* commutator vector fields:

$$\mathcal{Z} = \mathcal{Z}_g \sqcup \mathcal{Z}_b, \quad \mathcal{Z}_b = \{\partial_t, \partial_i \big| i = 1, 2, 3\}.$$

As shorthand notations, we use Z to denote an arbitrary vector field from $\mathcal{Z}$; similarly, we use Z_g and Z_b to denote vectors from $\mathcal{Z}_g$ and $\mathcal{Z}_b$ respectively. Geometrically, a good vector field Z_g is tangential to the outgoing light cone C_0, but a bad vector field Z_b is transversal to C_0.

1.4.2. *Main features of the proof.* We will discuss the main challenges of the problem and also the ideas to get around them.

- Relaxation in δ for the propagation estimates.

  In this discussion we focus on the behavior with respect to δ, and omit the behavior with respect to time and the higher order derivative for the moment. On the initial hypersurface Σ_1, it is easy to see that the data satisfy $\|\underline{L}\phi\|_{L^\infty(\Sigma_1)} \sim \delta^{-\frac{1}{2}}$ and $\|\slashed{\nabla}\phi\|_{L^\infty(\Sigma_1)} \sim \delta^{\frac{1}{2}}$. Up to a correct decay factor in t, we hope the size of $\underline{L}\phi$ and $\slashed{\nabla}\phi$ measured in δ can be propagated for later times, i.e. $\|\underline{L}\phi\|_{L^\infty(\Sigma_1)} \sim \delta^{-\frac{1}{2}}$ and $\|\slashed{\nabla}\phi\|_{L^\infty(\Sigma_1)} \sim \delta^{\frac{1}{2}}$ should be always true. Recall that the proof will be based on energy estimates. If we use $\underline{L}$ as a multiplier vector field and integrate in $D_{\underline{u},u}$ in the short pulse region, the energy on the left hand side of (1.10) is $\int_{\underline{C}_{\underline{u}}} |\underline{L}\phi|^2 + \int_{C_u} |\slashed{\nabla}\phi|^2$. Therefore, the expected propagation estimates suggest that $\int_{\underline{C}_{\underline{u}}} |\underline{L}\phi|^2 \lesssim 1$ and $\int_{C_u} |\slashed{\nabla}\phi|^2 \lesssim \delta$. So in view of the form of the energy, the disparity of the δ power for these two quantities only gives the desired bound for $\underline{L}\phi$, but not for $\slashed{\nabla}\phi$. This may lead to the failure of closing the bootstrap argument.

  To resolve this issue, we relax the bootstrap assumption on $\slashed{\nabla}\phi$ by assuming

$$\int_{C_u} |\slashed{\nabla}\phi|^2 \lesssim 1, \quad \int_{\underline{C}_{\underline{u}}} |\slashed{\nabla}\phi|^2 \lesssim \delta. \tag{1.11}$$

  In the bootstrap argument, we first recover the above weaker estimates (1.11), and then use the Sobolev inequalities (2.4)-(2.6) to recover the

L^∞-bound

$$\|\nabla\!\!\!/\phi\|_{L^\infty} \lesssim \delta^{\frac{1}{4}}. \tag{1.12}$$

Note that the bounds (1.11) and (1.12) are more relaxed than the expected profile $\nabla\!\!\!/\phi \sim \delta^{\frac{1}{2}}$.

- Relaxation in the decay factor in the short pulse region.

  According to the decay rate of linear waves, one may expect the decay $L\phi$ to be $\frac{1}{t^2}$ or $\frac{1}{\underline{u}^2}$ in the short pulse region. To gain the decay rate $\frac{1}{\underline{u}^2}$, it is natural to consider the multiplier $\underline{u}L$. However, this will cause a logarithmic divergence when we estimate the spacetime error integral of $\frac{\underline{u}}{r}\underline{L}Z^k\phi \cdot LZ^k\phi$. (See (2.36) when $\alpha = 1$.) The idea to resolve this is to relax the decay rate of $L\phi$ a little bit, i.e. instead of $\frac{1}{\underline{u}^2}$, we only prove that $L\phi$ decays like $\frac{1}{\underline{u}^{1+\alpha}}$ with $0 < \alpha < 1$. Therefore the estimate in (2.36) can be closed. This issue does not appear in the small data problem where we use the conformal Killing vectorfields ∂_t and $(t^2+r^2)\partial_t + 2tr\partial_r$ as multipliers, because the spacetime integrand $\frac{\underline{u}}{r}\underline{L}Z^k\phi \cdot LZ^k\phi$ do not appear due to the conformal invariance of ∂_t and $(t^2+r^2)\partial_t + 2tr\partial_r$.
- Smallness of the solution on C_δ.

  This is precisely the question that we will answer in Step 2 of the proof. As we discussed, in the short pulse region, we expect $\underline{L} \sim \delta^{-1}$. In particular, we expect that, for all the bad vector fields Z_b, we also have $Z_b \sim \delta^{-1}$. Therefore, for a given n, the restriction of ${Z_b}^n\phi$ on C_δ may be of size $\delta^{\frac{1}{2}-n}$. This is by no means small.

  The key point of the proof is the following observation: On the 2-sphere $S_{1-\delta,\delta}$, i.e. the initial sphere of C_δ, the data vanish completely since they are compactly supported on Σ_1 between $S_{1-\delta,\delta}$ and $S_{1,0}$. In particular, even the "bad" derivatives, which are supposed to be large, vanish initially at $S_{1-\delta,\delta}$. To get the smallness of ϕ, we will integrate along null geodesics on C_δ to trace all the information back to the data. In this way, we can show that up to an error of size $\delta^{\frac{1}{2}}$, all derivatives of ϕ are comparable to their initial values all over Σ_1.
- Estimates on nonlinearity.

  In the short pulse region we will prove that $\underline{L}\phi$ satisfies the pointwise estimates $\underline{L}\phi \sim \delta^{-\frac{1}{2}}$. Integrating $\underline{L}\phi$ along the incoming cone $\underline{C}_{\underline{u}}$, we obtain the pointwise estimate $\phi \sim \delta^{\frac{1}{2}}$ for ϕ in the short pulse region. In the small data region, we use the fact that the supercritical nonlinearity has high enough power that we have enough decay to close the a priori estimates.

1.4.3. *Outline of the paper.* The rest of the paper is organized as follows:

In Section 2, we establish *a priori* energy estimates for higher order derivatives of the solution in the short pulse region. As consequences, first of all, we can construct the solution in the short pulse region; Secondly, we can obtain a smallness estimate for the solution on C_δ, i.e. the inner boundary of the short pulse data region.

In Section 3, with a modified Klainerman-Sobolev inequality, we construct global solutions in the small data region.

2. Short pulse region

The goal of the current section is to construct the solution ϕ in the short pulse region. The construction relies on a priori energy estimates. We assume that the solution ϕ exists on spacetime domain $D_{\underline{u}^*,u^*}$. This domain is inside the short pulse region, i.e. $u^* \in (0,\delta)$ and $\underline{u}^* \in (1-u^*,+\infty)$.

We first introduce the energy norms. Let $u, u' \in (0,u^*)$ and $\underline{u}, \underline{u}' \in (1-u^*, \underline{u}^*)$. Let $C_u^{\underline{u}'}$ be the part of the cone C_u so that $1-u^* \leq \underline{u} \leq \underline{u}'$ and let $\underline{C}_{\underline{u}}^{u'}$ be the part of the cone $\underline{C}_{\underline{u}}$ so that $0 \leq u \leq u'$. Whenever there is no confusion, we will use C_u and $\underline{C}_{\underline{u}}$ instead of $C_u^{\underline{u}'}$ and $\underline{C}_{\underline{u}}^{u'}$. We use Σ_1 to denote the annulus region $\{(r,\theta) \mid 1-2\delta \leq r \leq 1\}$ on $\{t=1\}$ in the current section.

For a given $k \in \mathbb{Z}_{\geq 0}$, we introduce the following homogeneous norms:

$$E_k(u,\underline{u}) = \sum_{\substack{Z_g \in \mathcal{Z}_g, Z_b \in \mathcal{Z}_b \\ 0 \leq l \leq k}} \Big(\delta^{l-\frac{1}{2}} \|\nabla\!\!\!/ Z_b^l Z_g^{k-l}\phi\|_{L^2(\underline{C}_{\underline{u}}^u)} + \delta^{l-\frac{1}{2}} \| |\underline{u}|^{\frac{\alpha}{2}} L Z_b^l Z_g^{k-l}\phi\|_{L^2(C_u^{\underline{u}})} \Big),$$

$$\underline{E}_k(u,\underline{u}) = \sum_{\substack{Z_g \in \mathcal{Z}_g, Z_b \in \mathcal{Z}_b \\ 0 \leq l \leq k}} \Big(\delta^{l} \|\underline{L} Z_b^l Z_g^{k-l}\phi\|_{L^2(\underline{C}_{\underline{u}}^u)} + \delta^{l} \| |\underline{u}|^{\frac{\alpha}{2}} \nabla\!\!\!/ Z_b^l Z_g^{k-l}\phi\|_{L^2(C_u^{\underline{u}})} \Big).$$

We also introduce the inhomogeneous norms:

$$E_{\leq k}(u,\underline{u}) = \sum_{0 \leq j \leq k} E_j(u,\underline{u}), \quad \underline{E}_{\leq k}(u,\underline{u}) = \sum_{0 \leq j \leq k} \underline{E}_j(u,\underline{u}).$$

On the initial hypersurface Σ_1, we introduce the following initial energy norms:

$$E_{\leq n}(\Sigma_1) = \sum_{\substack{Z_g \in \mathcal{Z}_g, Z_b \in \mathcal{Z}_b \\ 0 \leq l \leq k, 0 \leq k \leq n}} \delta^l \|\underline{L} Z_b^l Z_g^{k-l}\phi\|_{L^2(\Sigma_1)} + \delta^{l-\frac{1}{2}} \|\nabla\!\!\!/ Z_b^l Z_g^{k-l}\phi\|_{L^2(\Sigma_1)} + \delta^{l-1} \|L Z_b^l Z_g^{k-l}\phi\|_{L^2(\Sigma_1)}.$$

According to the behavior of ϕ_0 and ϕ_1 on Σ_1 as well as the properties of commutation vectorfields, we have the following lemma:

LEMMA 2.1. *For all δ, we have*

$$E_{\leq n-1}(\Sigma_1) \lesssim I_n. \tag{2.1}$$

Here I_n is a constant depending only on n.

2.1. Main a priori estimates. This subsection is the central part of the paper. The goal is to bound $E_{\leq 3}(u,\underline{u})$ and $\underline{E}_{\leq 3}(u,\underline{u})$ on $D_{\underline{u}^*,u^*}$ where the solution ϕ is assumed to exist.

PROPOSITION 2.2. *There exists $\delta_0 > 0$, so that for all $\delta < \delta_0$, for all $u \in (0,u^*)$ and $\underline{u} \in (1-u^*, \underline{u}^*)$, we have*

$$E_{\leq 3}(u,\underline{u}) + \underline{E}_{\leq 3}(u,\underline{u}) \leq C(I_4), \tag{2.2}$$

where $C(I_4)$ is a constant depending only on I_4.

The proof of the proposition is based on a standard bootstrap argument. On $D_{\underline{u}^*,u^*}$, since we assume that ϕ exists, there is a large constant M, so that

$$E_{\leq 3}(u,\underline{u}) + \underline{E}_{\leq 3}(u,\underline{u}) \leq M, \tag{2.3}$$

for all $u \in (0,u^*)$ and $\underline{u} \in (1-u^*,\underline{u}^*)$. The large constant M may depend on ϕ itself. The purpose of the bootstrap argument is to show that, if δ is sufficiently small, then one can choose M in such a way that it depends only on I_4. Hence, we obtain the proof of (2.2)

2.1.1. *Preliminary estimates.* The goal of this subsection is to use the bootstrap assumption (2.3) to get estimates on lower order derivatives of ϕ (up to two derivatives).

We first recall the Sobolev inequalities on $S_{\underline{u},u}$, $\underline{C}_{\underline{u}}$ and C_u in the short pulse region. Recall that in the short pulse region, we have $|\underline{u}| \sim r$ provided δ is sufficiently small.

LEMMA 2.3. *Let ϕ be a C^6 function. On $S_{\underline{u},u}$, we have*

$$\|\phi\|_{L^\infty(S_{\underline{u},u})} \lesssim |\underline{u}|^{-1/2}\big(\|\phi\|_{L^4(S_{\underline{u},u})} + \|\Omega\phi\|_{L^4(S_{\underline{u},u})}\big), \tag{2.4}$$

$$\|\phi\|_{L^4(S_{\underline{u},u})} \lesssim |\underline{u}|^{-1/2}\big(\|\phi\|_{L^2(S_{\underline{u},u})} + \|\Omega\phi\|_{L^2(S_{\underline{u},u})}\big). \tag{2.5}$$

On $\underline{C}_{\underline{u}}$, if in addition we assume that $\phi \equiv 0$ on C_0, we have

$$\begin{aligned}
&\|\phi\|_{L^2(S_{\underline{u},u})} \lesssim \|\underline{L}\phi\|^{1/2}_{L^2(\underline{C}_{\underline{u}})}\|\phi\|^{1/2}_{L^2(\underline{C}_{\underline{u}})},\\
&\|\phi\|_{L^4(S_{\underline{u},u})} \lesssim |\underline{u}|^{-\frac{1}{2}}\|\underline{L}\phi\|^{1/2}_{L^2(\underline{C}_{\underline{u}})}\left(\|\phi\|^{1/2}_{L^2(\underline{C}_{\underline{u}})} + \|\Omega\phi\|^{1/2}_{L^2(\underline{C}_{\underline{u}})}\right),\\
&\|\phi\|_{L^\infty(S_{\underline{u},u})} \lesssim |\underline{u}|^{-1}\left(\|\underline{L}\phi\|_{L^2(\underline{C}_{\underline{u}})} + \|\underline{L}\Omega\phi\|_{L^2(\underline{C}_{\underline{u}})}\right)^{1/2}\\
&\qquad\cdot\left(\|\phi\|_{L^2(\underline{C}_{\underline{u}})} + \|\Omega\phi\|_{L^2(\underline{C}_{\underline{u}})} + \|\Omega^2\phi\|_{L^2(\underline{C}_{\underline{u}})}\right)^{1/2}.
\end{aligned} \tag{2.6}$$

Integrating the squared first inequality in (2.6) *along $\underline{C}_{\underline{u}}$, we have*

$$\|\phi\|_{L^2(\underline{C}_{\underline{u}})} \lesssim \delta\|\underline{L}\phi\|_{L^2(\underline{C}_{\underline{u}})}. \tag{2.7}$$

PROOF. Except the first inequality in (2.6), the proof of the above inequalities can be found in Chapter 5 and Chapter 10 of [**5**]. In particular, for (2.4), one refers to Lemma 5.2 in [**5**] on Page 156, for (2.5), one refers to Lemma 5.1 in [**5**] on Page 153, and for the second inequality in (2.6), one refers to the second inequality in Proposition 10.1 in [**5**] on Page 328. The third inequality in (2.6) can be derived from the first and the second in (2.6). Now we give a detailed proof for the first inequality in (2.6). Let $d\sigma$ be the standard measure on $S_{\underline{u},u}$, $d\sigma_{\mathbb{S}^2}$ be the standard measure on $\mathbb{S}^2$, and θ be the standard coordinate on $\mathbb{S}^2$. We have

$$\begin{aligned}
\|\phi\|^2_{L^2(S_{\underline{u},u})} &= \int_{S_{\underline{u},u}} |\phi|^2 d\sigma = \int_{\mathbb{S}^2} |\phi(\underline{u},u,\theta)|^2(\underline{u}-u)^2 d\sigma_{\mathbb{S}^2} \\
&= \int_0^u \underline{L}\left(\int_{\mathbb{S}^2} |\phi(\underline{u},u',\theta)|^2(\underline{u}-u')^2 d\sigma_{\mathbb{S}^2}\right) du' \\
&=2\int_0^u \int_{\mathbb{S}^2} \phi(\underline{u},u',\theta)(\underline{L}\phi)(\underline{u},u',\theta)(\underline{u}-u')^2 d\sigma_{\mathbb{S}^2} du' \\
&\quad + 2\int_0^u \int_{\mathbb{S}^2} |\phi(\underline{u},u',\theta)|^2(u'-\underline{u}) d\sigma_{\mathbb{S}^2} du' \\
&\leq 2\int_{\underline{C}_{\underline{u}}} \phi\cdot\underline{L}\phi d\sigma du \leq 2\|\phi\|_{L^2(\underline{C}_{\underline{u}})}\|\underline{L}\phi\|_{L^2(\underline{C}_{\underline{u}})}.
\end{aligned}$$

□

We also recall the Gronwall's inequality. Let $f(t)$ be a non-negative function defined on an interval I with initial point t_0. If f satisfies

$$\frac{d}{dt}f \leq a\cdot f + b$$

with two non-negative functions $a, b \in L^1(I)$, then for all $t \in I$, we have

$$f(t) \leq e^{A(t)}\big(f(t_0) + \int_{t_0}^t e^{-A(\tau)}b(\tau)d\tau\big)$$

where $A(t) = \int_{t_0}^t a(\tau)d\tau$. In particular, for the proof of the previous lemma, we used Gronwall's inequality in the last step.

We start to derive estimates and we treat $\underline{u}$ as a fixed constant. By virtue of null pair $(L, \underline{L})$, we rewrite the main system of equations (1.3) as

$$-L\underline{L}\phi + \not{\Delta}\phi + \frac{1}{r}(L\phi - \underline{L}\phi) = \pm|\phi|^{p-1}\phi. \tag{2.8}$$

We also need to commute derivatives with (2.8). Recall that, for all $Z \in \mathcal{Z}$ except for $Z = S$, we have $[\Box, Z] = 0$. Indeed, we have $[\Box, S] = 2\Box$. For all $k \geq 0$, we can commute k vectors $Z_1, Z_2, \cdots, Z_k \in \mathcal{Z}$ with (1.3) to obtain a semilinear wave equation for $Z_1Z_2\cdots Z_k\phi$. We use the shorthand notation $Z^k\phi$ to denote $Z_1Z_2\cdots Z_k\phi$, therefore, systematically we have

$$\Box Z^k\phi = \sum_{k_1+\ldots+k_p\leq k} Z^{k_1}\phi\ldots Z^{k_p}\phi. \tag{2.9}$$

We combine (2.6) and bootstrap assumption (2.3). We first have

$$\begin{aligned}
\|\not\nabla\phi\|_{L^4(S_{\underline{u},u})} &\lesssim |\underline{u}|^{-\frac12}\|\underline{L}\not\nabla\phi\|^{\frac12}_{L^2(\underline{C}_{\underline{u}})}\big(\|\not\nabla\phi\|^{\frac12}_{L^2(\underline{C}_{\underline{u}})} + |\underline{u}|^{\frac12}\|\not\nabla^2\phi\|^{\frac12}_{L^2(\underline{C}_{\underline{u}})}\big) \\
&\lesssim |\underline{u}|^{-\frac12}(|\underline{u}|^{-1}M)^{\frac12}\big((\delta^{\frac12}|\underline{u}|^{-\frac{\alpha}{2}}M)^{\frac12} + |\underline{u}|^{\frac12}(\delta^{\frac12}|\underline{u}|^{-\frac{2+\alpha}{2}}M)^{\frac12}\big).
\end{aligned}$$

Hence,

$$\|\not\nabla\phi\|_{L^4(S_{\underline{u},u})} \lesssim \delta^{\frac14}|\underline{u}|^{-1-\frac{\alpha}{4}}M. \tag{2.10}$$

Similarly, since in the bootstrap assumption (2.3), we have assumed bounds on four derivatives on ϕ, we can repeat the above argument to derive

$$\|\not\nabla\Omega\phi\|_{L^4(S_{\underline{u},u})} \lesssim \delta^{\frac14}|\underline{u}|^{-1-\frac{\alpha}{4}}M, \tag{2.11}$$

Combining (2.10) and (2.11), Sobolev inequality implies

$$\begin{aligned}\|\nabla\!\!\!/\phi\|_{L^\infty(S_{\underline{u},u})} &\lesssim |\underline{u}|^{-\frac{1}{2}}\big(\|\nabla\!\!\!/\phi\|_{L^4(S_{\underline{u},u})} + \|\nabla\!\!\!/\Omega\phi\|_{L^4(S_{\underline{u},u})}\big)\\ &\lesssim \delta^{\frac{1}{4}}|\underline{u}|^{-\frac{3}{2}-\frac{\alpha}{4}}M.\end{aligned} \tag{2.12}$$

By repeating the above argument, for $0 \leq l \leq k \leq 2$, we can also easily obtain

$$\|\nabla\!\!\!/ Z_b^l Z_g^{k-l}\phi\|_{L^4(S_{\underline{u},u})} \lesssim \delta^{\frac{1}{4}-l}|\underline{u}|^{-1-\frac{\alpha}{4}}M.$$

and for $0 \leq l \leq k \leq 1$

$$\|\nabla\!\!\!/ Z_b^l Z_g^{k-l}\phi\|_{L^\infty(S_{\underline{u},u})} \lesssim \delta^{\frac{1}{4}-l}|\underline{u}|^{-\frac{3}{2}-\frac{\alpha}{4}}M. \tag{2.13}$$

We turn to the bound of $L\phi$ in $L^\infty(S_{\underline{u},u})$. We have the following:

LEMMA 2.4. *Under the bootstrap assumption* (2.3), *we have*

$$\|L\phi\|_{L^\infty(S_{\underline{u},u})} \lesssim \delta^{\frac{1}{2}}|\underline{u}|^{-2}M. \tag{2.14}$$

PROOF. Let $a = \frac{1}{r}$ and $b = |\Delta\!\!\!/\phi| + \frac{1}{r}|\underline{L}\phi| + |\phi|^p$. (2.8) yields

$$\underline{L}|L\phi| \lesssim a|L\phi| + b. \tag{2.15}$$

We would like to integrate this equation directly along $\underline{L}$ to derive the pointwise bound on $L\phi$. Since $L\phi$ vanishes along C_0, in view of Gronwall's inequality, it suffices to control $\|a\|_{L^1_u L^\infty(S_{\underline{u},u})}$ and $\|b\|_{L^1_u L^\infty(S_{\underline{u},u})}$. The estimate on $\|a\|_{L^1_u L^\infty(S_{\underline{u},u})}$ is straightforward. We only give the estimates on $|\underline{L}\phi|$ appearing in b. The others can be estimated directly from (2.13). According to Sobolev inequalities (2.4) and (2.5), we have

$$\begin{aligned}\|\underline{L}\phi\|_{L^1_u L^\infty(S_{\underline{u},u})} &\lesssim |\underline{u}|^{-1}\sum_{0\leq j\leq 2}\|\Omega^j\underline{L}\phi\|_{L^1_u L^2(S_{\underline{u},u})}\\ &\lesssim |\underline{u}|^{-1}\delta^{\frac{1}{2}}\sum_{0\leq j\leq 2}\|\Omega^j\underline{L}\phi\|_{L^2_u L^2(S_{\underline{u},u})}\\ &\lesssim |\underline{u}|^{-1}\delta^{\frac{1}{2}}M,\end{aligned} \tag{2.16}$$

which implies

$$\|\phi\|_{L^\infty(S_{\underline{u},u})} \lesssim |\underline{u}|^{-1}\delta^{\frac{1}{2}}M \tag{2.17}$$

The above estimates and the the definition of a imply

$$\begin{aligned}\|a\|_{L^1_u L^\infty(S_{\underline{u},u})} &\lesssim |\underline{u}|^{-1}\delta^{\frac{1}{2}},\\ \|b\|_{L^1_u L^\infty(S_{\underline{u},u})} &\lesssim |\underline{u}|^{-2}\delta^{\frac{1}{2}}M.\end{aligned} \tag{2.18}$$

Therefore, Gronwall's inequality implies the desired estimate for $L\phi$. □

We now turn to the $L^\infty(S_{\underline{u},u})$ estimates on $\underline{L}\phi$ and ϕ. We have the following:

LEMMA 2.5. *Under the bootstrap assumption* (2.3), *we have*

$$\|\underline{L}\phi\|_{L^\infty(S_{\underline{u},u})} \lesssim \frac{\delta^{-\frac{1}{2}}}{|\underline{u}|}I_4 + \delta^{\frac{1}{4}}|\underline{u}|^{-1}M. \tag{2.19}$$

This gives an improved $L^\infty(S_{\underline{u},u})$*-estimate for* ϕ*:*

$$\|\phi\|_{L^\infty(S_{\underline{u},u})} \lesssim \frac{\delta^{\frac{1}{2}}}{|\underline{u}|}I_4 + \delta^{\frac{5}{4}}|\underline{u}|^{-1}M. \tag{2.20}$$

PROOF. We start with a computation of $L\big(\underline{u}^2(\underline{L}\phi)^2\big)$:

$$\begin{aligned} L\big(\underline{u}^2(\underline{L}\phi)^2\big) &= 2\underline{u}(\underline{L}\phi)^2 + 2\underline{u}^2(\underline{L}\phi)(L\underline{L}\phi) \\ &= 2\underline{u}(\underline{L}\phi)^2 + 2\underline{u}^2(\underline{L}\phi)\big[\Delta\!\!\!/\phi + \frac{1}{r}(L\phi - \underline{L}\phi) \mp |\phi|^{p-1}\phi\big] \\ &\lesssim |\underline{u}|^2|\underline{L}\phi|^2\left(\frac{2}{\underline{u}} - \frac{2}{r}\right) + |\underline{u}|^2|\underline{L}\phi|\left(|\Delta\!\!\!/\phi| + \frac{1}{r}|L\phi| + |\phi|^p\right). \end{aligned}$$

We make the following important observation: in the short pulse region, $|\frac{2}{r} - \frac{2}{\underline{u}}| \lesssim \frac{\delta}{|\underline{u}|^2}$. Therefore, if we define $y = |\underline{u}||\underline{L}\phi|$, according to the estimate obtained so far, if δ is sufficiently small, the previous computation yields

$$Ly^2 \lesssim \frac{\delta}{|\underline{u}|^2}y^2 + \frac{\delta^{\frac{1}{4}}}{|\underline{u}|^2}My.$$

We divide both sides of the equation by y, thus, we have

$$L\big(|\underline{u}||\underline{L}\phi|\big) \lesssim \left(\frac{\delta}{|\underline{u}|^2} + \frac{\delta^{\frac{1}{4}}}{|\underline{u}|^2}M\right)\big(|\underline{u}||\underline{L}\phi|\big) + \frac{\delta^{\frac{1}{4}}}{|\underline{u}|^2}M.$$

By integrating directly this equation, if δ is sufficiently small, we obtain

$$\big||\underline{u}||\underline{L}\phi|(\underline{u}, u, \theta) - C|1-u||\underline{L}\phi|(1-u, u, \theta)\big| \lesssim \delta^{\frac{1}{4}}M. \tag{2.21}$$

where the absolute constant C comes from the use of Gronwall's inequality. Therefore we obtain the desired estimate (2.19) and therefore (2.20). □

We remark that the derivation of (2.14) and (2.19) depends on not only on the bootstrap assumption (2.3) but also the main equation (1.3). We summarize the estimates derived so far as follows:

$$\begin{aligned} \|\nabla\!\!\!/ Z_b^l Z_g^{k-l}\phi\|_{L^4(S_{\underline{u},u})} &\lesssim \delta^{\frac{1}{4}-l}|\underline{u}|^{-1-\frac{\alpha}{4}}M, \quad 0 \le l \le k \le 1, \\ \|\nabla\!\!\!/\phi\|_{L^\infty(S_{\underline{u},u})} &\lesssim \delta^{\frac{1}{4}}|\underline{u}|^{-\frac{3}{2}-\frac{\alpha}{4}}M, \\ \|L\phi\|_{L^\infty(S_{\underline{u},u})} &\lesssim \delta^{\frac{1}{2}}|\underline{u}|^{-2}M, \\ \|LZ_b^l Z_g^{k-l}\phi\|_{L^2(S_{\underline{u},u})} &\lesssim \delta^{\frac{1}{2}-l}|\underline{u}|^{-1}M, \quad 0 \le l \le k \le 1, \\ \|\underline{L}\phi\|_{L^\infty(S_{\underline{u},u})} &\lesssim \frac{\delta^{-\frac{1}{2}}}{|\underline{u}|}I_4 + \delta^{\frac{1}{4}}|\underline{u}|^{-1}M, \\ \|\phi\|_{L^\infty(S_{\underline{u},u})} &\lesssim \frac{\delta^{\frac{1}{2}}}{|\underline{u}|}I_4 + \delta^{\frac{5}{4}}|\underline{u}|^{-1}M. \end{aligned} \tag{2.22}$$

REMARK 2.6. The bootstrap assumptions (2.3) involve relaxed estimates for $\nabla\!\!\!/\phi$. Roughly speaking, in (2.3) we expect the behavior of $\nabla\!\!\!/\phi$ with respect to δ and $\underline{u}$ is approximately $|\underline{u}|^{-1-\frac{\alpha}{2}}M$, i.e.

$$\|\nabla\!\!\!/\phi\|_{L^\infty(S_{\underline{u},u})} \sim |\underline{u}|^{-1-\frac{\alpha}{2}} \cdot \delta^0 \cdot M.$$

However, the estimates on $\nabla\!\!\!/\phi$ in (2.22) shows that, by affording two more derivatives (via Sobolev inequalities), we can improve the bound on $\nabla\!\!\!/\phi$: we get an extra $\delta^{\frac{1}{4}}$ factor and an extra $\underline{u}^{-\frac{1}{2}+\frac{\alpha}{4}}$ decay factor, i.e.,

$$\|\nabla\!\!\!/\phi\|_{L^\infty(S_{\underline{u},u})} \sim |\underline{u}|^{-\frac{3}{2}-\frac{\alpha}{4}} \cdot \delta^{\frac{1}{4}} \cdot M.$$

2.1.2. *Estimates on $E_{\leq 3}$ and $\underline{E}_{\leq 3}$.* Recall that for $Z \in \mathcal{Z}$ and $k \leq 3$, we have

$$\Box Z^k \phi = \sum_{k_1+\ldots+k_p \leq k} Z^{k_1}\phi \ldots Z^{k_p}\phi. \tag{2.23}$$

Let $l \leq k$ be the number of Z_b's appearing in Z^k, i.e. $Z^k = Z_b^l Z_g^{k-l}$. We use the vector field method outlined in the introduction to estimate $E_{\leq 3}$ and $\underline{E}_{\leq 3}$.

By (1.10) and (2.23), we replace ϕ by $Z^k\phi$ and take $X = \underline{L}$ to obtain

(2.24)

$$\int_{C_u} |\nabla\!\!\!/ Z^k\phi|^2 + \int_{\underline{C}_{\underline{u}}} |\underline{L}Z^k\phi|^2 = \int_{\Sigma_1} |\nabla\!\!\!/ Z^k\phi|^2 + |\underline{L}Z^k\phi|^2 - \iint_{D_{\underline{u},u}} \pm|\phi|^{p-1} Z^k\phi \underline{L}Z^k\phi$$
$$- \sum_{\substack{k_1+\ldots+k_p \leq k,\\ k_i<k}} \iint_{D_{\underline{u},u}} Z^{k_1}\phi \ldots Z^{k_p}\phi \underline{L}Z^k\phi + \iint_{D_{\underline{u},u}} \frac{1}{r}\underline{L}Z^k\phi \cdot LZ^k\phi.$$

We multiply both sides of the equation by δ^{2l} to normalize the contribution from the initial data to be close to 1, therefore, we obtain

(2.25)

$$\delta^{2l}\int_{C_u} |\nabla\!\!\!/ Z^k\phi|^2 + \delta^{2l}\int_{\underline{C}_{\underline{u}}} |\underline{L}Z^k\phi|^2 \lesssim I_3^2 + \delta^{2l}\left|\iint_{D_{\underline{u},u}} \phi^{p-1} Z^k\phi \underline{L}Z^k\phi\right|$$
$$+ \sum_{\substack{k_1+\ldots+k_p \leq k,\\ k_i<k}} \delta^{2l}\left|\iint_{D_{\underline{u},u}} Z^{k_1}\phi \ldots Z^{k_p}\phi \underline{L}Z^k\phi\right| + \delta^{2l}\left|\iint_{D_{\underline{u},u}} \frac{1}{r}\underline{L}Z^k\phi \cdot LZ^k\phi\right|.$$

We rewrite the right-hand side of the above inequality as

$$I_4^2 + S + T + W.$$

where S, T and W denote the three bulk integral terms in (2.25). We will bound S, T and W one by one.

For S, in view of the $L^\infty(S_{\underline{u},u})$ estimates on ϕ and (2.7), we have

$$\begin{aligned}
S &\leq \delta^{2l}\int_{1-u}^{\underline{u}}\int_0^u \left(\|\phi\|_{L^\infty(S_{\underline{u},u})}^{p-1}\|Z^k\phi\|_{L^2(S_{\underline{u}',u'})}\|\underline{L}Z^k\phi\|_{L^2(S_{\underline{u}',u'})}\right) du'd\underline{u}' \\
&\lesssim \delta^{2l}\int_{1-u}^{\underline{u}} \delta^{\frac{p-1}{2}}|\underline{u}'|^{-p+1}\int_0^u \left(\delta^{-1}\|Z^k\phi\|_{L^2(S_{\underline{u}',u'})}^2 + \delta\|\underline{L}Z^k\phi\|_{L^2(S_{\underline{u}',u'})}^2\right) du'd\underline{u}' \\
&\lesssim \delta^{2l}\int_{1-u}^{\underline{u}} \delta^{\frac{p-1}{2}}|\underline{u}'|^{-p+1}\left(\delta^{-1}\|Z^k\phi\|_{L^2(\underline{C}_{\underline{u}'})}^2 + \delta\|\underline{L}Z^k\phi\|_{L^2(\underline{C}_{\underline{u}'})}^2\right) d\underline{u}' \\
&\lesssim \int_{1-u}^{\underline{u}} \delta^{\frac{p+1}{2}}|\underline{u}'|^{-p+1}\left(\delta^{2l}\|\underline{L}Z^k\phi\|_{L^2(\underline{C}_{\underline{u}'})}^2\right)d\underline{u}'.
\end{aligned}$$

In view of the bootstrap assumption, we bound $\delta^{2l}\|\underline{L}Z^k\phi\|_{L^2(\underline{C}_{\underline{u}'})}^2$ by M^2. After an integration over $\underline{u}'$ on $[1-u, \underline{u}]$, we have

$$S \lesssim \delta^{\frac{p+1}{2}} M^2. \tag{2.26}$$

We now estimate the second term in (2.25), i.e. the estimates on T. We have

$$T \lesssim \sum_{\substack{k_1+..+k_p \leq k,\\ k_i<k}} \delta^{2l}\iint_{D_{\underline{u},u}} |Z^{k_1}\phi| \ldots |Z^{k_p}\phi||\underline{L}Z^k\phi|.$$

Without loss of generality, let $k_1 \geq k_i$ for $2 \leq i \leq p$. We bound the contribution from $Z^{k_1}\phi$ in L^2 while the contributions from $Z^{k_i}\phi, i \geq 2$ in L^∞. The rest of the argument is the same as in estimating S. This eventually yields

$$T \lesssim \delta^{\frac{p+1}{2}} M^2. \tag{2.27}$$

It remains to bound the third term W in (2.25). According to Young's inequality, we have

$$\begin{aligned} W \lesssim & \delta^{2l}\Big(\iint_{\mathcal{D}_{\underline{u},u}} \frac{\delta}{|\underline{u}'|^{2-\alpha}} |\underline{L}Z^k\phi|^2 + \iint_{\mathcal{D}_{\underline{u},u}} \frac{1}{\delta} ||\underline{u}'|^{\frac{\alpha}{2}} LZ^k\phi|^2\Big) \\ = & \int_{1-u}^{\underline{u}} \frac{1}{|\underline{u}'|^{2-\alpha}} \cdot \delta^{2l+1} \|\underline{L}Z^k\phi\|^2_{L^2(\underline{C}_{\underline{u}'})} d\underline{u}' + \delta^{2l-1} \int_0^u \||\underline{u}|^{\frac{\alpha}{2}} LZ^k\phi\|^2_{L^2(C_{u'})} du'. \end{aligned}$$

This leads to

$$W \lesssim \delta M^2. \tag{2.28}$$

By combining (2.25) with (2.26), (2.27) and (2.28), for sufficiently small δ, we obtain

$$\delta^{2l} \int_{C_u} |\nabla\!\!\!/ Z^k\phi|^2 + \delta^{2l} \int_{\underline{C}_{\underline{u}}} |\underline{L}Z^k\phi|^2 \lesssim I_4^2 + \delta M^2.$$

In other words, for all $0 \leq l \leq k \leq 2$, we have

$$\delta^l \|\nabla\!\!\!/ Z_b^l Z_g^{k-l}\phi\|_{L^2(C_u)} + \delta^l \|\underline{L}Z_b^l Z_g^{k-l}\phi\|_{L^2(\underline{C}_{\underline{u}})} \lesssim I_4 + \delta^{\frac{1}{2}} M. \tag{2.29}$$

To complete the estimates for $\underline{E}_{\leq 3}$ and $E_{\leq 3}$, it remains to bound $\delta^{l-\frac{1}{2}} \||\underline{u}|^{\frac{\alpha}{2}} LZ_b^l Z_g^{k-l}\phi\|_{L^2(C_u)}$ and $\delta^{l-\frac{1}{2}} \|\nabla\!\!\!/ Z_b^l Z_g^{k-l}\phi\|_{L^2(\underline{C}_{\underline{u}})}$. The starting point is similar to that of (2.24), but we need to change the multipliers to outgoing direction. More precisely, by (1.10) and (2.23), we replace ϕ by $Z^k\phi$ and take $X = \underline{u}^\alpha L$ to obtain

$$\begin{aligned} & \int_{C_u} |\underline{u}|^\alpha |LZ^k\phi|^2 + \int_{\underline{C}_{\underline{u}}} |\underline{u}|^\alpha |\nabla\!\!\!/ Z^k\phi|^2 = \int_{\Sigma_1} |\underline{u}|^\alpha \big(|\nabla\!\!\!/ Z^k\phi|^2 + |LZ^k\phi|^2\big) \\ & - \iint_{\mathcal{D}_{\underline{u},u}} |\underline{u}|^\alpha \pm |\phi|^{p-1} Z^k\phi LZ^k\phi \\ & - \sum_{\substack{k_1+\ldots+k_p \leq k, \\ k_i < k}} \iint_{D_{\underline{u},u}} |\underline{u}|^\alpha Z^{k_1}\phi \ldots Z^{k_p}\phi LZ^k\phi - \iint_{D_{\underline{u},u}} \frac{|\underline{u}|^\alpha}{r} \underline{L}Z^k\phi \cdot LZ^k\phi \\ & + \frac{\alpha}{2} \iint_{D_{\underline{u},u}} |\underline{u}|^{\alpha-1} |\nabla\!\!\!/ Z^k\phi|^2. \end{aligned} \tag{2.30}$$

We multiply both sides of the equation by δ^{2l-1} to normalize the contribution from the initial data to be close to 1. We remark that this normalization is with respect to the relaxed estimates on $\nabla\!\!\!/\phi$. The last term above is bounded as follows:

$$
\begin{aligned}
\iint_{D_{\underline{u},u}} |\underline{u}|^{\alpha-1}|\nabla\!\!\!/ Z^k\phi|^2 \leq & \frac{1}{2}\left(\iint_{D_{\underline{u},u}} |\underline{u}|^{2\alpha-2}|\nabla\!\!\!/ Z^k\phi|^2 + \iint_{D_{\underline{u},u}} |\nabla\!\!\!/ Z^k\phi|^2\right) \\
\leq & \frac{1}{2}\int_{1-u}^{\underline{u}} \underline{u}'^{\alpha-2}\left(\int_{\underline{C}_{\underline{u}'}} |\underline{u}'|^{\alpha}|\nabla\!\!\!/ Z^k\phi|^2\right) d\underline{u}' \\
& + \frac{1}{2}\int_0^u \left(\int_{C_{u'}} |\nabla\!\!\!/ Z^k\phi|^2\right) du'.
\end{aligned} \tag{2.31}
$$

The first term on the right hand side above is bounded using Gronwall, and the second term is bounded using (2.29). So omitting the last term in (2.30), we have

$$
\begin{aligned}
\delta^{2l-1}\int_{C_u} |\underline{u}|^{\alpha}|LZ^k\phi|^2 + \delta^{2l-1}\int_{\underline{C}_{\underline{u}}} |\underline{u}|^{\alpha}|\nabla\!\!\!/ Z^k\phi|^2 \lesssim & I_4^2 \\
& + \delta^{2l-1}\iint_{\mathcal{D}_{\underline{u},u}} |\underline{u}|^{\alpha}|\phi|^{p-1}|Z^k\phi||LZ^k\phi| \\
& + \sum_{\substack{k_1+\ldots+k_p\leq k,\\ k_i<k}} \delta^{2l-1}\iint_{D_{\underline{u},u}} |\underline{u}|^{\alpha}|Z^{k_1}\phi|\ldots|Z^{k_p}\phi||LZ^k\phi| \\
& + \delta^{2l-1}\left|\iint_{D_{\underline{u},u}} \frac{|\underline{u}|^{\alpha}}{r}\underline{L}Z^k\phi\cdot LZ^k\phi\right|.
\end{aligned} \tag{2.32}
$$

Note that here we used the fact $\underline{u}$ is of order 1 on Σ_1. Therefore the initial weighted energies are comparable to the usual initial energy. Later we will also use this fact and we will not explain this anymore. We remark that the data on Σ_1 is bounded by I_4^2 because the weight $\underline{u}^{\alpha}$ is of order 1 initially. We rewrite the right-hand side of the above inequality as

$$I_4^2 + S + T + W,$$

where S, T and W denote the three bulk integral terms in (2.32). We now bound S, T and W one by one. Again, T will be bounded in the same way as S, by singling out the top order factor, say, $Z^{k_1}\phi$ where $k_1 \geq k_i$ for $i \geq 2$. So we will only focus on S and W. To bound S, we first note that by improved L^∞-estimate for ϕ, we have, for sufficiently small δ,

$$\||\phi|^{p-1}\|_{L^\infty(S_{\underline{u},u})} \lesssim \delta^{\frac{p-1}{2}}|\underline{u}|^{-p+1}I_4^{p-1}. \tag{2.33}$$

So S is bounded by

(2.34)

$$
\begin{aligned}
S \lesssim & \delta^{2\delta-1}\delta^{\frac{p-1}{2}}\iint_{\mathcal{D}_{\underline{u},u}} \left(|\underline{u}'|^{-2p+2+\alpha}\|Z^k\phi\|^2_{L^2(S_{\underline{u}',u'})} + \|\underline{u}'^{\frac{\alpha}{2}}LZ^k\phi\|^2_{L^2(S_{\underline{u}',u'})}\right) d\underline{u}'du' \\
\lesssim & \delta^{2l-1}\delta^{\frac{p-1}{2}}\left(\int_{1-u}^{\underline{u}} |\underline{u}'|^{-2p+2+\alpha}\|Z^k\phi\|^2_{L^2(\underline{C}_{\underline{u}'})}d\underline{u}' + \int_0^u \|\underline{u}'^{\frac{\alpha}{2}}LZ^k\phi\|^2_{L^2(C_{u'})}du'\right) \\
\lesssim & \delta^{2l-1}\delta^{\frac{p-1}{2}}\left(\int_{1-u}^{\underline{u}} \delta|\underline{u}'|^{-2p+2+\alpha}\|\underline{L}Z^k\phi\|^2_{L^2(\underline{C}_{\underline{u}'})}d\underline{u}' + \int_0^u \|\underline{u}'^{\frac{\alpha}{2}}LZ^k\phi\|^2_{L^2(C_{u'})}du'\right).
\end{aligned}
$$

The last term above is treated using Gronwall while the first term is bounded using bootstrap assumption. So we obtain the following estimate for S:

$$S \lesssim \delta^{\frac{p-1}{2}} \left(M^2 + \int_0^u \delta^{2l-1} \| \underline{u}'^{\frac{\alpha}{2}} L Z^k \phi \|^2_{L^2(C_{u'})} du' \right). \tag{2.35}$$

It remains to control $W = \delta^{2l-1} \iint_{\mathcal{D}_{\underline{u},u}} \frac{|\underline{u}|^\alpha}{r} |LZ^k\phi||\underline{L}Z^k\phi|$. We proceed as follows

$$\begin{aligned} W &\lesssim \delta^{2l-1} \iint_{\mathcal{D}_{\underline{u},u}} (\delta^{-\frac{1}{2}} |\underline{u}'|^{\frac{\alpha}{2}} |LZ^k\phi|)(\delta^{\frac{1}{2}} |\underline{u}'|^{-\frac{2-\alpha}{2}} |\underline{L}Z^k\phi|) \\ &\overset{Young}{\lesssim} \delta^{2l-1} \Big(\iint_{\mathcal{D}_{\underline{u},u}} \frac{\delta}{|\underline{u}'|^{2-\alpha}} |\underline{L}Z^k\phi|^2 + \iint_{\mathcal{D}_{\underline{u},u}} \frac{1}{\delta} ||\underline{u}'|^{\frac{\alpha}{2}} LZ^k\phi|^2 \Big) \\ &= \delta^{2l} \int_{1-u}^{\underline{u}} \frac{1}{|\underline{u}'|^{2-\alpha}} \|\underline{L}Z^k\phi\|^2_{L^2(\underline{C}_{\underline{u}'})} d\underline{u}' + \delta^{2l-2} \int_0^u \| |\underline{u}|^{\frac{\alpha}{2}} LZ^k\phi\|^2_{L^2(C_{u'})} du'. \end{aligned} \tag{2.36}$$

The first term in the last line has already been controlled in (2.29). In view of the fact that $\alpha < 1$ (this is crucial to make the first factor integrable in $\underline{u}$!), for sufficiently small δ, we obtain

$$W \lesssim I_4^2 + \delta^{-1} \int_0^u \delta^{2l-1} \| |\underline{u}|^{\frac{\alpha}{2}} LZ^k\phi\|^2_{L^2(C_{u'})} du'. \tag{2.37}$$

By combining (2.32) with (2.35) and (2.37), for sufficiently small δ, we obtain

$$\begin{aligned} &\delta^{2l-1} \int_{C_u} |\underline{u}|^\alpha |LZ^k\phi|^2 + \delta^{2l-1} \int_{\underline{C}_{\underline{u}}} |\underline{u}|^\alpha |\nabla\!\!\!/ Z^k\phi|^2 \\ &\qquad \lesssim I_4^2 + \delta^{\frac{p-1}{2}} M^2 + \delta^{-1} \int_0^u \delta^{2l-1} \| |\underline{u}|^{\frac{\alpha}{2}} LZ^k\phi\|^2_{L^2(C_{u'})} du' \end{aligned}$$

The last term on the right-hand side can be absorbed by the Gronwall's inequality and the implicit constant in the estimate is changed. This finally proves that, for all $0 \leq l \leq k \leq 2$, we have

$$\delta^{l-\frac{1}{2}} \| |\underline{u}|^{\frac{\alpha}{2}} LZ_b^l Z_g^{k-l}\phi\|_{L^2(C_u)} + \delta^{l-\frac{1}{2}} \|\nabla\!\!\!/ Z_b^l Z_g^{k-l}\phi\|_{L^2(\underline{C}_{\underline{u}})} \lesssim C(I_4) + \delta^{\frac{p-1}{4}} M. \tag{2.38}$$

The estimates (2.29) and (2.38) together imply

$$\begin{aligned} &E_{\leq 3}(u,\underline{u}) + \underline{E}_{\leq 3}(u,\underline{u}) \leq C(I_4) + \delta M, \\ \Rightarrow\ &E_{\leq 3}(u,\underline{u}) + \underline{E}_{\leq 3}(u,\underline{u}) \leq C(I_4) \end{aligned} \tag{2.39}$$

for $\delta > 0$ sufficiently small.

2.2. Higher order estimates. Using the same argument as in proving Proposition 2.2, we obtain the higher order version of the a priori estimate in the short pulse region:

PROPOSITION 2.7. *Given a positive integer $3 \leq n \leq 5$, there exists $\delta_0 > 0$, so that for all $\delta < \delta_0$, for all $u \in (0, u^*)$ and $\underline{u} \in (1-u^*, \underline{u}^*)$, we have*

$$E_{\leq n}(u,\underline{u}) + \underline{E}_{\leq n}(u,\underline{u}) \leq C(I_{n+1}), \tag{2.40}$$

and

$$
(2.41)\qquad
\begin{aligned}
\|\slashed{\nabla} Z_b^l Z_g^{k-l}\|_{L^\infty(S_{\underline{u},u})} &\lesssim \delta^{\frac{1}{4}-l}|\underline{u}|^{-\frac{3}{2}-\frac{\alpha}{4}}C(I_{n+1}), \quad 0 \le l \le k \le n-2,\\
\|L Z_b^l Z_g^{k-l}\phi\|_{L^\infty(S_{\underline{u},u})} &\lesssim \delta^{\frac{1}{2}-l}|\underline{u}|^{-2}C(I_{n+1}), \quad 0 \le l \le k \le n-2,\\
\|\underline{L} Z_b^l Z_g^{k-l}\phi\|_{L^\infty(S_{\underline{u},u})} &\lesssim \delta^{-\frac{1}{2}-l}|\underline{u}|^{-1}C(I_{n+1}), \quad 0 \le l \le k \le n-2,\\
\|Z_b^l Z_g^{k-1}\phi\|_{L^\infty(S_{\underline{u},u})} &\lesssim \delta^{\frac{1}{2}-l}|\underline{u}|^{-1}C(I_{n+1}), \quad 0 \le l \le k \le n-2.
\end{aligned}
$$

where $C(I_{n+1})$ is a constant depending only on I_{n+1}.

REMARK 2.8. We present the detail for proving the a priori estimate in the short pulse region for $n = 3$ because this seems thc minimum regularity we need to close the estimate using the method in the present paper. The higher order estimate is used for proving the existence and decay estimate inside the region bounded by C_δ and Σ_1.

2.3. Improved estimates on C_δ. Recall that, given $3 \le n \le 5$, in the short pulse region, by Proposition 2.7 we have the following a priori L^∞ estimates on the solution ϕ (See Proposition 2.7):

$$
(2.42)\qquad
\begin{aligned}
\|\slashed{\nabla} Z_b^l Z_g^{k-l}\phi\|_{L^\infty(S_{\underline{u},u})} &\lesssim \delta^{\frac{1}{4}-l}|\underline{u}|^{-\frac{3}{2}-\frac{\alpha}{4}}C(I_{n+1}), \quad 0 \le l \le k \le n-2,\\
\|L Z_b^l Z_g^{k-l}\phi\|_{L^\infty(S_{\underline{u},u})} &\lesssim \delta^{\frac{1}{2}-l}|\underline{u}|^{-2}C(I_{n+1}), \quad 0 \le l \le k \le n-2,\\
\|\underline{L} Z_b^l Z_g^{k-l}\phi\|_{L^\infty(S_{\underline{u},u})} &\lesssim \delta^{-\frac{1}{2}-l}|\underline{u}|^{-1}C(I_{n+1}), \quad 0 \le l \le k \le n-2,\\
\|Z_b^l Z_g^{k-l}\phi\|_{L^\infty(S_{\underline{u},u})} &\lesssim \delta^{\frac{1}{2}-l}|\underline{u}|^{-1}C(I_{n+1}), \quad 0 \le l \le k \le n-2.
\end{aligned}
$$

The goal of this section is to improve these bounds for the solution on C_δ. More precisely, we will prove that

PROPOSITION 2.9. *On C_δ, for sufficiently small δ, we have*

$$
(2.43)\qquad
\begin{aligned}
\|\slashed{\nabla} Z_b^l Z_g^{k-l}\phi\|_{L^\infty(S_{\underline{u},\delta})} &\lesssim \delta^{\frac{1}{4}}|\underline{u}|^{-\frac{3}{2}-\frac{\alpha}{4}}C(I_{n+1}), \quad 0 \le l \le k \le n-2,\\
\|L Z_b^l Z_g^{k-l}\phi\|_{L^\infty(S_{\underline{u},\delta})} &\lesssim \delta^{\frac{1}{4}}|\underline{u}|^{-2}C(I_{n+1}), \quad 0 \le l \le k \le n-2,\\
\|\underline{L} Z_b^l Z_g^{k-l}\phi\|_{L^\infty(S_{\underline{u},\delta})} &\lesssim \delta^{\frac{1}{4}}|\underline{u}|^{-1}C(I_{n+1}), \quad 0 \le l \le k \le n-2,\\
\|Z_b^l Z_g^{k-l}\phi\|_{L^\infty(S_{\underline{u},u})} &\lesssim \delta^{\frac{1}{4}}|\underline{u}|^{-1}C(I_{n+1}), \quad 0 \le l \le k \le n-2.
\end{aligned}
$$

Since C_δ is also the outer boundary of the small data region (region I), the smallness (in terms of δ) of the solution stated in the proposition is indispensable for the construction of a global solution in the small data region. As we mentioned in the introduction, the proof relies on the following observation: On the $S_{1-\delta,\delta}$ or equivalently the lower boundary of C_δ, the data are *identically zero*. This is because that the data are compactly supported on Σ_1 between $S_{1-\delta,\delta}$ and $S_{1,0}$. Therefore, even for "bad" derivatives of ϕ, it is small at least initially. The idea of the proof is to integrate along the L direction to show that the smallness indeed propagates.

We use the induction argument on the pair (l, k) $(0 \le k \le n-2, 0 \le l \le k)$ to prove (2.43). First of all, we give an order on the set of such pairs: we say that $(l', k') < (l, k)$ if one of the following holds: (1) $k' < k$ or (2) $l' < l$, $k = k'$. We do the induction with respect to this order.

For $(l, k) = (0, 0)$, the bounds on $L\phi$, $\slashed{\nabla}\phi$ and ϕ are clear. It remains to prove that

$$\|\underline{L}\phi\|_{L^\infty(S_{\underline{u},\delta})} \lesssim \delta^{\frac{1}{4}}|\underline{u}|^{-1}C(I_n).$$

Recall that, in (2.21), we have obtained

$$\big||\underline{u}||\underline{L}\phi|(\underline{u},u,\theta)-C|1-u||\underline{L}\phi|(1-u,u,\theta)\big|\lesssim\delta^{\frac14}M.$$

In view of the higher order energy estimates derived in the previous subsection, the constant M should be replaced by $C(I_n)$. Let $u=\delta$, then the second term vanishes on the initial sphere $S_{1-\delta,\delta}$. This gives the desired estimates on $\|\underline{L}\phi\|_{L^\infty(S_{\underline{u},\delta})}$.

For $(l,k)=(0,k)$, we use the same argument as deriving (2.19) to obtain

$$\big||\underline{u}||\underline{L}Z_g^k\phi|(\underline{u},u,\theta)-C|1-u||\underline{L}Z_g^k\phi|(1-u,u,\theta)\big|\lesssim\delta^{\frac14-l}M. \tag{2.44}$$

From this, the desired estimates follow in a similar way.

We assume that for all $(l',k')<(l,k)$, we have

$$\begin{aligned}
\|\nabla\!\!\!/ Z_b^{l'}Z_g^{k'-l'}\phi\|_{L^\infty(S_{\underline{u},\delta})}&\lesssim\delta^{\frac14}|\underline{u}|^{-\frac32-\frac{\alpha}{4}}C(I_n),\\
\|LZ_b^{l'}Z_g^{k'-l'}\phi\|_{L^\infty(S_{\underline{u},\delta})}&\lesssim\delta^{\frac14}|\underline{u}|^{-2}C(I_n),\\
\|\underline{L}Z_b^{l'}Z_g^{k'-l'}\phi\|_{L^\infty(S_{\underline{u},\delta})}&\lesssim\delta^{\frac14}|\underline{u}|^{-1}C(I_n),\\
\|Z_b^{l'}Z_g^{k'-l'}\phi\|_{L^\infty(S_{\underline{u},\delta})}&\lesssim\delta^{\frac14}|\underline{u}|^{-1}C(I_n).
\end{aligned}$$

For (l,k), we now reduce the estimates to the above induction hypothesis. Because we have already proved the case for $(l,k)=(0,k)$, so we can assume in addition that $l\geq 1$.

We first bound $\nabla\!\!\!/ Z_b^lZ_g^{k-l}\phi$. In fact, we have

$$\begin{aligned}
|\nabla\!\!\!/ Z_b^lZ_g^{k-l}\phi|&\lesssim\frac{1}{|\underline{u}|}|\Omega Z_b^lZ_g^{k-l}\phi|\\
&\leq\frac{1}{|\underline{u}|}\Big(|Z_b\Omega Z_b^{l-1}Z_g^{k-l}\phi|+|[\Omega,Z_b]Z_b^{l-1}Z_g^{k-l}\phi|\Big).
\end{aligned}$$

Since $\Omega\in\mathcal{Z}_g$, we can use induction hypothesis (since reduced l), the first term is bounded by

$$\sum_{\partial\in\{L,\underline{L},\nabla\!\!\!/\}}\frac{1}{|\underline{u}|}|\partial Z_b^{l-1}Z_g^{k-l+1}\phi|\lesssim\delta^{\frac14}|\underline{u}|^{-2}C(I_{n+1}).$$

For the second term, notice that $[\Omega,Z_b]=Z_b$, therefore, we have decreased the number l by 1. According to the induction hypothesis, it is bounded by

$$\sum_{\partial\in\{L,\underline{L},\nabla\!\!\!/\}}\frac{1}{|\underline{u}|}|\partial Z_b^{l-1}Z_g^{k-l}\phi|\lesssim\delta^{\frac14}|\underline{u}|^{-2}C(I_{n+1}).$$

This gives the desired estimates for $\nabla\!\!\!/ Z_b^lZ_g^{k-l}\phi$.

For $Z_b^lZ_g^{k-l}\phi$, when $l>0$ we write it as

$$Z_b^lZ_g^{k-l}\phi=\sum_{\partial\in\{L,\underline{L},\nabla\!\!\!/\}}\partial Z_b^{l-1}Z_g^{(k-1)-(l-1)}\phi,$$

which is estimated by induction hypothesis. When $l=0$, the estimate is already established.

We turn to the bound on $LZ_b^lZ_g^{k-l}\phi$. Evidently, it is bounded by $\sum_{\partial\in\{L,\underline{L},\nabla\!\!\!/\}}|L\partial Z_b^{l-1}Z_g^{k-l}\phi|$. When $\partial=\nabla\!\!\!/$ in the sum, it can be bounded directly by the bound on $\nabla\!\!\!/ Z_b^{l-1}Z_g^{k-l+1}\phi$ derived above. Therefore, it suffices to bound $LLZ_b^{l-1}Z_g^{k-l}\phi$ and $L\underline{L}Z_b^{l-1}Z_g^{k-l}\phi$.

For $L\underline{L}Z_b^{l-1}Z_g^{k-l}\phi$, according to (2.23) (where we use $Z_b^{l-1}Z_g^{k-l}$ as commutator vector field), we have

$$\begin{aligned}&-L\underline{L}Z_b^{l-1}Z_g^{k-l}\phi+\not{\Delta}Z_b^{l-1}Z_g^{k-l}\phi\\&\qquad=\frac{1}{r}(LZ_b^{l-1}Z_g^{k-l}\phi-\underline{L}Z_b^{l-1}Z_g^{k-l}\phi)+\sum_{k_1+\ldots+k_p\leq k-1}Z^{k_1}\phi\ldots Z^{k_p}\phi.\end{aligned}$$

The second term on the left-hand side can be bounded by $\not{\nabla}Z_b^lZ_g^{k-l}\phi$. The terms on the right-hand side are all of lower degrees ($<k$) so that they are bounded by the induction hypothesis. Therefore, we have

$$\|L\underline{L}Z_b^{l-1}Z_g^{k-l}\phi\|_{L^\infty(S_{\underline{u},\delta})}\lesssim\delta^{\frac{1}{4}}|\underline{u}|^{-2}C(I_{n+1}),$$

For $LLZ_b^{l-1}Z_g^{k-l}\phi$, we use the following identity:

$$LLZ_b^{l-1}Z_g^{k-l}\phi=L\left(\frac{1}{\underline{u}}\left(SZ_b^{l-1}Z_g^{k-l}\phi-u\underline{L}Z_b^{l-1}Z_g^{k-l}\phi\right)\right).$$

The second term on the right-hand side can be bounded directly by the bound on $L\underline{L}Z_b^{l-1}Z_g^{k-l}\phi$ just derived. Therefore, it suffices to control the contribution from the first term, i.e.

$$\begin{aligned}L\left(\frac{1}{\underline{u}}\left(SZ_b^{l-1}Z_g^{k-l}\phi\right)\right)&=-\frac{1}{|\underline{u}|^2}SZ_b^{l-1}Z_g^{k-l}\phi+\frac{1}{\underline{u}}LSZ_b^{l-1}Z_g^{k-l}\phi\\&=\underbrace{-\frac{1}{|\underline{u}|^2}SZ_b^{l-1}Z_g^{k-l}\phi}_{A}+\underbrace{\frac{1}{\underline{u}}LZ_b^{l-1}Z_g^{k-l+1}\phi}_{B}.\end{aligned}$$

For A, by rewriting S as $\underline{u}L+u\underline{L}$, we can use induction hypothesis for $(l-1,k-1)$; for B, we can use induction hypothesis for $(l-1,k)$. Hence, we have obtained the desired estimates for $LZ_b^lZ_g^{k-l}\phi$.

Finally, to bound $\underline{L}Z_b^lZ_g^{k-l}\phi$, we use the equation

$$-L\underline{L}Z_b^lZ_g^{k-l}\phi+\not{\Delta}Z_b^lZ_g^{k-l}\phi=-\frac{1}{r}(LZ_b^lZ_g^{k-l}\phi-\underline{L}Z_b^lZ_g^{k-l}\phi)+\sum_{k_1+\ldots+k_p\leq k}Z^{k_1}\phi\ldots Z^{k_p}\phi.$$

We rewrite this as

$$\begin{aligned}&-L\underline{L}Z_b^lZ_g^{k-l}\phi-\frac{1}{r}\underline{L}Z_b^lZ_g^{k-l}\phi\\&=-\not{\Delta}Z_b^lZ_g^{k-l}\phi-\frac{1}{r}LZ_b^lZ_g^{k-l}\phi+\sum_{k_1+\ldots+k_p\leq k}Z^{k_1}\phi\ldots Z^{k_p}\phi.\end{aligned}$$

All the terms on the right-side have been controlled in previous steps. Therefore, it is straightforward to see that the right-hand side is bounded by $C(I_{n+1})|\underline{u}|^{-2}\delta^{\frac{1}{4}}$. We now can mimic the proof for (2.21) by defining $y=\underline{u}\underline{L}Z_b^lZ_g^{k-l}\phi$, this leads to

$$|\underline{L}Z_b^lZ_g^{k-l}\phi(\underline{u},\delta,\theta)-\frac{C(1-\delta)}{\underline{u}}\underline{L}Z_b^lZ_g^{k-l}\phi(1-\delta,\delta,\theta)|\lesssim\delta^{\frac{1}{4}}|\underline{u}|^{-1}C(I_{n+1}).$$

Taking into account of the vanishing property of $\underline{L}Z_b^lZ_g^{k-l}\phi$ on $S_{1-\delta,\delta}$, we complete the proof of Proposition 2.9.

REMARK 2.10. For applications in the next section, we only need a slightly weakened (in decay) version of the estimates from Proposition 2.9:

$$\begin{aligned} \|\nabla Z_b^l Z_g^{k-l}\phi\|_{L^\infty(S_{\underline{u},\delta})} + \|LZ_b^l Z_g^{k-l}\phi\|_{L^\infty(S_{\underline{u},\delta})} &\lesssim \delta^{\frac{1}{4}}|\underline{u}|^{-\frac{3}{2}}C(I_{n+1}), \\ \|\underline{L}Z_b^l Z_g^{k-l}\phi\|_{L^\infty(S_{\underline{u},\delta})} &\lesssim \delta^{\frac{1}{4}}|\underline{u}|^{-1}C(I_{n+1}), \end{aligned} \tag{2.45}$$

where $0 \leq l \leq k \leq n-2$.

2.4. Existence of short pulse Cauchy data. Now let us sketch a proof for the existence of the Cauchy data satisfying the requirement discussed in (1.4)-(1.5). In $\mathbb{R}^{3+1}$ let us consider the hypersurface H (denoted by the dash line in the picture) given by union of the of two pieces of spacelike hypersurfaces: $\Sigma_{1,1-2\delta} := \{(t,r)|t = 1, r \leq 1-2\delta\}$, $\Sigma_{1-\delta,1-\delta} := \{(t,r)|t = 1-\delta, r \geq 1-\delta\}$, and a piece of incoming null hypersurface $\underline{C}_{1-\delta}^{\delta} := \{(\underline{u}, u)|\underline{u} = 1-\delta, 0 \leq u \leq \delta\}$:

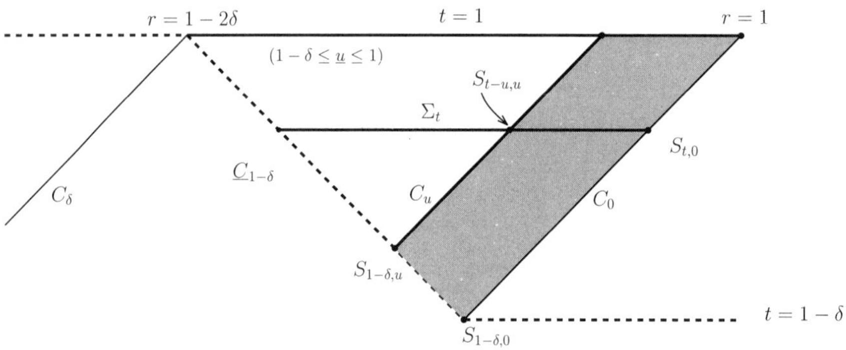

Let us consider the data on H for the main equation such that:

$$\begin{aligned} \phi &= 0 \quad \text{on} \quad \Sigma_{1,1-2\delta} \bigcup \Sigma_{1-\delta,1-\delta}, \\ \phi(1-\delta, u, \theta) &= \delta^{\frac{1}{2}}\phi_0\left(\frac{u}{\delta}, \theta\right) \quad \text{on} \quad \underline{C}_{1-\delta}^{\delta}. \end{aligned} \tag{2.46}$$

Here ϕ_0 is a C^∞ function defined on $(0,1) \times \mathbb{S}^2$ and compactly supported with respect to its first variable. Due to the specific form of the characteristic data (2.46), one can easily verify that

$$\begin{aligned} &\|(\partial_t - \partial_r)^k \nabla^l \phi_0\|_{L^\infty(\underline{C}_{1-\delta})} + \|(\partial_t - \partial_r)^{k-1}\nabla^l \phi_1\|_{L^\infty(\underline{C}_{1-\delta})} \lesssim \delta^{\frac{1}{2}-k}, \\ &\|(\partial_t + \partial_r)^k \nabla^l (\partial_t - \partial_r)^m \phi\|_{L^\infty(\underline{C}_{1-\delta})} \lesssim \delta^{\frac{1}{2}-m}. \end{aligned} \tag{2.47}$$

We will perform the estimates both in the upside-down triangle bounded by C_0, $\underline{C}_{1-\delta}$ and Σ_t as well as the grey region. The idea is exactly the same as we have seen in the proof of the main a priori estimate in the short pulse region. The only difference is that we replace the incoming null hypersurface by a Cauchy hypersurface here and the initial Cauchy hypersurface is replaced by an incoming null hypersurface. We first consider the spacetime domain D_t bounded by the hypersurfaces $\underline{C}_{1-\delta}$, C_0 and Σ_t where $1-\delta \leq t \leq 1$. So a portion of Σ_t where $2-2\delta-t \leq r \leq t$ is considered. Let us denote this portion by Σ_t^t. Applying divergence theorem in this domain we obtain the following energy estimate (using multiplier $\partial_t = \frac{1}{2}(L + \underline{L})$):

$$\begin{aligned}\int_{\Sigma_t^t} \left(|\partial_t Z^k \phi|^2 + |\nabla_x Z^k \phi|^2\right) &\leq \int_{\underline{C}_{1-\delta}^{t-1+\delta}} \left(|\underline{L} Z^k \phi|^2 + ||\nabla\!\!\!/ Z^k \phi|^2\right) \\ &+C \sum_{k_1+...+k_p \leq k} \left|\iint_{D_t} Z^{k_1}\phi...Z^{k_p}\phi \partial_t Z^k \phi\right|, \quad \text{for} \quad 0 \leq k \leq 5.\end{aligned} \tag{2.48}$$

We also consider the spacetime domain D_u bounded by C_0, C_u and Σ_t with $0 \leq u \leq \delta$, $1 - \delta \leq t \leq 1$. Applying divergence theorem in this domain we obtain the following energy estimate (using multiplier L):

$$\begin{aligned}\int_{C_u} |L Z^k \phi|^2 \leq \int_{\underline{C}_{1-\delta}^{t-1+\delta}} |\nabla\!\!\!/ Z^k \phi|^2 \\ + C \sum_{k_1+...+k_p \leq k} \left|\iint_{D_u} Z^{k_1}\phi...Z^{k_p}\phi L Z^k \phi\right|, \quad \text{for} \quad 0 \leq k \leq 5.\end{aligned} \tag{2.49}$$

The idea is to derive an a priori estimate in D_t up to $t = 1$, and in D_u up to $u = \delta$, using (2.48) and (2.49). Let us make the following bootstrap assumption:

$$\sum_{0 \leq k \leq 5} \delta^l \left(\|\partial_t Z^k \phi\|_{L^2(\Sigma_t^t)} + \|\nabla_x Z^k \phi\|_{L^2(\Sigma_t^t)}\right) + \sum_{0 \leq k \leq 5} \delta^{l-\frac{1}{2}} \|L Z^k \phi\|_{L^2(C_u^{1-u})} \leq M_d \tag{2.50}$$

for some $M_d > 0$. Using a continuity argument, one proves that M_d is actually an absolute constant depending only on ϕ_0. Since most of the a priori estimate is similar to what we have in this section, and moreover here we do not need to derive pointwise decay because we only solve the equation from $t = 1 - \delta$ to $t = 1$. Therefore the proof here is much easier than what we have done in this section and we only sketch some key ingredients. First we multiply both sides of (2.48) by δ^{2l}, with l being the number of Z_bs in Z^k. By the bootstrap assumption (2.50), the left hand side of (2.48) is bounded by M_d^2 up to a multiplication by an absolute constant. On the other hand, by (2.47), the first term on the right hand side of (2.48) is bounded by an absolute constant. Once we can prove that if δ is sufficiently small, the spacetime integral in (2.48) is also bounded by an absolute constant, the a priori estimate is proved (Similar argument applies to (2.49)). By finite speed of propagation, the solution ϕ is supported between C_δ and C_0, on Σ_1, ϕ is supported between $r = 1 - 2\delta$ and $r = 1$. This proves the existence of the Cauchy data we want as well as Lemma 2.1. Finally, let us briefly discuss how to control the spacetime integral in (2.48). First, it is easy to prove the following lemma using basic calculus:

LEMMA 2.11. *Let f be a C^∞ function supported in Σ_t^t. The following inequality holds:*

$$\begin{aligned}\|f\|_{L^2\left(S_{\frac{t+r}{2},\frac{t-r}{2}}\right)} &\leq C\delta^{\frac{1}{2}} \|\partial_r f\|_{L^2(\Sigma_t^t)}, \quad \|f\|_{L^2(\Sigma_t^t)} \leq C\delta \|\partial_r f\|_{L^2(\Sigma_t^t)}, \\ \|f\|_{L^2\left(S_{\frac{t+r}{2},\frac{t-r}{2}}\right)} &\leq C\|f\|_{L^2\left(S_{1-\delta,\frac{t-r}{2}}\right)} + C\|Lf\|_{L^2\left(C_{\frac{t-r}{2}}\right)}.\end{aligned} \tag{2.51}$$

where C is an absolute constant.

Using this lemma and (2.4)-(2.5), we have, based on (2.47), for $k \leq 9$,

$$\begin{aligned}\delta^l \|LZ^k\phi\|_{L^\infty\left(S_{\frac{t+r}{2},\frac{t-r}{2}}\right)} &\leq C\delta^{\frac{1}{2}} M_d,\\ \delta^l \|\not\nabla Z^k\phi\|_{L^\infty\left(S_{\frac{t+r}{2},\frac{t-r}{2}}\right)} &\leq C\delta^{\frac{1}{2}} M_d,\\ \delta^l \|\underline{L}Z^k\phi\|_{L^\infty\left(S_{\frac{t+r}{2},\frac{t-r}{2}}\right)} &\leq C\delta^{-\frac{1}{2}} M_d,\\ \delta^l \|Z^k\phi\|_{L^\infty\left(S_{\frac{t+r}{2},\frac{t-r}{2}}\right)} &\leq C\delta^{\frac{1}{2}} M_d.\end{aligned} \tag{2.52}$$

Plugging the pointwise estimate for $Z^k\phi$ into the spacetime integrals in (2.48) and (2.49), one sees that the spacetime integrals in (2.48) and (2.49) have more positive powers in δ than the left hand sides respectively. In other words, the spacetime integrals are lower order. Therefore, the bootstrap assumption (2.50) is recovered.

3. Small data region

In this section, we construct solutions in the entire small data region, i.e. region I. The approach is a modification of the classical approach with additional difficulties arising from the boundary C_δ.

3.1. Klainerman-Sobolev inequality revisited. We first introduce notations needed for the statement of the Klainerman-Sobolev inequality. We use Σ_t to denote the constant time slices in the small data region, i.e. for a fixed $t \in (1, +\infty)$,

$$\Sigma_t := \{(x,t) \big| t - r \geq \delta\}.$$

This is a ball of radius $t - \delta$. We recall that we use Σ_1 to denote the entire $t = 1$ hyperplane. Given a point $(t, x) \in \Sigma_t$ (assuming that $x \neq 0$), the point $\left(t, \frac{x}{|x|}(t-\delta)\right)$ is the unique point on the boundary of Σ_t (also on C_δ)which is the intersection of the boundary of Σ_t with the ray emanated from $(t, 0)$ and passing from (t, x). We now state the Klainerman-Sobolev inequality:

PROPOSITION 3.1. *For all $f \in C^\infty(\mathbb{R}^{3+1})$, $t > 1$ and a point (t, x) in the small data region, we have*

$$|f(t,x)| \lesssim \frac{1}{(1+|u|)^{1/2}} \left| f\left(t, \frac{x}{|x|}(t-\delta)\right)\right| + \frac{1}{(1+|\underline{u}|)(1+|u|)^{1/2}} \sum_{Z\in\mathcal{Z}, k\leq 3} \|Z^k f\|_{L^2(\Sigma_t)}. \tag{3.1}$$

The proof of this proposition can be found in [**12**].

3.2. A priori energy estimates. For a $t \in (1, +\infty)$, we still use $\Sigma_t = \{(x,t) \big| t - r \geq \delta\}$. to denote the constant time slices in the small data region. For $k \in \mathbb{Z}_{\geq 0}$ and $t > 1$, we introduce the following energy norms:

$$\begin{aligned}\widetilde{E}_k(t) &= \Big(\sum_{Z\in\mathcal{Z}} \int_{\Sigma_t} |\partial_t Z^k\phi|^2 + \sum_{j=1}^{3} |\partial_j Z^k\phi|^2 dx\Big)^{\frac{1}{2}},\\ \widetilde{E}_{\leq k}(t) &= \Big(\sum_{0\leq j\leq k} \widetilde{E}_j(t)^2\Big)^{\frac{1}{2}}.\end{aligned} \tag{3.2}$$

We use $D_{t,\delta}$ to denote the space-time region bounded by Σ_t, Σ_1 and C_δ. This region is obviously foliated by the constant time foliation $\{\Sigma_\tau \mid \tau \in [1,t]\}$ and this foliation is one of the foliations we use to derive energy estimates. The second foliation is the null foliation of outgoing null cones $\{C_u \mid u \in [\delta, t/2]\}$. This foliation is depicted as follows:

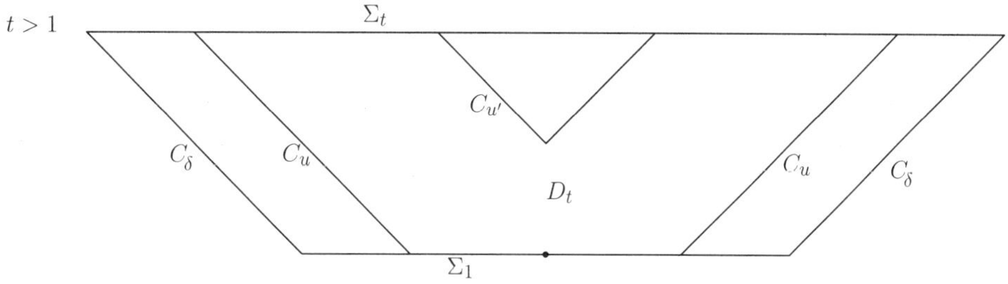

Whenever there is no confusion, we still use C_u to denote $C_u \cap D_t$. We use $D_{t,u}$ to denote the space-time region bounded by Σ_t, Σ_1 and C_u. This is a truncated solid light cone in $\mathbb{R}^{3+1}$. We use $\Sigma_{1,u}$ and $\Sigma_{t,u}$ to denote its bottom and top respectively. We remark that the bottom can be a single point.

Recall that (assuming that the solution ϕ exists up to time t), for $k \geq 0$ and $|\alpha| = k$, we have

$$\Box Z^k \phi = \sum_{k_1+\ldots+k_p \leq k} Z^{k_1}\phi \ldots Z^{k_p}\phi.$$

We multiply both sides by $\partial_t Z^k \phi$ and we then integrate over $D_{t,u}$. This leads to the following energy identity:

$$\begin{aligned}\int_{\Sigma_{t,u}} &|\partial_t Z^k \phi|^2 + \sum_{j=1}^{3} |\partial_j Z^k \phi|^2 \\ &= \int_{\Sigma_{1,u}} |\partial_t Z^k \phi|^2 + \sum_{j=1}^{3} |\partial_j Z^k \phi|^2 + \int_{C_u} |L Z^k \phi|^2 + |\nabla\!\!\!/ Z^k \phi|^2 \\ &\quad + \sum_{k_1+\ldots+k_p \leq k} \iint_{D_{t,u}} Z^{k_1}\phi \ldots Z^{k_p}\phi \partial_t Z^k \phi.\end{aligned}$$

Recall that we use $\partial \in \{L, \underline{L}, \nabla\!\!\!/\}$ to denote a generic derivative and use $\partial_g \in \{L, \nabla\!\!\!/\}$ to denote a "good" derivative. Therefore, by using $|\partial Z^k \phi|^2$ as a shorthand notation for $|\partial_t Z^k \phi|^2 + \sum_{j=1}^3 |\partial_j Z^k \phi|^2$ and using $|\partial_g Z^k \phi|^2$ as a shorthand notation for $|L Z^k \phi|^2 + |\nabla\!\!\!/ Z^k \phi|^2$, we have

$$\int_{\Sigma_{t,u}} |\partial Z^k \phi|^2 = \int_{\Sigma_{1,u}} |\partial Z^k \phi|^2 + \int_{C_u} |\partial_g Z^k \phi|^2 + \sum_{k_1+\ldots+k_p \leq k} \iint_{D_{t,u}} Z^{k_1}\phi \ldots Z^{k_p}\phi \partial_t Z^k \phi.$$

In applications, since the data prescribed on $\Sigma_{1,u}$ are trivial, we have

$$\int_{\Sigma_{t,u}} |\partial Z^k \phi|^2 = \int_{C_u} |\partial_g Z^k \phi|^2 + \sum_{k_1+\ldots+k_p \leq k} \iint_{D_{t,u}} Z^{k_1}\phi \ldots Z^{k_p}\phi \partial_t Z^k \phi. \tag{3.3}$$

Before we state the main estimates of the section, we first compute the energy flux $\int_{C_\delta} |\partial_g Z^k \phi|^2$ through the outermost cone C_δ. According to (2.43), for $k \leq n-2$, we have $|\partial_g Z^k\phi| \lesssim \delta^{\frac{1}{4}}|\underline{u}|^{-\frac{3}{2}-\frac{\alpha}{4}}C(I_{n+1})$, therefore,

$$\int_{C_\delta} |\partial_g Z^k\phi|^2 \lesssim \delta^{\frac{1}{2}}C(I_{n+1}), \tag{3.4}$$

where we still use $C(I_{n+1})$ to denote $C(I_{n+1})^2$ appearing in the previous section.

PROPOSITION 3.2. *Under the same assumptions as in the previous section, for sufficiently small δ, there exists a unique global future in time solution ϕ of (1.3) on the small data region, so that together with the solution constructed in the short pulse region, we have a unique future in time solution ϕ. Moreover, this solution ϕ on the small data region enjoys the following energy estimates:*

$$\widetilde{E}_{\leq 5}(t) \lesssim \delta^{\frac{1}{4}}C(I_6), \tag{3.5}$$

for all $t > 1$.

REMARK 3.3. The existence of solutions in the small data region follows from the *a priori* estimate (3.5). Since the argument is routine, we will not pursuit this point here.

We use a bootstrap argument to prove the proposition. We assume that the solution exists up to time t and for all $1 \leq t' \leq t$, we have

$$\widetilde{E}_{\leq 5}(t') \lesssim M\delta^{\frac{1}{4}}. \tag{3.6}$$

It suffices to show that we can indeed choose M so that it depends only on I_n. We first point out that we can derive L^∞ bound on $Z^m\phi$ for $m \leq 2$. According to Klainerman-Sobolev inequality, we have

$$\begin{aligned}
|\partial Z^m\phi(\tau,x)| &\lesssim \frac{1}{(1+|u|)^{1/2}}\left|\partial Z^m\phi\left(\tau,\frac{x}{|x|}(\tau-\delta)\right)\right| \\
&\quad + \frac{1}{(1+|\underline{u}|)(1+|u|)^{1/2}}\sum_{Z\in\mathcal{Z},l\leq 2+m}\|Z^l\partial Z\phi\|_{L^2(\Sigma_\tau)} \\
&\lesssim \frac{C(I_{n+1})}{(1+\underline{u})(1+u)^{1/2}}\delta^{\frac{1}{4}} + \frac{M}{(1+\underline{u})(1+|u|)^{1/2}}\delta^{\frac{1}{4}}.
\end{aligned}$$

In particular we choose $\partial = \underline{L}$ and integrating the above inequality along the incoming direction $\underline{L}$ to obtain:

$$\begin{aligned}
|Z^m\phi(\tau,x)| &\lesssim (1+|u|)^{1/2}\left|\partial Z^m\phi\left(\tau,\frac{x}{|x|}(\tau-\delta)\right)\right| \\
&\quad + \frac{(1+|u|)^{1/2}}{(1+|\underline{u}|)}\sum_{Z\in\mathcal{Z},l\leq 2+m}\|Z^l\partial Z\phi\|_{L^2(\Sigma_\tau)} \\
&\lesssim \frac{C(I_{n+1})(1+u)^{1/2}}{(1+\underline{u})}\delta^{\frac{1}{4}} + \frac{M(1+|u|)^{1/2}}{(1+\underline{u})}\delta^{\frac{1}{4}} \lesssim \frac{M\delta^{\frac{1}{4}}}{(1+|\underline{u}|)^{\frac{1}{2}}}
\end{aligned}$$

On the other hand, if $m > 2$, we estimate $\|Z^m\phi\|_{L^2(\Sigma_{t,u})}$ as follows:

$$\|Z^m\phi\|_{L^2(\Sigma_{t,\delta})} \lesssim (1+|\underline{u}|)\|\partial Z^m\phi\|_{L^2(\Sigma_{t,\delta})} \lesssim (1+|\underline{u}|)\widetilde{E}_{\leq 5}(t).$$

Substituting the above estimates for $\|Z^m\phi\|_{L^\infty(\Sigma_{t,\delta})}, \|Z^m\phi\|_{L^2(\Sigma_{t,\delta})}$ into the the energy identity (3.3), we obtain:

$$\begin{aligned}(3.7)\qquad &\sum_{k\le 5}\int_{\Sigma_{t,\delta}}|\partial Z^k\phi|^2 \lesssim C(I_6)\delta^{\frac{1}{2}} + \int_1^t \frac{\delta^{\frac{p-1}{4}}M^{p-1}}{(1+|\underline{u}|)^{\frac{p-3}{2}}}\sum_{k\le 5}\|\partial Z^k\phi\|^2_{L^2(\Sigma_{t',\delta})}dt',\\ \Rightarrow\quad &\widetilde{E}^2_{\le 5}(t) \lesssim C(I_6)\delta^{\frac{1}{2}} + \int_1^t \frac{\delta^{\frac{p-1}{4}}M^{p-1}}{(1+|\underline{u}|)^{\frac{p-3}{2}}}\widetilde{E}^2_{\le 5}(t')dt'.\end{aligned}$$

Since $\frac{p-3}{2} > 1$, using Gronwall and choosing $\delta > 0$ sufficiently small, we obtain

$$\widetilde{E}_{\le 5}(t) \lesssim \delta^{\frac{1}{4}}\sqrt{C(I_6)},$$

and this completes the bootstrap argument.

Acknowledgments

The author is grateful to two anonymous referees, who carefully read a previous version of this paper and suggested many valuable improvements and corrections. The author also would like to thank Shijun Zheng for interesting discussion and his encouragement to publish this result.

References

[1] H. Bahouri and P. Gérard, *High frequency approximation of solutions to critical nonlinear wave equations*, Amer. J. Math. **121** (1999), no. 1, 131–175. MR1705001

[2] H. Bahouri and J. Shatah, *Decay estimates for the critical semilinear wave equation* (English, with English and French summaries), Ann. Inst. H. Poincaré Anal. Non Linéaire **15** (1998), no. 6, 783–789, DOI 10.1016/S0294-1449(99)80005-5. MR1650958

[3] M. Beceanu and A. Soffer *Large Outgoing Solutions to Supercritical Wave Equations*, To appear on International Mathematics Research Notices.

[4] M. Beceanu and A. Soffer *Large Initial Data Global Well-Posedness for a Supercritical Wave Equation*, https://arxiv.org/abs/1602.08163

[5] D. Christodoulou, *The formation of black holes in general relativity*, EMS Monographs in Mathematics, European Mathematical Society (EMS), Zürich, 2009. MR2488976

[6] M. G. Grillakis, *Regularity and asymptotic behaviour of the wave equation with a critical nonlinearity*, Ann. of Math. (2) **132** (1990), no. 3, 485–509, DOI 10.2307/1971427. MR1078267

[7] C. E. Kenig and F. Merle, *Global well-posedness, scattering and blow-up for the energy-critical focusing non-linear wave equation*, Acta Math. **201** (2008), no. 2, 147–212, DOI 10.1007/s11511-008-0031-6. MR2461508

[8] C. E. Kenig and F. Merle, *Nondispersive radial solutions to energy supercritical non-linear wave equations, with applications*, Amer. J. Math. **133** (2011), no. 4, 1029–1065, DOI 10.1353/ajm.2011.0029. MR2823870

[9] S. Klainerman and I. Rodnianski, *On the formation of trapped surfaces*, Acta Math. **208** (2012), no. 2, 211–333, DOI 10.1007/s11511-012-0077-3. MR2931382

[10] J. Krieger and W. Schlag, *On the focusing critical semi-linear wave equation*, Amer. J. Math. **129** (2007), no. 3, 843–913, DOI 10.1353/ajm.2007.0021. MR2325106

[11] J. Krieger and W. Schlag, *Large global solutions for energy supercritical nonlinear wave equations on* $\mathbb{R}^{3+1}$, J. Anal. Math. **133** (2017), 91–131, DOI 10.1007/s11854-017-0029-0. MR3736488

[12] S. Miao, L. Pei and P. Yu *On Classical Global Solutions of Nonlinear Wave Equations with Large Data*, To appear on International Mathematics Research Notices.

[13] J. Rauch, *I. The* u^5 *Klein-Gordon equation. II. Anomalous singularities for semilinear wave equations*, Nonlinear partial differential equations and their applications. Collège de France Seminar, Vol. I (Paris, 1978/1979), Res. Notes in Math., vol. 53, Pitman, Boston, Mass.-London, 1981, pp. 335–364. MR631403

[14] J. Shatah and M. Struwe, *Well-posedness in the energy space for semilinear wave equations with critical growth*, Internat. Math. Res. Notices **7** (1994), 303ff., approx. 7 pp., DOI 10.1155/S1073792894000346. MR1283026

[15] M. Struwe, *Globally regular solutions to the u^5 Klein-Gordon equation*, Ann. Scuola Norm. Sup. Pisa Cl. Sci. (4) **15** (1988), no. 3, 495–513 (1989). MR1015805

[16] T. Tao, *Spacetime bounds for the energy-critical nonlinear wave equation in three spatial dimensions*, Dyn. Partial Differ. Equ. **3** (2006), no. 2, 93–110, DOI 10.4310/DPDE.2006.v3.n2.a1. MR2227039

[17] T. Tao, *Global regularity for a logarithmically supercritical defocusing nonlinear wave equation for spherically symmetric data*, J. Hyperbolic Differ. Equ. **4** (2007), no. 2, 259–265, DOI 10.1142/S0219891607001124. MR2329385

[18] J. Wang and P. Yu, *A large data regime for nonlinear wave equations*, J. Eur. Math. Soc. (JEMS) **18** (2016), no. 3, 575–622, DOI 10.4171/JEMS/597. MR3463418

[19] J. Wang and P. Yu, *Long time solutions for wave maps with large data*, J. Hyperbolic Differ. Equ. **10** (2013), no. 2, 371–414, DOI 10.1142/S0219891613500136. MR3078077

[20] S. Yang, *Global solutions of nonlinear wave equations with large data*, Selecta Math. (N.S.) **21** (2015), no. 4, 1405–1427, DOI 10.1007/s00029-014-0176-4. MR3397453

Bâtiment des Mathématiques, École Polytechnique Fédérale de Lausanne, CH-1015 Lausanne, Switzerland

Current address: School of Mathematics and Statistics, Wuhan University, Wuhan 430072, China

Email address: `shuang.miao@epfl.ch`

Contemporary Mathematics
Volume **725**, 2019
https://doi.org/10.1090/conm/725/14560

The nonlinear Schrödinger equation with an inverse-square potential

Jason Murphy

ABSTRACT. We discuss recent results on the scattering/blow-up dichotomy below the ground state threshold for the focusing nonlinear Schrödinger equation with an inverse-square potential and a nonlinearity that is mass-supercritical and energy-subcritical.

1. Introduction

Nonlinear Schrödinger equations (NLS) with power-type nonlinearities comprise a class of PDE of wide mathematical and physical interest. The most basic equations of this type take the form

$$i\partial_t u = -\Delta u + \mu|u|^p u, \tag{1.1}$$

where $u : \mathbb{R}_t \times \mathbb{R}_x^d \to \mathbb{C}$, $p > 0$, and $\mu \in \{\pm 1\}$ (giving the defocusing and focusing cases, respectively). One may also consider such equations in the presence of an external potential, in which case $-\Delta$ is replaced by a Schrödinger operator of the form $-\Delta + V$. In this article, we consider the case of an inverse-square potential $V(x) = a|x|^{-2}$.

More precisely, we consider the Schrödinger operator

$$\mathcal{L}_a = -\Delta + a|x|^{-2}, \quad a > -(\tfrac{d-2}{2})^2, \tag{1.2}$$

in dimensions $d \geq 3$, defined via the Friedrichs extension with domain $C_c^\infty(\mathbb{R}^d\backslash\{0\})$. The restriction on a comes from the sharp Hardy inequality—it guarantees that $\mathcal{L}_a$ is a positive operator. The operator $\mathcal{L}_a$ appears in a variety of physical settings and is an interesting borderline case from the mathematical point of view.

The addition of an external potential to (1.1) often breaks symmetries of the underlying equation, resulting in new analytic challenges. The inverse-square potential breaks space-translation symmetry; however, one feature of this particular model is that it retains the scaling symmetry

$$u(t,x) \mapsto \lambda^{\frac{2}{p}} u(\lambda^2 t, \lambda x). \tag{1.3}$$

This suggests that one cannot generally rely on perturbative techniques, as the potential and the Laplacian are of equal strength at every scale.

2010 *Mathematics Subject Classification.* Primary 35Q55.

The work discussed in this article was supported in part by NSF DMS-1400706, while the author was a postdoctoral fellow at the University of California, Berkeley.

Two cases of (1.1) have received a great deal of attention, singled out by the fact that in these cases the equation has a conserved quantity that is invariant under the scaling symmetry of the equation. These are the mass- and energy-critical NLS, corresponding to $p = \frac{4}{d}$ and $p = \frac{4}{d-2}$ (in dimensions $d \geq 3$), respectively. The intercritical case refers to $p \in (\frac{4}{d}, \frac{4}{d-2})$.

For the NLS with an inverse-square potential, scattering in the defocusing energy-critical case was established in [**12**], while scattering/blowup dichotomies below the ground state threshold in the focusing intercritical setting were established in [**9**, **10**]. These parallel results previously obtained for the standard NLS (see e.g. [**1**, **3**–**8**]). In fact, a wealth of refined results have been established for (1.1); accordingly, there are many directions of research to be pursued for the inverse-square potential.

To treat the nonlinear equation with an inverse-square potential firstly requires a good linear theory (e.g. Strichartz estimates [**2**]), as well as a set of harmonic analysis tools adapted to the inverse-square potential (see [**12**]). With these ingredients, many of the strategies developed to treat (1.1) (e.g. the concentration compactness approach to induction on energy) can be adapted to attack problems involving the inverse-square potential.

The results discussed in this article, which appeared originally in [**9**, **10**], concern the long-time behavior of solutions to the intercritical focusing NLS with an inverse square potential:

$$i\partial_t u = \mathcal{L}_a u - |u|^p u, \tag{1.4}$$

where $u : \mathbb{R}_t \times \mathbb{R}^d_x \to \mathbb{C}$, $\frac{4}{d} < p < \frac{4}{d-2}$ and $d \geq 3$.

For $a \in (-(\frac{d-2}{2})^2, 0]$, equation (1.4) admits a global but non-scattering solution of the form $u(t) = e^{it}Q_a$, where Q_a (the 'ground state') solves the elliptic problem

$$-\mathcal{L}_a Q_a - Q_a + |Q_a|^p Q_a = 0. \tag{1.5}$$

The functions Q_a may be constructed as optimizers to the Gagliardo–Nirenberg inequality

$$\|f\|_{L^{p+2}}^{p+2} \leq C_a \|f\|_{L^2}^{\frac{4-(d-2)p}{2}} \|f\|_{\dot{H}^1_a}^{\frac{dp}{2}}, \tag{1.6}$$

where $\dot{H}^1_a$ is the homogeneous Sobolev space defined in terms of the operator $\mathcal{L}_a$[1]. For $a > 0$, the sharp constant C_a is equal to C_0; however, no optimizers exist[2].

The functions Q_a, which lead to global but non-scattering solutions to (1.4), provide a natural threshold below which one can prove a simple scattering/blow-up dichotomy.

THEOREM 1.1 (Scattering/blowup dichotomy, [**9**, **10**]). *Suppose $d \geq 3$, $\frac{4}{d} < p < \frac{4}{d-2}$, and $a > -(\frac{d-2}{2})^2$ are as in (1.9) below, and let $u_0 \in H^1(\mathbb{R}^d)$.*

There exists a unique maximal-lifespan solution u to (1.4) with $u|_{t=0} = u_0$. If u_0 is below the ground state threshold, in the sense that

$$M(u_0)^{\frac{4-p(d-2)}{dp-4}} E_a(u_0) < M(Q_{a\wedge 0})^{\frac{4-p(d-2)}{dp-4}} E_{a\wedge 0}(Q_{a\wedge 0}), \tag{1.7}$$

[1]For $a > -(\frac{d-2}{2})^2$, the spaces $\dot{H}^1_a$ and $\dot{H}^1$ are equivalent by the sharp Hardy inequality.

[2]This is due to the lack of compactness coming from space translation. When $a \leq 0$, the compactness can be restored via radial rearrangements and the Riesz rearrangement inequality.

where M and E_a are defined in (1.10) *and* $a \wedge 0 = \min\{a, 0\}$, *then the following dichotomy holds:*

(i) *If*

$$\|u_0\|_{L^2}^{\frac{4-p(d-2)}{dp-4}} \|u_0\|_{\dot H^1_a} < \|Q_{a\wedge 0}\|_{L^2}^{\frac{4-p(d-2)}{dp-4}} \|Q_{a\wedge 0}\|_{\dot H^1_{a\wedge 0}}, \tag{1.8}$$

then u is global in time and scatters in both time directions; that is, there exist solutions $v_\pm$ to the linear equation $i\partial_t v_\pm = \mathcal{L}_a v_\pm$ such that

$$\lim_{t\to\pm\infty} \|u(t) - v_\pm(t)\|_{H^1} = 0.$$

(ii) *If*

$$\|u_0\|_{L^2}^{\frac{4-p(d-2)}{dp-4}} \|u_0\|_{\dot H^1_a} > \|Q_{a\wedge 0}\|_{L^2}^{\frac{4-p(d-2)}{dp-4}} \|Q_{a\wedge 0}\|_{\dot H^1_{a\wedge 0}},$$

and u_0 is radial or $xu_0 \in L^2$, then u blows up in finite time in both time directions.

REMARK 1.2. The precise set of (d, p, a) under consideration is the following:

$$\begin{cases} a > -(\frac{d-2}{2})^2 & \text{for} \quad d = 3 \quad \text{and} \quad \frac{4}{3} < p \le 2, \\ a > -(\frac{d-2}{2})^2 + (\frac{d-2}{2} - \frac{1}{p})^2 & \text{for} \quad 3 \le d \le 6 \quad \text{and} \quad \frac{2}{d-2} \vee \frac{4}{d} < p < \frac{4}{d-2}, \end{cases} \tag{1.9}$$

where $x \vee y = \max\{x, y\}$. These constraints come from the H^1 local well-posedness theory and are related to the issue of equivalence of Sobolev spaces defined in terms of $\mathcal{L}_a$ and the standard Sobolev spaces. We discuss the local theory in Section 2.

REMARK 1.3. The conserved mass and energy of a solution u to (1.4) are defined by

$$M(u) = \int_{\mathbb{R}^d} |u|^2\,dx, \quad E_a(u) = \int_{\mathbb{R}^d} \tfrac{1}{2}|\nabla u|^2 + \tfrac{a}{2|x|^2}|u|^2 - \tfrac{1}{p+2}|u|^{p+2}\,dx, \tag{1.10}$$

respectively. Note that $M(u_0), |E_a(u_0)| < \infty$ for $u_0 \in H^1$ (cf. (1.6) above). The power appearing in (1.7) is chosen to make the product invariant under the scaling symmetry (1.3).

REMARK 1.4. The analogue of Theorem 1.1 for equation (1.1) was originally established in [**4**, **6**–**8**], and the proof of Theorem 1.1 follows a similar overall strategy. However, because of the broken space-translation symmetry, it is important to understand the sense in which (1.1) may be 'embedded' into (1.4) in certain limiting regimes; in particular, one needs to use the analogous result for (1.1) as a black box in order to treat the equation with broken symmetry. For a further discussion, see Lemma 3.4.

In the rest of this article, we describe the proof of Theorem 1.1. The starting point is the following virial identity, which follows from a direct computation using (1.4): Let u be a solution to (1.4) and

$$V(t) := \int |u(t,x)|^2|x|^2\,dx.$$

Then

$$\begin{aligned} V'(t) &= 4\mathrm{Im}\int \bar u \nabla u \cdot x\,dx, \\ V''(t) &= 8\|u(t)\|_{\dot H^1_a}^2 - \tfrac{4pd}{p+2}\|u(t)\|_{L^{p+2}}^{p+2}. \end{aligned} \tag{1.11}$$

Combining this identity with the sharp Gagliardo–Nirenberg inequality (1.6), one deduces that in scenario (i) of Theorem 1.1 one has $V''(t) > 0$, while in scenario (ii) one has $V''(t) < 0$. In particular, in scenario (i), one may expect the solution to spread out and scatter ($V(t) \to \infty$), while in scenario (ii), one may expect the solution to concentrate and blow up in finite time ($V(t) \to 0$ in finite time).

In fact, using the virial identity to establish blow-up in scenario (ii) is fairly standard, and so for the remainder of the article we will focus on proving scattering in scenario (i).

Because we work with merely H^1 data, one cannot implement the virial identity directly—indeed, none of the quantities appearing in the identity are necessarily finite. We therefore follow the concentration compactness approach to induction on energy: we reduce the problem of proving scattering to the problem of precluding the existence of a special type of non-scattering solution. In particular, we reduce to the problem of precluding solutions satisfying (1.7) and (1.8) that have a pre-compact orbit in H^1; for such solutions, a localized version of the virial identity may be implemented in order to reach a contradiction (see Section 3.1).

The main problem therefore reduces to proving the existence of compact 'minimal blowup solutions' under the assumption that Theorem 1.1(i) fails. We will discuss this problem in Section 3; it is here that the broken translation symmetry present in (1.4) plays its most important role.

2. Linear and local theory

In order to treat the nonlinear problem (1.4), it is important to have (i) good estimates for the underlying linear equation and (ii) harmonic analysis tools adapted to the Schrödinger operator $\mathcal{L}_a$.

Strichartz estimates for the linear propagator $e^{-it\mathcal{L}_a}$ were established in [**2**]:

THEOREM 2.1 (Strichartz estimates, [**2**]). *Let $a > -(\frac{d-2}{2})^2$ and $d \geq 3$. Let (q,r) and $(\tilde{q},\tilde{r})$ be admissible pairs, i.e.*

$$2 \leq q, \tilde{q} \leq \infty \quad \text{and} \quad \tfrac{2}{q} + \tfrac{d}{r} = \tfrac{2}{\tilde{q}} + \tfrac{d}{\tilde{r}} = \tfrac{d}{2},$$

with $(q,\tilde{q}) \neq (2,2)$. Suppose $u : I \times \mathbb{R}^d \to \mathbb{C}$ solves

$$(i\partial_t - \mathcal{L}_a)u = F.$$

Then for any $t_0 \in I$, the following estimate holds:

$$\|u\|_{L_t^q L_x^r(I\times\mathbb{R}^d)} \lesssim \|u(t_0)\|_{L_x^2(\mathbb{R}^d)} + \|F\|_{L_t^{\tilde{q}'} L_x^{\tilde{r}'}(I\times\mathbb{R}^d)}.$$

One constructs solutions to (1.4) via a fixed point argument and Strichartz estimates; in particular, solutions satisfy the Duhamel formula

$$u(t) = e^{-it\mathcal{L}_a}u_0 + i\int_0^t e^{-i(t-s)\mathcal{L}_a}(|u|^p u)(s)\,ds.$$

One closes the estimates in spaces of the form $C_t H_a^1 \cap L_t^q H_a^{1,r}$, where (q,r) is an admissible pair and $H_a^{s,r}$ denotes the Sobolev space defined in terms of $\mathcal{L}_a$, i.e.

$$\|f\|_{H_a^{s,r}(\mathbb{R}^d)} = \|(1+\mathcal{L}_a)^{\frac{s}{2}} f\|_{L^r(\mathbb{R}^d)}.$$

In particular, one needs to estimate $\sqrt{\mathcal{L}_a}(|u|^p u)$ in some dual Strichartz space. For this purpose, it is useful to substitute $\sqrt{\mathcal{L}_a}$ with ∇, use the chain rule, and subsequently exchange ∇ with $\sqrt{\mathcal{L}_a}$ again. For this to work, however, one needs to know that Sobolev spaces defined in terms of $\mathcal{L}_a$ are equivalent to those defined in terms of Δ. The sharp range of exponents for which this is the case was worked out in [**11**]:

LEMMA 2.2 (Equivalence of Sobolev spaces, [**11**]). *Let $d \geq 3$, $a > -(\frac{d-2}{2})^2$, and*

$$\rho = \tfrac{d-2}{2} - \big[\big(\tfrac{d-2}{2}\big)^2 + a\big]^{\frac{1}{2}}.$$

Let $0 < s < 2$. If $1 < p < \infty$ satisfies $\frac{s+\rho}{d} < \frac{1}{p} < \min\{1, \frac{d-\rho}{d}\}$, then

$$\| |\nabla|^s f\|_{L^p_x} \lesssim_{d,p,s} \|(\mathcal{L}_a)^{\frac{s}{2}} f\|_{L^p_x} \quad \text{for all} \quad f \in C_c^\infty(\mathbb{R}^d \backslash \{0\}).$$

If $\max\{\frac{s}{d}, \frac{\rho}{d}\} < \frac{1}{p} < \min\{1, \frac{d-\rho}{d}\}$, then

$$\|(\mathcal{L}_a)^{\frac{s}{2}} f\|_{L^p_x} \lesssim_{d,p,s} \| |\nabla|^s f\|_{L^p_x} \quad \text{for all} \quad f \in C_c^\infty(\mathbb{R}^d \backslash \{0\}).$$

Using these ideas, one can establish an H^1 well-posedness theory for (1.4), at least for (d, p, a) obeying (1.9). With

$$q_0 = \tfrac{p(d+2)}{2},$$

we have the following result.

THEOREM 2.3 (Local well-posedness). *Let $t_0 \in \mathbb{R}$, $u_0 \in H^1$, and suppose (d, p, a) satisfy (1.9).*

- *There exist $T = T(\|u_0\|_{H^1}) > 0$ and a unique solution u to (1.1) on $(t_0 - T, t_0 + T)$ with $u(t_0) = u_0$. In particular, if u remains uniformly bounded in H^1 throughout its lifespan, then u extends to a global solution.*
- *Furthermore, there exists $\eta_0 > 0$ so that if*

  $$\|e^{-i(t-t_0)\mathcal{L}} u_0\|_{L^{q_0}_{t,x}((t_0,\infty)\times\mathbb{R}^d)} < \eta \quad \text{for some} \quad 0 < \eta < \eta_0,$$

  then u is forward global and obeys

  $$\|u\|_{L^{q_0}_{t,x}((t_0,\infty)\times\mathbb{R}^d)} \lesssim \eta.$$

  The analogous statement holds backward in time and on all of $\mathbb{R}$.
- *Finally, for any $\psi \in H^1$ there exists a solution to (1.1) that scatters to ψ as $t \to \infty$, and the analogous statement holds backward in time.*

Using persistence of regularity arguments, one can also show that if $u(t_0) \in H^1$ and u obeys $L^{q_0}_{t,x}$-bounds on some interval $I \ni t_0$, then in fact $(1+\mathcal{L}_a)^{\frac{1}{2}} u$ is bounded in every Strichartz norm on I. Consequently, if u remains uniformly bounded in $L^{q_0}_{t,x}$ throughout its lifespan, then u is global and scatters.

Related to the local theory is the stability theory for (1.4), which concerns approximate solutions $\tilde{u}$ to (1.4) and will be important in Section 3.

THEOREM 2.4 (Stability). *Let (d, p, a) satisfy (1.9). Let $\tilde{u}$ solve*

$$i\partial_t \tilde{u} = \mathcal{L}_a \tilde{u} - |\tilde{u}|^p \tilde{u} + e$$

on an interval I for some function e. Suppose

$$\|u_0\|_{H^1} + \|\tilde{u}(t_0)\|_{H^1} \leq E, \quad \|\tilde{u}\|_{L^{q_0}_{t,x}(I\times\mathbb{R}^d)} \leq L.$$

There exists $\varepsilon_0(E, L) > 0$ so that if $0 < \varepsilon < \varepsilon_0$ and

$$\|u_0 - \tilde{u}(t_0)\|_{H^1} + \| |\nabla|^{s_c} e\|_{N(I)} < \varepsilon,$$

where $s_c = \frac{d}{2} - \frac{2}{p}$ *and* N *is a sum of dual Strichartz spaces, then there exists a solution* u *to* (1.4) *with* $u(t_0) = u_0$ *satsifying*

$$\|(\mathcal{L}_a)^{\frac{s_c}{2}}[u - \tilde{u}]\|_{S(I)} \lesssim \varepsilon, \quad \|(1 + \mathcal{L}_a)^{\frac{1}{2}} u\|_{S(I)} \lesssim_{E,L} 1$$

for any Strichartz space S.

REMARK 2.5. The parameter s_c is the critical regularity associated to (1.4). For $\frac{4}{d} < p < \frac{4}{d-2}$, we have $0 < s_c < 1$.

2.1. Harmonic analysis adapted to $\mathcal{L}_a$. In the following, it will be important to have a set of harmonic analysis tools adapted to $\mathcal{L}_a$. Such a toolkit was developed in [**11**], where the central ingredient was a Mihklin-type multiplier theorem for functions of $\mathcal{L}_a$.

We make use of Littlewood–Paley projections defined via the heat kernel:

$$P_N^a := e^{-\mathcal{L}_a/N^2} - e^{-4\mathcal{L}_a/N^2} \quad \text{for} \quad N \in 2^{\mathbb{Z}}.$$

In order to state results, it is convenient to define

$$\tilde{q} := \begin{cases} \infty & \text{if } a \geq 0, \\ \frac{d}{\rho} & \text{if } -(\frac{d-2}{2})^2 < a < 0, \end{cases} \quad \text{where} \quad \rho = \tfrac{d-2}{2} - \big[\big(\tfrac{d-2}{2}\big)^2 + a\big]^{\frac{1}{2}}.$$

We write $\tilde{q}'$ for the dual exponent to $\tilde{q}$. We summarize the tools we need in the following:

LEMMA 2.6 (Harmonic analysis tools, [**11**]). *For* $\tilde{q}' < q \leq r < \tilde{q}$,

$$f = \sum_{N \in 2^{\mathbb{Z}}} P_N^a f \quad \text{as elements of} \quad L_x^r.$$

Furthermore, we have the following Bernstein estimates:

(i) *The operators* P_N^a *are bounded on* L_x^r.
(ii) *The operators* P_N^a *map* L_x^q *to* L_x^r, *with norm* $\mathcal{O}(N^{\frac{d}{q} - \frac{d}{r}})$.
(iii) *For any* $s \in \mathbb{R}$,

$$N^s \|P_N^a f\|_{L_x^r} \sim \big\|(\mathcal{L}_a)^{\frac{s}{2}} P_N^a f\big\|_{L_x^r}.$$

Finally, for $0 \leq s < 2$, *we have the square function estimate:*

$$\Big\|\Big(\sum_{N \in 2^{\mathbb{Z}}} N^{2s} |P_N^a f|^2\Big)^{\frac{1}{2}}\Big\|_{L_x^r} \sim \|(\mathcal{L}_a)^{\frac{s}{2}} f\|_{L_x^r}.$$

3. Construction and exclusion of minimal blowup solutions

As described in the introduction, the most important step in the proof of Theorem 1.1(i) is to show that if the result fails, there exists a minimal blowup solution with certain compactness properties. In this section, we sketch the proof of this fact. At the end of the section, we also sketch the proof that such solutions cannot exist, completing the sketch of the proof of Theorem 1.1(i).

We first define

$$L(\mathcal{E}) := \sup\big\{\|u\|_{L_{t,x}^{q_0}(I \times \mathbb{R}^d)}\big\}, \quad q_0 = \tfrac{p(d+2)}{2},$$

where the supremum is taken over all maximal-lifespan solutions $u : I \times \mathbb{R}^d$ such that

$$M(u)^{\frac{4-p(d-2)}{dp-4}} E_a(u) \leq \mathcal{E} \quad \text{and} \quad \|u(t)\|_{L_x^2}^{\frac{4-p(d-2)}{dp-4}} \|u(t)\|_{\dot{H}_a^1} < \mathcal{K}_a$$

for some $t \in I$. Recall from Theorem 2.3 that uniform boundedness of the $L^{q_0}_{t,x}$-norm throughout the maximal lifespan of a solution implies scattering. Recall also that $L(\mathcal{E})$ is finite for all $\mathcal{E}$ sufficiently small. Thus if Theorem 1.1(i) fails, there exists some critical $\mathcal{E}_c$ such that

$$0 < \mathcal{E}_c < M(Q_{a\wedge 0})^{\frac{4-p(d-2)}{dp-4}} E_{a\wedge 0}(Q_{a\wedge 0})$$

and

$$L(\mathcal{E}) < \infty \quad \text{for} \quad \mathcal{E} < \mathcal{E}_c, \quad L(\mathcal{E}) = \infty \quad \text{for} \quad \mathcal{E} > \mathcal{E}_c.$$

The main result of this section is the following:

THEOREM 3.1 (Existence of minimal blowup solutions). *Suppose Theorem 1.1(i) fails. Then there exists a global solution v to (1.4) satisfying*

$$M(v) = 1, \quad E_a(v) = \mathcal{E}_c, \quad \|v\|_{L^{q_0}_{t,x}((-\infty,0)\times\mathbb{R}^d)} = \|v\|_{L^{q_0}_{t,x}((0,\infty)\times\mathbb{R}^d)} = \infty.$$

Moreover, the orbit $\{v(t)\}_{t\in\mathbb{R}}$ is precompact in H^1.

In fact, Theorem 3.1 can be deduced from the following lemma:

LEMMA 3.2. *Let (d,p,a) satisfy (1.9). Suppose u_n is a sequence of solutions to (1.4) such that*

$$M(u_n) \equiv 1, \quad E_a(u_n) \nearrow \mathcal{E}_c,$$

and suppose t_n are such that

$$\|u_n(t_n)\|_{L^2}^{\frac{4-p(d-2)}{dp-4}} \|u_n(t_n)\|_{\dot H^1_a} < \|Q_{a\wedge 0}\|_{L^2}^{\frac{4-p(d-2)}{dp-4}} \|Q_{a\wedge 0}\|_{\dot H^1_{a\wedge 0}}, \tag{3.1}$$

$$\lim_{n\to\infty} \|u_n\|_{L^{q_0}_{t,x}(\{t>t_n\}\times\mathbb{R}^d)} = \lim_{n\to\infty} \|u_n\|_{L^{q_0}_{t,x}(\{t<t_n\}\times\mathbb{R}^d)} = \infty. \tag{3.2}$$

Then $\{u_n(t_n)\}$ converges along a subsequence in H^1.

PROOF OF THEOREM 3.1. Assume that Theorem 1.1(i) fails. Applying a rescaling, we may find a sequence of solutions u_n and times t_n as in Lemma 3.2. Let v be the solution to (1.1) with initial data given by the subsequential limit of $u_n(t_n)$. Another application of Lemma 3.2 to $v(\tau_n)$ (for an arbitrary sequence τ_n) demonstrates the precompactness. □

We turn to a discussion of the proof of Lemma 3.2. To begin, we remark that the solutions u_n are global. In fact, using

$$M(u_n)^{\frac{4-p(d-2)}{dp-4}} E_a(u_n) < M(Q_{a\wedge 0})^{\frac{4-p(d-2)}{dp-4}} E_{a\wedge 0}(Q_{a\wedge 0})$$

and (3.1), the sharp Gagliardo–Nirenberg inequality implies that (3.1) holds throughout the lifespan of u_n. Thus the u_n remain uniformly bounded in H^1 and hence (by Theorem 2.3) are global.

The proof of Lemma 3.2 proceeds in several steps.

Step 1. After translating so that $t_n \equiv 0$, the first step is to expand the H^1-bounded sequence $\{u_n(0)\}$ in a linear profile decomposition, which provides a way of quantifying any possible lack of compactness in the sequence.

LEMMA 3.3 (Linear profile decomposition). *Passing to a subsequence, there exist $J^* \in \{0,1,2,\ldots,\infty\}$, $\{\phi^j\}_{j=1}^{J^*} \subset H^1$, and $\{(t_n^j, x_n^j)\}_{j=1}^{J^*} \subset \mathbb{R}\times\mathbb{R}^d$ such that for each J,*

$$u_n(0) = \sum_{j=1}^{J} \phi_n^j + r_n^J, \tag{3.3}$$

where $\phi_n^j = [e^{it_n^j \mathcal{L}_a^{n_j}} \phi^j](\cdot - x_n^j)$, *with* $\mathcal{L}_a^{n_j} := -\Delta + \frac{a}{|x+x_n^j|^2}$. *The remainder* r_n^J *satisfies*

$$(e^{-it_n^J \mathcal{L}_a} r_n^J)(x + x_n^J) \rightharpoonup 0 \quad \textit{weakly in} \quad H^1,$$

$$\lim_{J\to J^*} \limsup_{n\to\infty} \|e^{-it\mathcal{L}_a} r_n^J\|_{L_{t,x}^{q_0}(\mathbb{R}\times\mathbb{R}^d)} = 0.$$

The parameters (t_n^j, x_n^j) *are asymptotically orthogonal: for any* $j \neq k$,

$$\lim_{n\to\infty} \left(|t_n^j - t_n^k| + |x_n^j - x_n^k|\right) = \infty. \tag{3.4}$$

Furthermore, for each j, *we may assume that either* $t_n^j \to \pm\infty$ *or* $t_n^j \equiv 0$, *and either* $|x_n^j| \to \infty$ *or* $x_n^j \equiv 0$. *Finally, we have the following decoupling:*

$$\lim_{n\to\infty} \Big\{ \|(\mathcal{L}_a)^{\frac{s}{2}} u_n(0)\|_{L_x^2}^2 - \sum_{j=1}^J \|(\mathcal{L}_a)^{\frac{s}{2}} \phi_n^j\|_{L_x^2}^2 - \|(\mathcal{L}_a)^{\frac{s}{2}} r_n^J\|_{L_x^2}^2 \Big\} = 0, \;\; s \in \{0, 1\},$$

$$\lim_{n\to\infty} \Big\{ \|u_n(0)\|_{L_x^{p+2}}^{p+2} - \sum_{j=1}^J \|\phi_n^j\|_{L_x^{p+2}}^{p+2} - \|r_n^J\|_{L_x^{p+2}}^{p+2} \Big\} = 0.$$

Using the energy decoupling and the sharp Gagliardo–Nirenberg inequality, one can verify that each profile carries positive energy. Note that with this decomposition, the proof of Lemma 3.2 boils down to proving $J = 1$, $t_n^1 \equiv 0$, $x_n^1 \equiv 0$, and $r_n^1 \to 0$ strongly in H^1. Before proceeding to this, we say a few words about the proof of Lemma 3.3.

To prove Lemma 3.3, one successively extracts bubbles of Strichartz-norm concentration from the sequence $\{u_n(0)\}$, each with a well-defined position in space-time and scale. The starting point is a refined Strichartz estimate of the following form:

$$\|e^{-it\mathcal{L}_a} u_n(0)\|_{L_{t,x}^{q_0}(\mathbb{R}\times\mathbb{R}^d)} \lesssim \|u_n(0)\|_{\dot H_a^{s_c}}^{\theta} \sup_{N\in 2^{\mathbb{Z}}} \|e^{-it\mathcal{L}_a} P_N^a u_n(0)\|_{L_{t,x}^{q_0}(\mathbb{R}\times\mathbb{R}^d)}^{1-\theta},$$

which allows one to identify a scale N_n^{-1} responsible for $L_{t,x}^{q_0}$-norm concentration. This estimate is proved by employing harmonic analysis tools adapted to $\mathcal{L}_a$, specifically the square function estimate and Bernstein estimates of Lemma 2.6. After identifying a scale, one can use Hölder's inequality, Strichartz estimates, and Bernstein estimates to identify a location in space-time (τ_n, x_n) where concentration occurs.

Because the sequence is H^1-bounded, one can prove upper and lower bounds for the scale N_n^{-1}; thus without loss of generality $N_n \equiv 1$ (say). One can also arrange that $x_n \to x_\infty \in \mathbb{R}^d \backslash \{0\}$ or $|x_n| \to \infty$, and that $\tau_n \to \tau_\infty \in [-\infty, \infty]$. The first profile ϕ^1 is obtained as the weak limit of the sequence

$$[e^{-it_n\mathcal{L}_a} u_n(0)](x + x_n),$$

where $t_n \equiv 0$ if $\tau_\infty \in \mathbb{R}$ and $\tau_n = t_n$ otherwise. After removing this first bubble from the sequence, one repeats the argument to extract the second bubble, and so on. One can then prove that the sequence of profiles, space-time positions, and remainders obtained in this way satisfy the conclusions of Lemma 3.3. At a technical level, one encounters some issues related to the fact that translation does not commute with $\mathcal{L}_a$, resulting in the need to prove some convergence properties for certain linear operators.

Step 2. Having written $u_n(0)$ in the form (3.3), the next step is to construct nonlinear profiles associated to each ϕ^j. If $x_n^j \equiv 0$ and $t_n^j \equiv 0$, we let v^j be the solution to (1.4) with $v^j(0) = \phi^j$; if instead $t_n^j \to \pm\infty$ we use Theorem 2.3 to find a solution that scatters to $e^{-it\mathcal{L}_a}\phi^j$ as $t \to \pm\infty$. In either case, we then set $v_n^j(t,x) = v^j(t + t_n^j, x)$.

If $|x_n^j| \to \infty$, however, we encounter a significant obstacle: because translation symmetry is broken in (1.4), we cannot simply construct the solution v^j with data ϕ^j and then define v_n^j via translation by x_n^j. Indeed, v_n^j would fail to be a solution.

The following result shows that we may construct nonlinear solutions corresponding to profiles with $|x_n^j| \to \infty$, provided they are below the *Euclidean* ground state. Furthermore, these solutions are global and scatter.

LEMMA 3.4 (Embedding of nonlinear profiles). *Let (d,p,a) satisfy (1.9). Suppose $t_n \equiv 0$ or $t_n \to \pm\infty$ and assume $|x_n| \to \infty$. Let ϕ in H^1 and define*

$$\phi_n(x) = [e^{-it_n\mathcal{L}_a^n}\phi](x - x_n), \quad \mathcal{L}_a^n := -\Delta + \tfrac{a}{|x+x_n|^2}.$$

Suppose

$$M(\phi)^{\frac{4-p(d-2)}{dp-4}} E_0(\phi) < M(Q_0)^{\frac{4-p(d-2)}{dp-4}} E_0(Q_0),$$
$$\|\phi\|_{L^2}^{\frac{4-p(d-2)}{dp-4}} \|\phi\|_{\dot{H}^1} < \|Q_0\|_{L^2}^{\frac{4-p(d-2)}{dp-4}} \|Q_0\|_{\dot{H}^1},$$

in the case $t_n \equiv 0$ and

$$\tfrac{1}{2}\|\phi\|_{L^2}^{\frac{2[4-p(d-2)]}{dp-4}} \|\phi\|_{\dot{H}^1}^2 < M(Q_0)^{\frac{4-p(d-2)}{dp-4}} E_0(Q_0)$$

in the case $t_n \to \pm\infty$. Then there exists a global solution v_n to (1.4) with $v_n(0) = \phi_n$ satisfying $\|(1+\mathcal{L}_a)^{\frac{1}{2}} v_n\|_{S(\mathbb{R})} \lesssim 1$ for any Strichartz norm S.

The key to Lemma 3.4 is to observe that as $|x| \to \infty$, the effect of the potential becomes increasingly weak. In particular, we expect that the desired solution v_n should approximately solve the *free* NLS. Indeed, roughly speaking, the idea of the proof of Lemma 3.4 is as follows:

Using the assumption that ϕ is below the Euclidean ground state, one can use the results of [**1, 4, 7**] to construct scattering solutions $\tilde{v}_n$ to the free NLS (1.1) with data ϕ_n. Relying on convergence results for various linear operators (related to the sense in which $\mathcal{L}_a^n$ 'converges' to $-\Delta$ when $|x_n| \to \infty$), one can prove that the $\tilde{v}_n$ approximately solve (1.4). Thus, using the stability result (Theorem 2.4), one can construct a true solution v_n to (1.4) with data ϕ_n; furthermore, this solution inherits good space-time bounds from $\tilde{v}_n$.

It is an important but simple observation that the ground state thresholds for (1.4) are smaller than the thresholds for (1.1). In particular, because $\mathcal{E}_c$ is below the threshold appearing in Theorem 1.1, we know that all of the profiles in the decomposition of $u_n(0)$ are in fact below the Euclidean threshold; thus we may use Lemma 3.4 to construct nonlinear solutions associated to profiles with $|x_n^j| \to \infty$.

Step 3. The third step is to establish that $J^* = 1$, that is, there can be only one profile. Equivalently, this means that one profile captures all of the critical energy. This is achieved by a contradiction argument. If there were multiple profiles, then each would carry strictly less than the critical energy $\mathcal{E}_c$. In particular, by the definition of $\mathcal{E}_c$, the corresponding nonlinear profiles v_n^j would scatter and obey global space-time bounds. One then considers the following function:

$$u_n^J(t) = \sum_{j=1}^{J} v_n^j(t) + e^{-it\mathcal{L}_a} r_n^J.$$

This function has the following properties: (i) u_n^J agrees with the true solutions u_n at time $t=0$, (ii) u_n^J obeys global space-time bounds, and (iii) u_n^J approximately solves (1.4). Indeed, (i) follows by construction. Along with the space-time bounds obeyed by the v_n^j, the key ingredient to establishing (ii) and (iii) is the orthogonality (3.4) satisfied by the profiles. Indeed, orthogonality is essential if one hopes to (approximately) solve a nonlinear equation by a linear combination of solutions. Using properties (i)–(iii), the stability result Theorem 2.4 implies uniform space-time bounds for the solutions u_n, yielding a contradiction to (3.2).

Step 4. We have reduced the decomposition (3.3) to the form

$$u_n(0) = [e^{it_n \mathcal{L}_a^n} \phi](\cdot - x_n) + r_n, \quad \mathcal{L}_a^n = -\Delta + \tfrac{a}{|x+x_n|^2}.$$

To complete the proof of Lemma 3.2, we need to show $t_n \equiv 0$, $x_n \equiv 0$, and $r_n \to 0$ in H^1. In fact, by the energy decoupling one can see already that $r_n \to 0$ in $\dot{H}^1$. To see that $x_n \equiv 0$, we consider the alternative, namely $|x_n| \to \infty$. In this case, an application of the embedding result Lemma 3.4 and the stability result Theorem 2.4 would imply uniform space-time bounds for the u_n, contradicting (3.2). Similarly, if $t_n \to \pm\infty$, then an application of Theorem 2.4 (comparing u_n with the linear solutions $e^{-it\mathcal{L}_a} u_n(0)$) would contradict (3.2). Finally, to establish $r_n \to 0$ in L^2, it suffices to show $\|\phi\|_{L^2} = 1$. In fact, if $\|\phi\|_{L^2} < 1$, then the definition of $\mathcal{E}_c$ would imply that ϕ leads to a scattering solution, and Theorem 2.4 would once again lead to a contradiction to (3.2). Thus, we complete the sketch of the proof of Lemma 3.2, and hence of Theorem 3.1.

3.1. Preclusion of minimal blowup solutions. To complete our discussion of the proof of Theorem 1.1(i), it remains to describe how one can preclude the existence of solutions as in Theorem 3.1. As mentioned in the introduction, the key is to use a localized version of the virial identity, as follows.

Suppose toward a contradiction that v is a solution as in Theorem 3.1 and let $\varepsilon > 0$. By precompactness, there exists R large enough that

$$\int_{|x|>R} |v(t,x)|^2 + |\nabla v(t,x)|^2 + |v(t,x)|^{p+2}\, dx < \varepsilon$$

uniformly in $t \in \mathbb{R}$. As v is below the ground state threshold, one can use the sharp Gagliardo–Nirenberg inequality and precompactness to deduce

$$8\|u(t)\|_{\dot{H}_a^1}^2 - \tfrac{4dp}{p+2}\|u(t)\|_{L^{p+2}}^{p+2} \gtrsim \|u(t)\|_{\dot{H}_a^1}^2 \gtrsim 1$$

uniformly in $t \in \mathbb{R}$. For ε sufficiently small, one can combine these last two bounds with identities related to (1.11) to deduce that

$$\partial_t^2 \int a_R(x)|u(t,x)|^2\, dx \gtrsim 1,$$

$$\left| \partial_t a_R(x)|u(t,x)|^2\, dx \right| \lesssim R,$$

uniformly in $t \in \mathbb{R}$, where $a_R(x)$ is a smooth function equal to $|x|^2$ for $|x| \le R$ and constant for $|x| \ge 2R$. Integrating over any interval of the form $[0,T]$ and using the fundamental theorem of calculus, one can deduce $T \lesssim R$ (uniformly in T). Choosing T sufficiently large now leads to the desired contradiction and completes the proof of Theorem 1.1(i).

References

[1] T. Akahori and H. Nawa, *Blowup and scattering problems for the nonlinear Schrödinger equations*, Kyoto J. Math. **53** (2013), no. 3, 629–672, DOI 10.1215/21562261-2265914. MR3102564

[2] N. Burq, F. Planchon, J. G. Stalker, and A. S. Tahvildar-Zadeh, *Strichartz estimates for the wave and Schrödinger equations with the inverse-square potential*, J. Funct. Anal. **203** (2003), no. 2, 519–549, DOI 10.1016/S0022-1236(03)00238-6. MR2003358

[3] J. Colliander, M. Keel, G. Staffilani, H. Takaoka, and T. Tao, *Global well-posedness and scattering for the energy-critical nonlinear Schrödinger equation in* $\mathbb{R}^3$, Ann. of Math. (2) **167** (2008), no. 3, 767–865, DOI 10.4007/annals.2008.167.767. MR2415387

[4] T. Duyckaerts, J. Holmer, and S. Roudenko, *Scattering for the non-radial 3D cubic nonlinear Schrödinger equation*, Math. Res. Lett. **15** (2008), no. 6, 1233–1250, DOI 10.4310/MRL.2008.v15.n6.a13. MR2470397

[5] B. Dodson and J. Murphy, *A new proof of scattering below the ground state for the 3D radial focusing cubic NLS*, Proc. Amer. Math. Soc. **145** (2017), no. 11, 4859–4867, DOI 10.1090/proc/13678. MR3692001

[6] D. Fang, J. Xie, and T. Cazenave, *Scattering for the focusing energy-subcritical nonlinear Schrödinger equation*, Sci. China Math. **54** (2011), no. 10, 2037–2062, DOI 10.1007/s11425-011-4283-9. MR2838120

[7] C. D. Guevara, *Global behavior of finite energy solutions to the d-dimensional focusing nonlinear Schrödinger equation*, Appl. Math. Res. Express. AMRX **2** (2014), 177–243. MR3266698

[8] J. Holmer and S. Roudenko, *A sharp condition for scattering of the radial 3D cubic nonlinear Schrödinger equation*, Comm. Math. Phys. **282** (2008), no. 2, 435–467, DOI 10.1007/s00220-008-0529-y. MR2421484

[9] J. Lu, C. Miao, and J. Murphy, *Scattering in* H^1 *for the intercritical NLS with an inverse-square potential*, J. Differential Equations **264** (2018), no. 5, 3174–3211, DOI 10.1016/j.jde.2017.11.015. MR3741387

[10] R. Killip, J. Murphy, M. Visan, and J. Zheng, *The focusing cubic NLS with inverse-square potential in three space dimensions*, Differential Integral Equations **30** (2017), no. 3-4, 161–206. MR3611498

[11] R. Killip, C. Miao, M. Visan, J. Zhang, and J. Zheng, *Sobolev spaces adapted to the Schrödinger operator with inverse-square potential*, Math. Z. **288** (2018), no. 3-4, 1273–1298, DOI 10.1007/s00209-017-1934-8. MR3778997

[12] R. Killip, C. Miao, M. Visan, J. Zhang, and J. Zheng, *The energy-critical NLS with inverse-square potential*, Discrete Contin. Dyn. Syst. **37** (2017), no. 7, 3831–3866, DOI 10.3934/dcds.2017162. MR3639442

Department of Mathematics, University of California, Berkeley, Berkeley, California 94720; and Department of Mathematics and Statistics, Missouri University of Science and Technology, Rolla, Missouri 65409

Email address: `jason.murphy@mst.edu`

Contemporary Mathematics
Volume **725**, 2019
https://doi.org/10.1090/conm/725/14550

The harmonic map heat flow on conic manifolds

Yuanzhen Shao and Changyou Wang

ABSTRACT. In this article, we study the harmonic map heat flow from a manifold with conic singularities to a closed manifold. In particular, we prove the short time existence and uniqueness of solutions as well as the existence of global solutions into manifolds with nonpositive sectional curvature. These results are established in virtue of the maximal regularity theory on manifolds with conic singularities.

1. Introduction

The construction of an $(n+1)$-dimensional conic manifold M starts from a C^∞-compact manifold $(\tilde{\mathsf{M}}, \tilde{g})$ with boundary. Let $(\mathbb{B}, g_{\mathbb{B}}) = (\partial\tilde{\mathsf{M}}, \tilde{g}_{\partial\tilde{\mathsf{M}}})$. Note that $\mathbb{B}$ does not need to be connected. We equip $\mathsf{M} = \tilde{\mathsf{M}} \setminus \mathbb{B}$ with a smooth metric g such that in a closed collar neighbourhood $(0,1] \times \mathbb{B}$ of the boundary

$$g(x,y) = dx^2 + x^2 g_{\mathbb{B}}(y), \quad (x,y) \in (0,1] \times \mathbb{B}.$$

Here x is a boundary defining function of $\tilde{\mathsf{M}}$. Outside $(0,1] \times \mathbb{B}$, g is equivalent to $\tilde{g}$. The resulting C^∞-manifold (M, g) is called a *conic manifold*, and the set $\{x = 0\}$ is called the set of *conic singularities*. See Figure (1).

In their seminal work [**9**], J. Eells and J. H. Sampson initiated the study of the harmonic map heat flow between closed manifolds. In our article, a closed manifold always refers to one that is compact and without boundary.

The goal of this article is to study the harmonic map heat flow from an $(n+1)$-dimensional conic manifold (M, g) to an m-dimensional smooth closed manifold (N, h). We will assume that (N, h) is isometrically embedded into a Euclidean space $\mathbb{R}^L$, and thus can be viewed as a Riemannian submanifold of $\mathbb{R}^L$.

For each $u : \mathsf{M} \to \mathsf{N}$, the harmonic map heat flow aims at finding $u : [0,T) \times \mathsf{M} \to \mathsf{N}$ solving the following equation:

$$\begin{cases} \partial_t u - \Delta_g u = A_g(u)(\nabla u, \nabla u) & \text{on} \quad \mathsf{M}_T; \\ \qquad u(0) = u_0 & \text{on} \quad \mathsf{M}, \end{cases} \tag{1.1}$$

where Δ_g and ∇ denote the Laplace-Beltrami operator on (M, g) and the gradient operator on M, respectively, $A(\cdot)$ is the second fundamental form of N in $\mathbb{R}^L$, $A_g(u)(\nabla u, \nabla u) = g^{\alpha\beta} A(u)(\frac{\partial u}{\partial x_\alpha}, \frac{\partial u}{\partial x_\beta})$, and $\mathsf{M}_T := (0,T) \times \mathsf{M}$ for some $T > 0$.

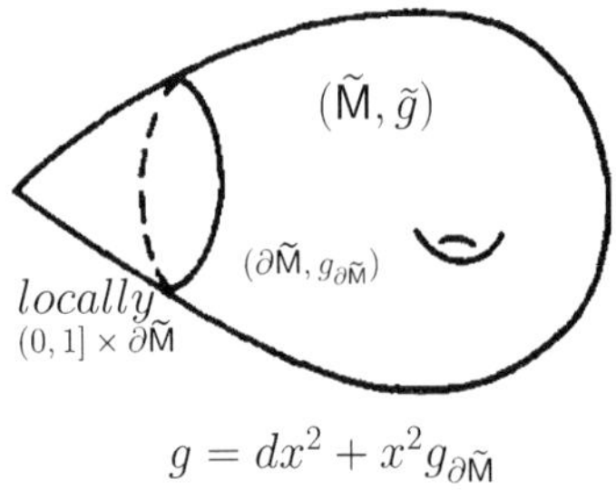

FIGURE 1. A manifold (M, g) with a conical singularity

The harmonic map heat flow is the negative L_2-gradient flow of the Dirichlet energy:

$$E(u) = \frac{1}{2}\int_{\mathsf{M}} |\nabla u|_g^2 dv_g, \tag{1.2}$$

where $|\nabla u|_g^2 = |\nabla u|^2_{g^*\otimes h(u)} = g^{\alpha\beta}h_{ij}(u)\frac{\partial u^i}{\partial x_\alpha}\frac{\partial u^j}{\partial x_\beta}$ with g^* being the induced metric on the cotangent bundle $T^*\mathsf{M}$ by g, and $(u^i)_{i=1,\cdots,m}$ and $(x_\alpha)_{\alpha=1,\cdots,n+1}$ denote the local coordinates of N and M. In addition, v_g stands for the volume element of (M, g).

The critical points of (1.2) are called harmonic maps. A problem of particular interest in geometric analysis is that, given any smooth $u : \mathsf{M} \to \mathsf{N}$, can we deform u into a harmonic map that is homotopic to u?

The harmonic map heat flow (1.1) is aiming at studying the above question by deforming u_0 continuously along the flow to find the critical points of (1.2).

When both the target manifold N and the domain manifold M are closed, J. Eells and J. H. Sampson proved in **[9]** that any smooth initial date $u_0 \in C^\infty(\mathsf{M}, \mathsf{N})$ admits a unique local in time smooth solution to (1.1). However, a certain curvature condition on the target manifold N is essential for their proof for the existence of global smooth solutions. In **[17]**, F. Lin and the second author of this article have obtained an alternative condition on the existence of global smooth solutions, which is different from the curvature condition on N. We would like to refer the reader to **[18]** for a thorough survey of the problem between compact manifolds.

R. Hamilton studied the harmonic map heat flow on compact manifolds with smooth boundary in his book **[12]**. The study of (1.1) on complete and noncompact domain manifolds was initiated by R. Schoen and S.-T Yau **[25]**. In **[25]**, the authors showed that any C^1-map with finite total energy from a complete noncompact M into a complete manifold N with nonpositive sectional curvature can be deformed into a harmonic map along the flow (1.1). Later, many authors considered this problem under various conditions on the complete noncompact domain manifold M and the initial datum u_0. Interested readers may refer to **[10, 14–16, 26, 27]** for more details of the harmonic map heat flow on complete noncompact domain manifolds.

As far as we know, our paper is the first one on the study of the harmonic map heat flow on domain manifolds with singularities. The main tool in our paper is the maximal L_p-regularity theory for nonlinear parabolic equations. Our main results are the following two theorems.

THEOREM 1.1. *Suppose that* (M, g) *is an* $(n+1)$*-dimensional conic manifold and* (N, h) *is an* m*-dimensional smooth closed manifold. Assume that either*

- *when* $n = 1, 2$*, it holds* $\frac{4}{p} < \gamma < 2$*,* $p > 6$*, and*
$$\gamma < \min\{\sqrt{1-\lambda_1} - 1, \sqrt{4-n} - 1\},$$
where λ_1 *is the first non-zero eigenvalue of* $\Delta_{\mathbb{B}}$*; or*
- $0 < \gamma < \frac{n+1}{2}$*,* $p > n+3$ *and* $\gamma + 2 - \frac{4}{p} > \frac{n+1}{2}$*, and*
  - *when* $n \geq 3$*,* $\sqrt{(\frac{n-1}{2})^2 - \lambda_1} > \gamma + 1$*,*
  - *when* $n = 2$ *and* $\lambda_1 < -2$*,* $\gamma \in (\frac{1}{2}, \frac{3}{2})$ *is close enough to* $\frac{1}{2}$ *such that* $\sqrt{\frac{1}{4} - \lambda_1} > \gamma + 1 > \frac{3}{2}$*,*
  - *when* $n = 1$ *and* $\lambda_1 < -1$*, we choose* $\gamma > 0$ *so small that* $\sqrt{-\lambda_1} > \gamma + 1$*.*

Let
$$u_0 = w_0 + \mathsf{p} \quad \textit{with} \quad \mathsf{p} \in \mathsf{N} \quad \textit{and} \quad w_0 \in \mathcal{H}_p^{2-\frac{2}{p}+\delta, \gamma+2-\frac{2}{p}+\delta}(\mathsf{M}, \mathsf{N} - \mathsf{p})$$
for an arbitrary positive constant δ*. Then* (1.1) *with initial condition* u_0 *has a unique solution*
$$u \in L_p(J_T, \mathcal{H}_p^{2,\gamma+2}(\mathsf{M}, \mathbb{R}^L) \oplus \mathbb{R}^L) \cap H_p^1(J_T, \mathcal{H}_p^{0,\gamma}(\mathsf{M}, \mathsf{N}))$$
on $J_T = [0, T)$ *for some* $T > 0$*. Moreover,* $u \in C^\infty((0,T) \times \mathsf{M}, \mathsf{N})$*.*

The space $\mathcal{H}_p^{s,\gamma}(\mathsf{M}, \mathbb{R}^L)$ can be viewed as a Bessel potential space with weight, called Mellin-Sobolev space, which will be introduced in Section 2.1.

For the global existence, we have the following result.

THEOREM 1.2. *Suppose that the sectional curvature of* N *is nonpositive. Assume that*

- *when* $n = 1, 2$*, it holds* $\frac{4}{p} < \gamma < \frac{n+1}{2}$*,* $p > 6$*, and*
$$\gamma < \min\{\sqrt{1-\lambda_1} - 1, \sqrt{4-n} - 1\},$$
where λ_1 *is the first non-zero eigenvalue of* $\Delta_{\mathbb{B}}$*; or*
- $0 < \gamma < \frac{n+1}{2}$*,* $p > n+3$ *and* $\gamma + 2 - \frac{4}{p} > \frac{n+1}{2}$*, and*
  - *when* $n \geq 3$*,* $\sqrt{(\frac{n-1}{2})^2 - \lambda_1} > \gamma + 1$*,*
  - *when* $n = 2$ *and* $\lambda_1 < -2$*,* $\gamma \in (\frac{1}{2}, \frac{3}{2})$ *is close enough to* $\frac{1}{2}$ *such that* $\sqrt{\frac{1}{4} - \lambda_1} > \gamma + 1 > \frac{3}{2}$*,*
  - *when* $n = 1$ *and* $\lambda_1 < -1$*, we choose* $\gamma > 0$ *so small that* $\sqrt{-\lambda_1} > \gamma + 1$*.*

Let
$$u_0 = w_0 + \mathsf{p} \quad \textit{with} \quad \mathsf{p} \in \mathsf{N} \quad \textit{and} \quad w_0 \in \mathcal{H}_p^{2-\frac{2}{p}+\delta, \gamma+2-\frac{2}{p}+\delta}(\mathsf{M}, \mathsf{N} - \mathsf{p})$$
for an arbitrary positive constant δ*; in addition* $\|\nabla w_0\|_\infty < \infty$*. Then* (1.1) *with initial condition* u_0 *has a unique global solution*
$$u \in L_p([0,T), \mathcal{H}_p^{2,\gamma+2}(\mathsf{M}, \mathbb{R}^L) \oplus \mathbb{R}^L) \cap H_p^1([0,T), \mathcal{H}_p^{0,\gamma}(\mathsf{M}, \mathsf{N}))$$
for any $T > 0$*. Moreover,* $u \in C^\infty((0,\infty) \times \mathsf{M}, \mathsf{N})$*.*

We would like to point out that in Theorem 1.2, the necessity of replacing Condition (A) in Theorem 1.1 by the stronger Condition (A') stems from the simple fact that any nonzero constant function belongs to $\mathcal{H}_p^{0,\gamma}(\mathsf{M})$ for $p > 1$ iff $\gamma < \frac{n+1}{2}$.

One of the essential difficulties in the analysis of the Laplace-Beltrami operator on conic manifolds is that there is no canonical choice for a closed extension of this operator, as noticed first by J. Brüning and R. Seeley [**4**]. Indeed, the domains of the minimal and the maximal extensions of Δ_g differ by a non-trivial finite dimensional space. Functions in this finite dimensional space admit certain asymptotic behaviors as $x \to 0^+$, as we will see in Section 3.2. We would like to refer the readers to [**7**, **13**, **22**–**24**] for more details.

This article is organized as follows. In Sections 2.1 and 2.2, we will introduce the precise definitions and fundamental properties of the function spaces on conic manifolds. In Section 2.3, we will establish the point-wise multiplication and Nemyskii operator theory for the function spaces defined in Sections 2.1 and 2.2. Section 3 provides a review of some necessary results on the maximal L_p-regularity theory and closed extensions of the Laplace-Beltrami operator on conic manifolds. In Section 4, we state our choice for the close extensions of the Laplace-Beltrami operator and prove the local existence and uniqueness of solutions to the harmonic map heat flow by means of the results presented in Sections 2.3 and 3. In Section 5, we derive a gradient estimate for the solution to the harmonic map heat flow; and then we utilize the maximal regularity theory to obtain some higher order estimates, which gives us the benefit of establishing a global solution.

NOTATIONS. For any two Banach spaces X, Y, $X \doteq Y$ means that they are equal in the sense of equivalent norms. The notation $\mathcal{L}^j(X,Y)$ means the set of all bounded j-linear maps from X to Y. In particular, we use the notation $\mathcal{L}(X,Y) = \mathcal{L}^1(X,Y)$. Moreover, $\mathcal{L}\mathrm{is}(X,Y)$ stands for the subset of $\mathcal{L}(X,Y)$ consisting of all bounded linear isomorphisms from X to Y.

Given any Banach space X and a manifold $\mathscr{M}$, let $\|\cdot\|_{k,\infty;c}$ and $\|\cdot\|_{s,\gamma;p}$ denote the norms of the X-valued Banach spaces $\mathcal{C}_g^k(\mathscr{M},X)$ and $\mathcal{H}_p^{s,\gamma}(\mathscr{M},X)$, respectively. We will introduce these spaces in Section 2. Meanwhile, $\|\cdot\|_{k,\infty}$ and $\|\cdot\|_{s,p}$ stand for the usual norms of the spaces $BC^k(\mathscr{M},X)$ and $H_p^s(\mathscr{M},X)$.

If the letter X is omitted in the definitions of these spaces, e.g., $\mathcal{C}_g(\mathscr{M})$, it means that the corresponding space is $\mathbb{C}$-valued.

$\mathbb{R}_+$ denotes $(0,\infty)$ and $I = (0,1]$. In addition, $\mathbb{N}_0 = \mathbb{N} \cup \{0\}$.

Throughout the rest of this paper, unless stated otherwise, we always assume that

$$\boxed{s \geq 0,\ k \in \mathbb{N}_0,\ 1 < p < \infty \text{ and } \gamma \in \mathbb{R}.}$$

2. Function spaces on conic manifolds

2.1. Mellin Sobolev spaces. In this subsection, we describe a scale of weighted Sobolev spaces $\mathcal{H}_p^{s,\gamma}(\mathsf{M})$. These weighted Sobolev spaces are widely used in the analysis on conic manifolds. See [**13**, **22**, **23**].

We pick a cut-off function ω on $[0,1)$, which means $\omega \in C^\infty([0,1),[0,1])$ with $\omega \equiv 1$ near 0 and $\omega \equiv 0$ near 1.

For $k \in \mathbb{N}_0$, $\mathcal{H}_p^{k,\gamma}(\mathsf{M})$ is the space of all functions $u \in H_{p,loc}^k(\mathsf{M})$ such that near the conic singularities, or more precisely, in $I \times \mathbb{B}$

$$x^{\frac{n+1}{2}-\gamma}(x\partial_x)^j \partial_y^\alpha(\omega u) \in L_p((0,1] \times \mathbb{B}, \frac{dx}{x}dy), \quad j + |\alpha| \leq k,$$

where $(x, y) \in I \times \mathbb{B}$ and $\alpha \in \mathbb{N}_0^n$. Here ∂_y^α can be considered as the derivatives in local coordinates of $\mathbb{B}$, and we will use this slight abuse of notation in the sequel.

We also put

$$\mathcal{H}_p^{\infty,\gamma}(\mathsf{M}) := \bigcap_{k\in\mathbb{N}_0} \mathcal{H}_p^{k,\gamma}(\mathsf{M}).$$

To understand the motivation of this somewhat unusual definition, let us consider the flat cone $\mathsf{M} = I \times S^n$ in $\mathbb{R}^{n+1}$, where S^n is the n-sphere. Taking polar coordinates in M, then $\mathcal{H}_2^{0,0}(\mathsf{M})$ coincides with the usual $L_2(B_1)$ space (B_1 is the closed unit ball in $\mathbb{R}^{n+1}$).

For arbitrary $\gamma \in \mathbb{R}$, define the map

$$S_\gamma : C_c^\infty(I \times \mathbb{B}) \to C_c^\infty(\overline{\mathbb{R}}_+ \times \mathbb{B}) : u(x, y) \mapsto e^{(\gamma - \frac{n+1}{2})t} u(e^{-t}, y).$$

Then for any $s \geq 0$ and $\gamma \in \mathbb{R}$, $\mathcal{H}_p^{s,\gamma}(\mathsf{M})$ is the space of all distributions on M such that

$$\|u\|_{\mathcal{H}_p^{s,\gamma}(\mathsf{M})} = \|S_\gamma(\omega u)\|_{H_p^s(\overline{\mathbb{R}}_+ \times \mathbb{B})} + \|(1-\omega)u\|_{H_p^s(\mathsf{M})} < \infty.$$

It is understood that ω is extended to be zero outside $(0, 1) \times \mathbb{B}$.

In the sequel, we always assume that $\overline{\mathbb{R}}_+ \times \mathbb{B}$ is equipped with the product metric $\bar{g} = dt^2 + g_{\mathbb{B}}$, and $I \times \mathbb{B}$ is equipped with $g = dx^2 + x^2 g_{\mathbb{B}}$.

LEMMA 2.1.
(i) *For all* $s \geq 0$, $S_\gamma \in \mathcal{L}\mathrm{is}(\mathcal{H}_p^{s,\gamma}(I \times \mathbb{B}), H_p^s(\overline{\mathbb{R}}_+ \times \mathbb{B}))$.
(ii) *Let* $1 < q < \infty$. *Assume that* $s_1 - \frac{n+1}{p} > s_0 - \frac{n+1}{q}$ *and* $\gamma_1 \geq \gamma_0$. *Then the embedding*

$$\mathcal{H}_p^{s_1,\gamma_1}(\mathsf{M}) \hookrightarrow \mathcal{H}_q^{s_0,\gamma_0}(\mathsf{M}).$$

is compact.

PROOF. (i) easily follows from the definition of Mellin-Sobolev spaces.

(ii) First it is easy to see that $\mathcal{H}_p^{s_1,\gamma_1}(\mathsf{M}) \hookrightarrow \mathcal{H}_p^{s_1,\gamma_0}(\mathsf{M})$. Then (ii) is a direct consequence of (i) and the classical embedding results for Bessel potential spaces. □

For $0 < \theta < 1$, we denote by $[\cdot,\cdot]_\theta$ the complex interpolation method, cf. [**1**, Example I.2.4.2], and by $(\cdot,\cdot)_{\theta,q}$ with $1 \leq q \leq \infty$ the real interpolation method, cf. [**1**, Example I.2.4.1].

It was shown in [**23**, Lemma 3.7] that, for Mellin-Sobolev spaces of the same weight, any complex interpolation will lead to another Mellin-Sobolev space.

LEMMA 2.2. *Suppose that* $0 \leq s_0 < s_1$ *and* $\gamma \in \mathbb{R}$. *For any* $0 < \theta < 1$,

$$[\mathcal{H}_p^{s_0,\gamma}(\mathsf{M}), \mathcal{H}_p^{s_1,\gamma}(\mathsf{M})]_\theta \doteq \mathcal{H}_p^{s,\gamma}(\mathsf{M})$$

where $s = (1-\theta)s_0 + \theta s_1$.

Whenever it causes no confusion, in the sequel, we will denote by $\mathbb{C}$ the space of all constant functions on M.

LEMMA 2.3. *Suppose that* $0 \leq s_0 < s_1$ *and* $\gamma_0, \gamma_1 \in \mathbb{R}$. *For any* $\varepsilon > 0$ *and* $0 < \theta < 1$,

$$\mathcal{H}_p^{s+\varepsilon,\gamma+\varepsilon}(\mathsf{M}) \hookrightarrow (\mathcal{H}_p^{s_0,\gamma_0}(\mathsf{M}), \mathcal{H}_p^{s_1,\gamma_1}(\mathsf{M}))_{\theta,p} \hookrightarrow \mathcal{H}_p^{s-\varepsilon,\gamma-\varepsilon}(\mathsf{M}),$$

where $s = (1-\theta)s_0 + \theta s_1$ *and* $\gamma = (1-\theta)\gamma_0 + \theta\gamma_1$, *and for any* $s \geq 0$

$$\begin{aligned}\mathcal{H}_p^{s+2\theta+\varepsilon,\gamma+2\theta+\varepsilon}(\mathsf{M}) \oplus \mathbb{C} &\hookrightarrow (\mathcal{H}_p^{s,\gamma}(\mathsf{M}), \mathcal{H}_p^{s+2,\gamma+2}(\mathsf{M}) \oplus \mathbb{C})_{\theta,p} \\ &\hookrightarrow \mathcal{H}_p^{s+2\theta-\varepsilon,\gamma+2\theta-\varepsilon}(\mathsf{M}) \oplus \mathbb{C}.\end{aligned}$$

The second statement holds as long as $\gamma + 2\theta - \varepsilon > \frac{n+1}{2}$.

PROOF. The first embeddings follow from [**6**, Lemma 5.4] and [**23**, Lemma 3.6]. The second embeddings were proved in [**23**, Lemma 5.2]. The extra assumption $\gamma + 2\theta - \varepsilon > \frac{n+1}{2}$ is to guarantee the sums of the weighted Mellin- Sobolev spaces with $\mathbb{C}$ are direct. ☐

2.2. Conic BC^k-spaces. In this subsection, we will first give a brief introduction to the usual BC^k-spaces on

$$(\mathscr{M}, g_{\mathscr{M}}) \in \{(\tilde{\mathsf{M}}, \tilde{g}), (\overline{\mathbb{R}}_+ \times \mathbb{B}, dt^2 + g_{\mathbb{B}}), (\Omega, dx_1^2 + \cdots + dx_L^2)\},$$

where Ω is an open subset of some Euclidean space $\mathbb{R}^L$.

The space $BC^k(\mathscr{M})$ is defined by

$$BC^k(\mathscr{M}) := \{u \in C^k(\mathscr{M}) : \|u\|_{k,\infty} := \sum_{i=0}^{k} \|\nabla^i_{g_{\mathscr{M}}} u\|_\infty < \infty\}.$$

We also set $BC^\infty(\mathscr{M}) := \bigcap_{k\in\mathbb{N}_0} BC^k(\mathscr{M})$.

Similarly, we can also define conic BC^k-spaces on (M, g) with respect to the conic metric g. More precisely, the space $\mathcal{C}_g^k(\mathsf{M})$ consists of all the $k-$times continuously differentiable functions u such that

$$\|u\|_{k,\infty;c} := \|(1-\omega)u\|_{k,\infty} + \sum_{j+|\alpha|\leq k} \|(\partial_x)^j(\frac{1}{x}\partial_y)^\alpha(\omega u)\|_\infty < \infty.$$

We set $\mathcal{C}_g(\mathsf{M}) := \mathcal{C}_g^0(\mathsf{M})$ and

$$\mathcal{C}_g^\infty(\mathsf{M}) := \bigcap_{k\in\mathbb{N}_0} \mathcal{C}_g^k(\mathsf{M}).$$

REMARK 2.4. In the above definitions of Mellin-Sobolev spaces and conic BC^k-spaces, those functions are the usual Bessel potential or BC^k-spaces outside $I \times \mathbb{B}$, and the conic metric $g = dx^2 + x^2 g_{\mathbb{B}}$ only plays a role near the conic singularities. We can always decompose a function u into $u = \omega u + (1-\omega)u$. Therefore, in the sequel, for a function in Mellin-Sobolev spaces or conic BC^k-spaces, we will only analyze it inside $I \times \mathbb{B}$.

Recall that $S_\gamma u(t,y) = e^{(\gamma - \frac{n+1}{2})t} u(e^{-t}, y)$. It is not a hard job to check that

$$\|S_{\frac{n+1}{2}}(\omega u)\|_{k,\infty} \leq \|(\omega u)\|_{k,\infty;c} \leq \|S_{\frac{n+1}{2}+k}(\omega u)\|_{k,\infty}$$

for any $u \in \mathcal{C}_g^k(\mathsf{M})$. The rightmost term can be infinite.

This equality gives rise to the following two propositions.

PROPOSITION 2.5.

$$S_{\frac{n+1}{2}} \in \mathcal{L}(\mathcal{C}_g^k(I \times \mathbb{B}), BC^k(\overline{\mathbb{R}}_+ \times \mathbb{B})), \quad k \in \mathbb{N}_0.$$

PROPOSITION 2.6. *Suppose that* $s > k + \frac{n+1}{p}$ *and* $\gamma \geq \frac{n+1}{2} + k$. *Then*

$$\mathcal{H}_p^{s,\gamma}(\mathsf{M}) \hookrightarrow \mathcal{C}_g^k(\mathsf{M}).$$

PROOF. By definition, we have $\mathcal{H}_p^{s,\gamma}(\mathsf{M}) \hookrightarrow \mathcal{H}_p^{s,\frac{n+1}{2}+k}(\mathsf{M})$. If $u \in \mathcal{H}_p^{s,\frac{n+1}{2}+k}(\mathsf{M})$, then it follows from the assumption $s > k + \frac{n+1}{p}$ and the conventional Sobolev embedding theorem, cf. [**2**, Theorem 14.2(ii)], that

$$\|S_{\frac{n+1}{2}+k}(\omega u)\|_{BC^k(\overline{\mathbb{R}}_+ \times \mathbb{B})} \le C \|S_{\frac{n+1}{2}+k}(\omega u)\|_{H_p^s(\overline{\mathbb{R}}_+ \times \mathbb{B})}.$$

Now applying the conventional Sobolev embedding theorem once more to $(1-\omega)u$ will establish the desired result. □

REMARK 2.7. It follows from [**21**, Corollary 2.5] that if $s > \frac{n+1}{p}$, we actually have the estimate $|u(x,y)| \le C x^{\gamma - \frac{n+1}{2}} \|u\|_{s,\gamma;p}$.

2.3. The Nemyskii operator in Mellin Sobolev spaces. One of the crucial steps in showing the well-posedness and regularity of the harmonic map heat flow (1.1) is to obtain the necessary regularity for the operator

$$[u \mapsto A_g(u)(\nabla u, \nabla u)].$$

To this end, the objective of this subsection is to establish a Nemyskii operator theorem for Mellin-Sobolev spaces defined in Section 2.1. However, in order not to cause too much distraction and to keep this article in a reasonable length, we will not state the optimal result, but only confine ourselves to the least necessary level.

For any Banach space E, one can generalize the definitions of Mellin-Sobolev and conic BC^k-spaces introduced in Sections 2.1 and 2.2 to E-valued spaces without difficulty. We will denote these generalized spaces by $\mathcal{H}_p^{s,\gamma}(\mathsf{M}, E)$ and $\mathcal{C}_g^k(\mathsf{M}, E)$, respectively.

Suppose X_i with $i = 0, 1, 2$ are finite dimensional Banach spaces. We call

$$\bullet : (v_1, v_2) \mapsto v_1 \bullet v_2 = v_0, \qquad v_i \in X_i,$$

a Banach multiplication from $X_1 \times X_2 \to X_0$ if it satisfies

$$\|v_0\|_{X_0} \le c \|v_1\|_{X_1} \|v_2\|_{X_2}.$$

THEOREM 2.8. *Let $\bullet$ be a Banach multiplication from $X_1 \times X_2 \to X_0$. Then $[(v_1, v_2) \mapsto v_1 \bullet v_2]$ is a continuous bilinear map from*

(i) *$\mathcal{H}_p^{s,\gamma}(\mathsf{M}, X_1) \times \mathcal{H}_p^{s,\gamma'}(\mathsf{M}, X_2)$ to $\mathcal{H}_p^{s,\gamma+\gamma'-\frac{n+1}{2}}(\mathsf{M}, X_0)$ if $s > \frac{n+1}{p}$. In particular, if in addition $\gamma \ge \frac{n+1}{2}$, $\mathcal{H}_p^{s,\gamma}(\mathsf{M})$ is a Banach algebra.*

(ii) *$\mathcal{H}_p^{s,\gamma}(\mathsf{M}, X_1) \times \mathcal{C}_g^k(\mathsf{M}, X_2)$ to $\mathcal{H}_p^{s,\gamma}(\mathsf{M}, X_0) \times 0)$ if $k \ge s$.*

PROOF. As we have pointed out in Section 2.2, only the part of a function in $I \times B$ will be considered. For notional brevity, let $\vartheta = \gamma + \gamma' - \frac{n+1}{2}$. For any $u \in \mathcal{H}_p^{s,\gamma}(\mathsf{M}, X_1)$ and $v \in \mathcal{H}_p^{s,\gamma'}(\mathsf{M}, X_2)$, we have

$$\begin{aligned}
\|u \bullet v\|_{s,\vartheta;p} &\le c \|S_\vartheta(u \bullet v)(t,y)\|_{s,p} \\
&= c \|e^{(\vartheta - \frac{n+1}{2})t} u(e^{-t}, y) \bullet v(e^{-t}, y)\|_{s,p} \\
&= c \|e^{(\gamma - \frac{n+1}{2})t} u(e^{-t}, y) \bullet e^{(\gamma' - \frac{n+1}{2})t} v(e^{-t}, y)\|_{s,p} \\
&\le c \|e^{(\gamma - \frac{n+1}{2})t} u(e^{-t}, y)\|_{H_p^s(\bar{R}_+ \times \mathbb{B}, X_1)} \|e^{(\gamma' - \frac{n+1}{2})t} v(e^{-t}, y)\|_{H_p^s(\bar{R}_+ \times \mathbb{B}, X_2)} \\
&\le c \|u\|_{\mathcal{H}_p^{s,\gamma}(I \times \mathbb{B}, X_1)} \|v\|_{\mathcal{H}_p^{s,\gamma'}(I \times \mathbb{B}, X_2)}. \qquad (2.1)
\end{aligned}$$

Step (2.1) follows from an easy generalization of the conventional pointwise multiplication theorem on compact manifolds to the half cylinder $\overline{\mathbb{R}}_+ \times \mathbb{B}$.

This proves (i). One can check directly via the definitions of Mellin-Sobolev and conic BC^k-spaces that Statement (ii) holds for $s \in \mathbb{N}_0$ and $s \leq k$. For non-integer s, the assertion follows from the interpolation theory and Lemma 2.2. $\square$

In the rest of this section, we assume that $\mathcal{O} \subset \mathbb{R}^L$ for some $L \in \mathbb{N}$ is open, $\mathscr{M}$ is a manifold, and E is a Banach space. Let $\phi : \mathcal{O} \to E$. Then the Nemyskii operator, or the so called substitution operator, Φ induced by ϕ is defined by

$$\Phi : \mathcal{O}^{\mathscr{M}} \to E^{\mathscr{M}}, \quad u \mapsto \phi(u).$$

We use $\mathrm{Rng}(f)$ to denote the range of an $\mathbb{R}^L$-valued function f. For a set X of $\mathbb{R}^L$-valued functions, similarly, we set

$$\mathrm{Rng}(X) = \bigcup_{f \in X} \mathrm{Rng}(f).$$

PROPOSITION 2.9. *Suppose that $p > n+1$ and $\gamma \geq \frac{n+1}{2}$. Let $\phi \in BC^{\nu}(\mathcal{O}, E)$ with $[s] \leq \nu \in \mathbb{N}_0$ and X be an open subset of $\mathcal{H}_p^{s,\gamma}(\mathsf{M}, \mathbb{R}^L) \oplus \mathbb{R}^L$ such that* $\mathrm{Rng}(X) \subset \mathcal{O}$. *Then for any $\vartheta < \frac{n+1}{2}$*

$$\Phi(X) \subset \mathcal{H}_p^{[s],\vartheta}(\mathsf{M}, E).$$

PROOF. Let $u \in X$. When $s \in [0,1)$, since $\|\Phi(u)\|_\infty < \infty$, for any $\vartheta < \frac{n+1}{2}$ we obviously have

$$\Phi(u) \in \mathcal{H}_p^{0,\vartheta}(\mathsf{M}, E).$$

For $s \geq 1$, note that

$$\partial^j \phi \in BC^{\nu - j}(\mathcal{O}, \mathcal{L}^j(\mathbb{R}^L, E)), \quad j \leq \nu.$$

The assertion now follows from the definition of Mellin-Sobolev spaces, Theorem 2.8 and the chain rule for higher order derivatives, that is,

$$\frac{\partial^l}{\partial z^l}\phi(u) = \sum_{k_1 + 2k_2 + \cdots + lk_l = l} \frac{l!}{k_1! k_2! \cdots k_l!} \partial^k \phi(u) \left(\frac{\partial_z u}{1!}\right)^{k_1} \left(\frac{\partial_z^2 u}{2!}\right)^{k_2} \cdots \left(\frac{\partial_z^l u}{l!}\right)^{k_l}$$

where $z \in \{x, y_1, \cdots, y_n\}$, $l \leq [s]$ and $k = k_1 + k_2 + \cdots + k_l$. $\square$

THEOREM 2.10. *Suppose that $k \in \mathbb{N}_0$, $s > \frac{n+1}{p}$, $p > n+1$ and $\gamma \geq \frac{n+1}{2}$. Let $\phi \in BC^{\nu+k+1}(\mathcal{O}, E)$ for some $[s] \leq \nu \in \mathbb{N}_0$ and X be an open subset of $\mathcal{H}_p^{s,\gamma}(\mathsf{M}, \mathbb{R}^L) \oplus \mathbb{R}^L$ such that* $\mathrm{Rng}(X) \subset \mathcal{O}$. *Then for any $\vartheta < \frac{n+1}{2}$*

$$\Phi \in C^{k+1-}(X, \mathcal{H}_p^{[s],\vartheta}(\mathsf{M}, E)).$$

PROOF. For each $u \in X$, by the openness of $\mathcal{O}$ and Proposition 2.6, we can find a bounded convex neighbourhood U of u in $\mathcal{H}_p^{s,\gamma}(\mathsf{M}, \mathbb{R}^L) \oplus \mathbb{R}^L$ such that $U \subset X$.

(i) We first consider the case $k = 0$. Pick $u, v \in U$. Then

$$\Phi(u) - \Phi(v) = m(u,v)(u - v),$$

where

$$m(u,v) = \int_0^1 \partial\phi(v + \tau(u - v))\, d\tau.$$

Since $\partial\phi \in BC^{\nu}(\mathcal{O}, \mathcal{L}(\mathbb{R}^L, E))$, by Proposition 2.9 and its proof, one can find a constant $C = C(U) > 0$ such that

$$\|\partial\phi(w)\|_{\mathcal{H}_p^{[s],\vartheta}(\mathsf{M}, \mathcal{L}(\mathbb{R}^L, E))} \leq C, \quad w \in U.$$

So we can infer from $\gamma \geq \frac{n+1}{2}$, Proposition 2.6 and Theorem 2.8 that

$$\|\Phi(u) - \Phi(v)\|_{\mathcal{H}_p^{[s],\vartheta}(\mathsf{M},E)} \leq C\|u - v\|_{\mathcal{H}_p^{s,\gamma}(\mathsf{M},\mathbb{R}^L)\oplus\mathbb{R}^L}.$$

In particular, when $s < 1$, we apply both Proposition 2.6 and Theorem 2.8(ii) in the above inequality. When $s \geq 1$, applying only $\gamma \geq \frac{n+1}{2}$ and Theorem 2.8(i) is sufficient. This proves the asserted statement for $k = 0$.

(ii) Suppose that $k = 1$. For any $h \in \mathcal{H}_p^{s,\gamma}(\mathsf{M}, \mathbb{R}^L) \oplus \mathbb{R}^L$ with $\|h\|_{\mathcal{H}_p^{s,\gamma}(\mathsf{M},\mathbb{R}^L)\oplus\mathbb{R}^L}$ small enough,

$$\Phi(u + h) - \Phi(u) - \partial\phi(u)h = \int_0^1 [\partial\phi(u + \tau h) - \partial\phi(u)]h \, d\tau \tag{2.2}$$

holds point-wise on M.

Because $\partial\phi \in BC^{1+\nu}(\mathcal{O}, \mathcal{L}(\mathbb{R}^L, E))$, by Step (i), we know that the Nemyskii operator B induced by $\partial\phi$ satisfies

$$B \in C^{1-}(X, \mathcal{H}_p^{[s],\vartheta}(\mathsf{M}, \mathcal{L}(\mathbb{R}^L, E)).$$

Using (2.2), one can obtain

$$\begin{aligned}
&\|\Phi(u + h) - \Phi(u) - \partial\phi(u)h\|_{\mathcal{H}_p^{[s],\vartheta}(\mathsf{M},E)} \\
\leq& \int_0^1 \|[\partial\phi(u + \tau h) - \partial\phi(u)]h\|_{\mathcal{H}_p^{[s],\vartheta}(\mathsf{M},E)} \, d\tau \\
\leq& \int_0^1 \|\partial\phi(u + \tau h) - \partial\phi(u)\|_{\mathcal{H}_p^{[s],\vartheta}(\mathsf{M},\mathcal{L}(\mathbb{R}^L,E))} \, d\tau \|h\|_{\mathcal{H}_p^{s,\gamma}(\mathsf{M},\mathbb{R}^L)\oplus\mathbb{R}^L} \\
\leq& C\|h\|^2_{\mathcal{H}_p^{s,\gamma}(\mathsf{M},\mathbb{R}^L)\oplus\mathbb{R}^L}
\end{aligned}$$

via Proposition 2.6 and Theorem 2.8 as in Step (i). This implies that

$$\Phi \in C^{2-}(X, \mathcal{H}_p^{[s],\vartheta}(\mathsf{M}, E))$$

with $\partial\Phi(u)h = \partial\phi(u)h$.

(iii) The general case $k > 1$ follows now by an induction argument. □

3. The Laplace-Beltrami operator

In this section, we will state some known results about the Laplace-Beltrami operator on conic manifolds, which will serve as the theoretical basis for the analysis of the harmonic map heat flow in the next section.

The Laplace-Beltrami operator Δ_g induced by the conic metric g is a second order differential operator, which near the conic singularities, i.e. inside $I \times \mathbb{B}$, can be written as

$$\Delta_g = \frac{1}{x^2}[(x\partial_x)^2 + (n - 1)(x\partial_x) + \Delta_{\mathbb{B}}],$$

where $\Delta_{\mathbb{B}}$ is the Laplace-Beltrami operator on $\mathbb{B}$ with respect to $g_{\mathbb{B}}$.

This operator acts in a natural way on the scale of weighted Mellin-Sobolev spaces defined in Section 2:

$$\Delta_g \in \mathcal{L}(\mathcal{H}_p^{s+2,\gamma+2}(\mathsf{M}), \mathcal{H}_p^{s,\gamma}(\mathsf{M})),$$

for all $s \geq 0$, $\gamma \in \mathbb{R}$ and $1 < p < \infty$, as a bounded differential operator.

The conormal symbol of Δ_g is

$$\sigma_{\mathsf{M}}(\Delta_g)(z) := z^2 - (n - 1)z + \Delta_{\mathbb{B}}, \quad z \in \mathbb{C}.$$

In particular, $\sigma_{\mathsf{M}}(\Delta_g) \in \mathcal{A}(\mathbb{C}, \mathcal{L}(H_p^{s+2}(\mathbb{B}), H_p^s(\mathbb{B})))$, where $\mathcal{A}(\mathbb{C}, E)$ denotes the space of holomorphic E-valued functions on $\mathbb{C}$ for any Banach space E. The definition of the conormal symbols is inspired by the fact that

$$M(x\partial_x f)(z) = zMf(z),$$

where the Mellin transform is defined by

$$Mf(z) = \int_0^\infty x^{z-1} f(x)\, dx, \qquad z \in \mathbb{C}.$$

We would like to refer the reader to [**13**] for more details of the Mellin transform and the conormal symbols.

3.1. $\mathscr{R}$-sectorial operators and L_p-maximal regularity. Maximal regularity theory has proved itself a very important tool in the study of nonlinear parabolic equations. In combination with a fixed point argument and the implicit function theorem, the theory of maximal regularity can be used to establish well-posedness, regularity and stability of solutions. In this subsection, we will introduce several important concepts and theorems in maximal regularity theory. The reader may refer to the treatises [**1, 8, 20**] for a thorough survey of the theory.

For $\theta \in (0, \pi]$, the open sector with angle 2θ is denoted by

$$\Sigma_\theta := \{\omega \in \mathbb{C} \setminus \{0\} : |\arg \omega| < \theta\}.$$

DEFINITION 3.1. Let X be a complex Banach space, and $\mathcal{A}$ be a densely defined closed linear operator in X with dense range. $\mathcal{A}$ is called sectorial if $\Sigma_\theta \subset \rho(-\mathcal{A})$ for some $\theta > 0$ and

$$\sup\{\|\mu(\mu + \mathcal{A})^{-1}\|_{\mathcal{L}(X)} : \mu \in \Sigma_\theta\} < \infty.$$

The class of sectorial operators in X is denoted by $\mathcal{S}(X)$. The spectral angle $\phi_{\mathcal{A}}$ of $\mathcal{A}$ is defined by

$$\phi_{\mathcal{A}} := \inf\{\phi : \Sigma_{\pi-\phi} \subset \rho(-\mathcal{A}), \sup_{\mu \in \Sigma_{\pi-\phi}} \|\mu(\mu + \mathcal{A})^{-1}\|_{\mathcal{L}(X)} < \infty.\}.$$

DEFINITION 3.2. Suppose that $\mathcal{A} \in \mathcal{S}(X)$. Then $\mathcal{A}$ is said to admit bounded imaginary powers if $\mathcal{A}^{is} \in \mathcal{L}(X)$ for each $s \in \mathbb{R}$, and there exists some constant $C > 0$ such that

$$\|\mathcal{A}^{is}\|_{\mathcal{L}(X)} \leq C, \quad |s| \leq 1.$$

The class of such operators is denoted by $\mathcal{BIP}(X)$. The power angle $\theta_{\mathcal{A}}$ of $\mathcal{A}$ is defined by

$$\theta_{\mathcal{A}} := \overline{\lim}_{|s| \to \infty} \frac{1}{|s|} \log \|\mathcal{A}^{is}\|_{\mathcal{L}(X)}.$$

DEFINITION 3.3. Let X and Y be two Banach spaces. A family of operators $\mathcal{T} \in \mathcal{L}(X, Y)$ is called $\mathscr{R}$-bounded, if there is a constant $C > 0$ and $p \in [1, \infty)$ such that for each $N \in \mathbb{N}$, $T_j \in \mathcal{T}$ and $x_j \in X$ and for all independent, symmetric, $\{-1, 1\}$-valued random variables ε_j on a probability space $(\Omega, \mathcal{M}, \mu)$ the inequality

$$\|\sum_{j=1}^N \varepsilon_j T_j x_j\|_{L_p(\Omega, Y)} \leq C \|\sum_{j=1}^N \varepsilon_j x_j\|_{L_p(\Omega, X)}$$

is valid. The smallest such C is called the $\mathscr{R}$-bound of $\mathcal{T}$. We denote it by $\mathscr{R}(\mathcal{T})$.

DEFINITION 3.4. Suppose that $\mathcal{A} \in \mathcal{S}(X)$. Then $\mathcal{A}$ is called $\mathscr{R}$-sectorial if there exists some $\phi > 0$ such that

$$\mathscr{R}_{\mathcal{A}}(\phi) := \mathscr{R}\{\mu(\mu + \mathcal{A})^{-1} : \mu \in \Sigma_\phi\} < \infty.$$

The $\mathscr{R}$-angle $\phi^R_{\mathcal{A}}$ is defined by

$$\phi^R_{\mathcal{A}} := \inf\{\theta \in (0, \pi) : \mathscr{R}_{\mathcal{A}}(\pi - \theta) < \infty\}.$$

The class of $\mathscr{R}$-sectorial operators in X is denoted by $\mathcal{RS}(X)$.

DEFINITION 3.5. A Banach space X is said to belong to the class $\mathcal{HT}$ if the Hilbert Transform defined by

$$H(f)(t) := \lim_{\epsilon \to 0} \int_{|s|>\epsilon} f(t - s)\frac{ds}{\pi s}, \quad t \in \mathbb{R},\ f \in C_0(\mathbb{R}, X)$$

can be extended to a bounded linear operator on $L_p(\mathbb{R}, X)$ for some $p \in (1, \infty)$.

In particular, Mellin-Sobolev spaces defined in Section 2.1 are of class $\mathcal{HT}$. Suppose that X is a Banach space of class $\mathcal{HT}$. Then by [**8**, formula (2.15), Remark 3.2(1), Theorem 4.5], we obtain the inclusions

$$\mathcal{BIP}(X) \subset \mathcal{RS}(X) \subset \mathcal{S}(X), \tag{3.1}$$

and the inequalities

$$\theta_{\mathcal{A}} \geq \phi^R_{\mathcal{A}} \geq \phi_{\mathcal{A}}. \tag{3.2}$$

DEFINITION 3.6. Assume that $X_1 \overset{d}{\hookrightarrow} X_0$ is some densely embedded Banach couple. Suppose that $\mathcal{A} \in \mathcal{S}(X_0)$ with $\mathrm{dom}(\mathcal{A}) = X_1$. The Cauchy problem

$$\begin{cases} \partial_t u(t) + \mathcal{A}u(t) = f(t) & t \geq 0 \\ u(0) = u_0 \end{cases} \tag{3.3}$$

has maximal L_p-regularity if for any

$$(f, u_0) \in L_p(\mathbb{R}_+, X_0) \times (X_0, X_1)_{1-1/p,p},$$

(3.3) has a unique solution

$$u \in L_p(\mathbb{R}_+, X_1) \cap H^1_p(\mathbb{R}_+, X_0).$$

We denote this by

$$\mathcal{A} \in \mathcal{MR}_p(X_0).$$

Note that Definition 3.6 basically tells us that an operator $\mathcal{A} \in \mathcal{MR}_p(X_0)$ iff the map

$$(\partial_t + \mathcal{A}, \gamma_0) \in \mathcal{L}\mathrm{is}(L_p(\mathbb{R}_+, X_1) \cap H^1_p(\mathbb{R}_+, X_0), L_p(\mathbb{R}_+, X_0) \times (X_0, X_1)_{1-1/p,p}), \tag{3.4}$$

where γ_0 is the temporal trace operator.

The class $\mathcal{MR}_p(X_0)$ is closely related to the class of $\mathscr{R}$-sectorial operators.

PROPOSITION 3.7. [**8**, Theorem 4.4] *Assume that $X_1 \overset{d}{\hookrightarrow} X_0$ is some densely embedded Banach couple. Suppose that $\mathcal{A} \in \mathcal{S}(X_0)$ with $\mathrm{dom}(\mathcal{A}) = X_1$, and X_0 belongs to the class $\mathcal{HT}$. Then the Cauchy problem (3.3) has maximal L_p-regularity for all $1 < p < \infty$ iff*

$$\mathcal{A} \in \mathcal{RS}(X_0)$$

with $\phi^R_{\mathcal{A}} < \pi/2$.

Maximal regularity theory is a powerful tool in the theory of nonlinear parabolic equations. To illustrate its power, let us consider the following abstract evolution equation

$$\begin{cases} \partial_t u + \mathcal{A}u = F(u), & t > 0; \\ u(0) = u_0, \end{cases} \tag{3.5}$$

in X_0. We have the following existence and uniqueness result for equation (3.5).

THEOREM 3.8. [**5**, Theorem 2.1] *Let $1 < p < \infty$ and $X_1 \overset{d}{\hookrightarrow} X_0$ be a densely embedded pair of Banach spaces. Assume that $\mathcal{A} \in \mathcal{MR}_p(X_0)$ with* $\mathrm{dom}(\mathcal{A}) = X_1$. *Setting $X_{1-1/p} := (X_0, X_1)_{1-1/p,p}$, suppose that $U \subset X_{1-1/p}$ is open and that F satisfies*

$$F \in C^{1-}(U, X_0).$$

Then for every $u_0 \in U$, there exist $T = T(u_0) > 0$ and a unique solution of (3.5) *on $J = [0, T)$ with*

$$u \in L_p(J, X_1) \cap H^1_p(J, X_0).$$

3.2. Closed extensions of the Laplace-Beltrami operator on conic manifolds. In this subsection, we will quote some well-established results on the closed extensions of the conic Laplace-Beltrami operator. More details of these results can be found in [**22–24**].

If we consider Δ_g as an unbounded operator on $\mathcal{H}^{s,\gamma}_p(\mathsf{M})$ with domain $C^\infty_c(\mathsf{M})$, denote its closure by $\Delta_{g,\min} = \Delta^\gamma_{g,s,\min}$, and its maximal closed extension by $\Delta_{g,\max} = \Delta^\gamma_{g,s,\max}$, where

$$\mathrm{dom}(\Delta_{g,\max}) = \{u \in \mathcal{H}^{s,\gamma}_p(\mathsf{M}) : \Delta_g u \in \mathcal{H}^{s,\gamma}_p(\mathsf{M})\}.$$

We have

$$\mathrm{dom}(\Delta_{g,\min}) = \mathrm{dom}(\Delta_{g,\max}) \cap \bigcap_{\varepsilon>0} \mathcal{H}^{s+2,\gamma+2-\varepsilon}_p(\mathsf{M}).$$

In particular, $\mathrm{dom}(\Delta_{g,\min}) = \mathcal{H}^{s+2,\gamma+2}_p(\mathsf{M})$ iff $\sigma_{\mathsf{M}}(\Delta_g)(z)$ is invertible for all z satisfying $\mathrm{Re}z = \frac{n+1}{2} - \gamma - 2$. The reader may refer to [**24**, Proposition 5.1] for the details of this result.

We denote by $0 = \lambda_0 > \lambda_1 > \cdots$ the distinct eigenvalues of $\Delta_{\mathbb{B}}$ and by $E_0, E_1, \cdots$ the corresponding eigenspaces. Then the non-bijectivity points of $\sigma_{\mathsf{M}}(\Delta_g)$ are exactly

$$q^\pm_j = \frac{n-1}{2} \pm \sqrt{(\frac{n-1}{2})^2 - \lambda_j}, \quad j \in \mathbb{N}_0.$$

Note that $q^+_j = (n-1) - q^-_j$ and $q^-_0 = 0$. From the discussion in the previous paragraph, in case $q^\pm_j \neq \frac{n+1}{2} - \gamma - 2$ for all $j \in \mathbb{N}_0$, it holds that

$$\mathrm{dom}(\Delta_{g,\min}) = \mathcal{H}^{s+2,\gamma+2}_p(\mathsf{M}).$$

For $q^\pm_j$ with $j \neq 0$, we define the function spaces

$$\mathcal{E}_{q^\pm_j} = \omega x^{-q^\pm_j} \otimes E_j = \{\omega(x) x^{-q^\pm_j} e_j(y) : e_j \in E_j\}.$$

Recall that x is the singularity (boundary) defining function. When $j = 0$, we put

$$\mathcal{E}_{q^\pm_0} = \begin{cases} \omega x^{q^\pm_0} \otimes E_0 & n > 1; \\ \omega \otimes E_0 + \omega \log x \otimes E_0 & n = 1. \end{cases}$$

We will also introduce the set I_γ defined by

$$I_\gamma = \{q_j^\pm : j \in \mathbb{N}_0\} \cap (\frac{n+1}{2} - \gamma - 2, \frac{n+1}{2} - \gamma).$$

As a conclusion from [**11**, Theorem 3.6] and [**24**, Proposition 5.1], we have the following proposition concerning the maximal domain of Δ_g.

PROPOSITION 3.9. *Suppose that $q_j^\pm \neq \frac{n+1}{2} - \gamma - 2$ for all $j \in \mathbb{N}_0$. Then*

$$\mathrm{dom}(\Delta_{g,\max}) = \mathcal{H}_p^{s+2,\gamma+2}(\mathsf{M}) \oplus \bigoplus_{q_j^\pm \in I_\gamma} \mathcal{E}_{q_j^\pm}.$$

Given a subspace $\underline{\mathcal{E}}_{q_j^\pm}$ of $\mathcal{E}_{q_j^\pm}$, we associate with it a subspace $\underline{\mathcal{E}}_{q_j^\pm}^\perp$ according to the following rules:

(i) Suppose that either $q_j^\pm \neq 0$ or $n > 1$. If $\underline{\mathcal{E}}_{q_j^\pm} = \omega x^{-q_j^\pm} \otimes \underline{E}_j$ for some subspace $\underline{E}_j \subset E_j$, then we define

$$\underline{\mathcal{E}}_{q_j^\pm}^\perp = \omega x^{-q_j^\mp} \otimes \underline{E}_j^\perp,$$

where $\underline{E}_j^\perp$ is the orthogonal complement of $\underline{E}_j$ in E_j with respect to the $L_2(\mathbb{B})$-inner product.

(ii) When $q_0^\pm = 0$ and $n = 1$, define $\underline{\mathcal{E}}_0^\perp = \{0\}$ if $\underline{\mathcal{E}}_0 = \mathcal{E}_0$, $\underline{\mathcal{E}}_0^\perp = \mathcal{E}_0$ if $\underline{\mathcal{E}}_0 = \{0\}$, $\underline{\mathcal{E}}_0^\perp = \underline{\mathcal{E}}_0$ if $\underline{\mathcal{E}}_0 = \omega \otimes E_0$.

Note that $\underline{\mathcal{E}}_{q_j^\pm}^\perp$ is a subspace of $\mathcal{E}_{q_j^\mp}$.

DEFINITION 3.10. We define the extension $\underline{\Delta}_g = \underline{\Delta}_{g,s}^\gamma$ of Δ_g with the domain

$$\mathrm{dom}(\underline{\Delta}_g) = \mathcal{H}_p^{s+2,\gamma+2}(\mathsf{M}) \oplus \bigoplus_{q_j^\pm \in I_\gamma} \underline{\mathcal{E}}_{q_j^\pm}$$

with $\underline{\mathcal{E}}_{q_j^\pm}$ chosen as follows

(i) if $q_j^\pm \in I_\gamma \cap I_{-\gamma}$, then $\underline{\mathcal{E}}_{q_j^\pm}^\perp = \underline{\mathcal{E}}_{q_j^\mp}$;

(ii) if $\gamma \geq 0$ and $q_j^\pm \in I_\gamma \setminus I_{-\gamma}$, then $\underline{\mathcal{E}}_{q_j^\pm}^\perp = \mathcal{E}_{q_j^\pm}$;

(iii) if $\gamma \leq 0$ and $q_j^\pm \in I_\gamma \setminus I_{-\gamma}$, then $\underline{\mathcal{E}}_{q_j^\pm}^\perp = \{0\}$.

In the sequel, we will confine ourselves to the extension $\underline{\Delta}_g$. This is a closed extension of Δ_g. Indeed by Proposition 3.9 an extension of Δ_g is closed iff its domain is of the form

$$\mathcal{H}_p^{s+2,\gamma+2}(\mathsf{M}) \oplus \mathcal{E},$$

where $\mathcal{E}$ is a subspace of $\bigoplus_{q_j^\pm \in I_\gamma} \mathcal{E}_{q_j^\pm}$, as long as $q_j^\pm \neq \frac{n+1}{2} - \gamma - 2$ for all $j \in \mathbb{N}_0$.

With all these preparations, we are ready to state the following result on the Laplace-Beltrami operator based on [**23**, Theorem 3.11].

THEOREM 3.11. *Assume $s \geq 0$, $|\gamma| < \frac{1}{2}\mathrm{dim}(\mathsf{M})$. Then the extension $\underline{\Delta}_g$ defined in Definition 3.10 satisfies*

$$c - \underline{\Delta}_g \in \mathcal{MR}_p(\mathcal{H}_p^{s,\gamma}(\mathsf{M}))$$

for sufficiently large $c > 0$.

PROOF. Under the given conditions, [**23**, Theorem 3.11] states that

$$c - \underline{\Delta}_g \in \mathcal{BIP}(\mathcal{H}_p^{s,\gamma}(\mathsf{M}))$$

with power angle $\theta_{c-\underline{\Delta}_g} < \pi/2$, for $c > 0$ large enough. By (3.1), the $\mathscr{R}$-angle of $c - \underline{\Delta}_g$ is smaller than $\pi/2$. Now the assertion follows from Proposition 3.7. □

4. Short time existence of the harmonic map heat flow on conic manifolds

Throughout, suppose that u is a harmonic map heat flow from an $n+1$-dimensional conic manifold (M, g) to an m-dimensional smooth closed manifold (N, h).

Recall that the harmonic map heat flow for $u : \mathsf{M}_T \to \mathsf{N}$ is

$$\begin{cases} \partial_t u - \Delta_g u = A_g(u)(\nabla u, \nabla u) & \text{on} \quad \mathsf{M}_T; \\ u(0) = u_0 & \text{on} \quad \mathsf{M}, \end{cases} \tag{4.1}$$

where the initial condition $u_0 : \mathsf{M} \to \mathsf{N}$, and $\mathsf{M}_T := (0, T) \times \mathsf{M}$ for some $T > 0$.

By the Nash embedding theorem, we may assume that (N, h) is isometrically embedded into a Euclidean space $\mathbb{R}^L$. Since N is smooth and closed, we can always find a a-tubular neighbourhood

$$\mathsf{T}_a := \{z \in \mathbb{R}^L : \text{dist}(z, \mathsf{N}) < a\}$$

for sufficiently small $a > 0$ such that within which the nearest point projection π_{N}, that is the map satisfying

$$|z - \pi_{\mathsf{N}}(z)| = \text{dist}(z, \mathsf{N}) \qquad z \in \mathsf{T}_a,$$

is a smooth function onto N. Without confusion, we also use $\pi_{\mathsf{N}} : \mathbb{R}^L \to \mathbb{R}^L$ to denote a smooth extension of the nearest point projection π_{N} from T_a to N.

We will first consider u as a function from M to $\mathbb{R}^L$. With this relaxation, let us first study an alternative problem to (4.1)

$$\begin{cases} \partial_t u - \Delta_g u = \phi(u)(\nabla u, \nabla u)_{g^*} & \text{on} \quad \mathsf{M}_T; \\ u(0) = u_0 & \text{on} \quad \mathsf{M}, \end{cases} \tag{4.2}$$

with initial value $u_0 : \mathsf{M} \to \mathsf{N}$, where $\phi = -\partial^2 \pi_{\mathsf{N}}$ and

$$\phi(u)(\nabla u, \nabla u)_{g^*} = g^{\alpha\beta}\phi(u)\Big(\frac{\partial u}{\partial x_\alpha}, \frac{\partial u}{\partial x_\alpha}\Big).$$

In order to apply Theorems 3.8 and 3.11 to (4.2), we will impose the following conditions.

(A) When $n = 1, 2$, we assume that

$$4/p < \gamma < 2, \quad p > 6,$$

and $\gamma < \min\{\sqrt{1 - \lambda_1} - 1, \sqrt{4 - n} - 1\}$.

or

(B0) $0 < \gamma < \frac{n+1}{2}$, $p > n + 3$ and $\gamma + 2 - \frac{4}{p} > \frac{n+1}{2}$.

(B1) When $n = 1$ and $\lambda_1 < -1$, we choose $\gamma > 0$ so small that

$$-\sqrt{-\lambda_1} < -1 - \gamma.$$

(B2) When $n = 2$ and $\lambda_1 < -2$, $\gamma \in (1/2, 3/2)$ is close enough to $1/2$ such that

$$q_1^- < -\frac{1}{2} - \gamma < -1.$$

(B3) When $n \geq 3$,

$$q_1^- < \frac{n+1}{2} - \gamma - 2.$$

REMARKS 4.1. (i) Although (B0)-(B2) have some overlapping with (A), they can render us some flexibility in choosing the parameter p when the first eigenvalue λ_1 of $\mathbb{B}$ satisfies the given condition.

(ii) The functions of Assumptions (B0)-(B3) are to fulfill the conditions in Theorem 3.11 and to obtain enough regularity for the map

$$[u \mapsto \phi(u)(\nabla u, \nabla u)_{g^*}].$$

The latter only uses (B0) and will be clarified right after this remark. We will give some explanation for the former here.

(B0) ensures that $|\gamma| < \dim(\mathsf{M})/2$. We will show that, for any $s \geq 0$,

$$\mathrm{dom}\underline{\Delta}_{g,s}^{\gamma} = \mathcal{H}_p^{s+2,\gamma+2}(\mathsf{M}) \oplus \mathbb{C}$$

as long as (B0) and (Bj) are met when $j = \min\{3, n\}$ for $j \in \mathbb{N}$.

Case 1: Suppose that $n \geq 3$. The third condition in (B0) implies

$$\gamma + 2 > \frac{n+1}{2} \Longleftrightarrow \frac{n+1}{2} - \gamma - 2 < 0,$$

and thus $q_0^+ = n - 1 \geq 2 > \frac{n+1}{2} - \gamma$. Combining with (B3), this yields

$$I_\gamma = \{0\} = \{q_0^-\}.$$

On the other hand, $\frac{n+1}{2} + \gamma - 2 > 0$ implies that

$$q_0^- \in I_\gamma \setminus I_{-\gamma}.$$

By Definition 3.10(ii), we conclude

$$\underline{\mathcal{E}}_{q_0^-} = \mathcal{E}_{q_0^-} = \mathbb{C}.$$

Case 2: Suppose that $n = 2$. Due to (B2),

$$\{q_0^-\} \in I_\gamma \setminus I_{-\gamma}$$

and $q_0^+ = 1 > \frac{3}{2} - \gamma$. Hence Definition 3.10(ii) again yields

$$\underline{\mathcal{E}}_{q_0^-} = \mathcal{E}_{q_0^-} = \mathbb{C}.$$

Case 3: Suppose that $n = 1$. Note that in this case $q_0^\pm = 0$ and $q_1^\pm = \pm\sqrt{-\lambda_1}$. (B0) implies

$$q_0^\pm \in I_\gamma \cap I_{-\gamma}.$$

At the same time, (B1) guarantees that

$$q_1^- < -1 - \gamma < 1 - \gamma < 1 < q_1^+,$$

and thus

$$I_\gamma = \{0\} = \{q_0^\pm\}.$$

Now it follows from Definition 3.10(i) that

$$\underline{\mathcal{E}}_0^\pm = \underline{\mathcal{E}}_0 = \mathbb{C}.$$

All the discussions in (ii) imply that

$$\mathrm{dom}\underline{\Delta}^{\gamma}_{g,s} = \mathcal{H}^{s+2,\gamma+2}_p(\mathsf{M}) \oplus \mathbb{C}, \quad s \geq 0.$$

Therefore, Theorem 3.11 will apply as long as (B0) and (Bj) are assumed when $j = \min\{3, n\}$.

(iii) As we can see from (ii), the extra conditions on λ_1 in (B1) and (B2) are coming from the need to avoid containing q_1^- and q_0^+ in the interval $(\frac{n+1}{2} - \gamma - 2, \frac{n+1}{2} - \gamma)$ such that

$$\mathrm{dom}\underline{\Delta}^{\gamma}_{g,s} = \mathcal{H}^{s+2,\gamma+2}_p(\mathsf{M}) \oplus \mathbb{C}.$$

If either q_1^- or q_0^+ is contained in $(\frac{n+1}{2} - \gamma - 2, \frac{n+1}{2} - \gamma)$, the asymptotic behavior of $\mathrm{dom}\underline{\Delta}^{\gamma}_{g,s}$ will become much more complicated, as indicated by Definition 3.10. This in turn will create an essential difficulty in computing

$$(\mathcal{H}^{s,\gamma}_p(\mathsf{M}), \mathrm{dom}\underline{\Delta}^{\gamma}_{g,s})_{1-1/p,p},$$

and as we learn from Definition 3.6, the above interpolation space is the space of initial data of (4.1).

However, when $n \geq 3$, we have $q_0^+ = n - 1 \geq 2$. This provides us with enough room to avoid both q_1^- and q_0^+ in $(\frac{n+1}{2} - \gamma - 2, \frac{n+1}{2} - \gamma)$.

In the sequel, we will always assume that (B0) and (Bj) hold when $j = \min\{3, n\}$.

For notational brevity, for any $\vartheta \in \mathbb{R}$, we put

$$\vartheta_{\varepsilon,p} = \vartheta + 2 - 2/p - \varepsilon.$$

One can compute inside $I \times \mathbb{B}$ that

$$(\nabla u, \nabla v)_{g^*} = \partial_x u \partial_x v + \frac{1}{x^2}(\nabla_{g_{\mathbb{B}}} u, \nabla_{g_{\mathbb{B}}} v)_{g^*_{\mathbb{B}}},$$

where $g^*_{\mathbb{B}}$ is the induced cotangent metric by $g_{\mathbb{B}}$. Since for any $s \geq 0$ and $\vartheta \in \mathbb{R}$

$$[u \mapsto \partial_x u] \in \mathcal{L}(\mathcal{H}^{s+1,\vartheta+1}_p(\mathsf{M}), \mathcal{H}^{s,\vartheta}_p(\mathsf{M}))$$

and

$$[u \mapsto \frac{1}{x}\nabla_{g_{\mathbb{B}}} u] \in \mathcal{L}(\mathcal{H}^{s+1,\vartheta+1}_p(\mathsf{M}), \mathcal{H}^{s,\vartheta}_p(\mathsf{M}, T^*\mathbb{B})),$$

we can obtain the following lemma in virtue of Theorem 2.8(i).

LEMMA 4.2. *If (B0) is assumed and let $\varepsilon > 0$ so small that*

$$1 - \frac{n+3}{p} - \varepsilon > 0, \tag{4.3}$$

then for any $s \geq 0$

$$[(u, v) \mapsto (\nabla u, \nabla v)_{g^*}]$$

is a continuous bilinear map from

$$(\mathcal{H}^{s_{\varepsilon,p},\gamma_{\varepsilon,p}}_p(\mathsf{M}) \oplus \mathbb{C}) \times (\mathcal{H}^{s_{\varepsilon,p},\gamma_{\varepsilon,p}}_p(\mathsf{M}) \oplus \mathbb{C}) \to \mathcal{H}^{s_{\varepsilon,p}-1, 2\gamma_{\varepsilon,p}-2-\frac{n+1}{2}}_p(\mathsf{M}).$$

Recall that $\phi = \partial^2 \pi_{\mathsf{N}}$ and $\pi_{\mathsf{N}} \in C^\infty(\mathsf{T}_a, \mathsf{N})$. By possibly further shrinking a, we have

$$\phi \in BC^\infty(\mathsf{T}_a, \mathcal{L}^2(\mathbb{R}^L, \mathbb{R}^L)).$$

Assume that the conditions of Lemma 4.2 are satisfied. In view of (B0), we can find $\varepsilon > 0$ so small that both (4.3) and

$$\gamma_{\varepsilon,p} \geq \frac{n+1}{2} \tag{4.4}$$

are fulfilled.

Let $\mathsf{p} \in \mathsf{N} \subset \mathbb{R}^L$. For every $v = w + \mathsf{p}$ with $w \in \mathcal{H}_p^{s_{\varepsilon,p},\gamma_{\varepsilon,p}}(\mathsf{M}, \mathsf{N} - p)$ satisfying (4.3) and (4.4), by Proposition 2.6, we can find a bounded neighbourhood $U = U(v)$ of v in $\mathcal{H}_p^{s_{\varepsilon,p},\gamma_{\varepsilon,p}}(\mathsf{M}, \mathbb{R}^L) \oplus \mathbb{R}^L$ such that

$$\mathrm{Rng}(U) \subset \mathsf{T}_a.$$

Then we can apply Theorem 2.10 and infer that the Nemyskii operator Φ induced by ϕ satisfies

$$\Phi \in C^\infty(U, \mathcal{H}^{[s_{\varepsilon,p}],\vartheta}(\mathsf{M}, \mathcal{L}^2(\mathbb{R}^L, \mathbb{R}^L))$$

for any $\vartheta < \frac{n+1}{2}$.

Making use of Theorem 2.8(i) and Lemma 4.2, direct computations show that

$$[u \mapsto \phi(u)(\nabla u, \nabla u)_{g^*}] \in C^\infty(U, \mathcal{H}^{s_{\varepsilon,p}-1, 2\gamma_{\varepsilon,p}-2+\vartheta-n-1}(\mathsf{M}, \mathbb{R}^L)). \tag{4.5}$$

In view of the third condition in (B0) and by choosing $\varepsilon > 0$ sufficiently small and ϑ close enough to $\frac{n+1}{2}$, we can always have

$$2\gamma_{\varepsilon,p} - 2 + \vartheta - n - 1 > \gamma. \tag{4.6}$$

Now we put

$$E_0^s = \mathcal{H}_p^{s,\gamma}(\mathsf{M}, \mathbb{R}^L), \qquad E_1^s = \mathcal{H}_p^{s+2,\gamma+2}(\mathsf{M}, \mathbb{R}^L) \oplus \mathbb{R}^L,$$

and

$$E_{1-1/p}^s = (\mathcal{H}_p^{s,\gamma}(\mathsf{M}, \mathbb{R}^L), \mathcal{H}_p^{s+2,\gamma+2}(\mathsf{M}, \mathbb{R}^L) \oplus \mathbb{R}^L)_{1-1/p,p}.$$

Due to Lemma 2.3, we have the embedding

$$E_{1-1/p}^s \hookrightarrow \mathcal{H}^{s_{\varepsilon,p},\gamma_{\varepsilon,p}}(\mathsf{M}, \mathbb{R}^L) \oplus \mathbb{R}^L.$$

Denote by $\iota : E_{1-1/p}^s \to \mathcal{H}^{s_{\varepsilon,p},\gamma_{\varepsilon,p}}(\mathsf{M}, \mathbb{R}^L) \oplus \mathbb{R}^L$ the inclusion map. Let $\mathcal{O} = \mathcal{O}(v) := \iota^{-1}(U)$ be an open subset in $E_{1-1/p}^s$. We thus infer from (4.5) and (4.6) that

$$[u \mapsto \phi(u)(\nabla u, \nabla u)_{g^*}] \in C^\infty(\mathcal{O}, E_0^s). \tag{4.7}$$

Careful readers may have noticed that the regularity in (4.7) is much stronger than what is asked by Theorem 3.8, but later we will find out that (4.7) is indeed necessary for obtaining a smooth solution of the harmonic map heat flow.

Now we can use (4.7) and apply Theorem 3.8 to prove the local well-posedness of (4.2).

THEOREM 4.3. *Assume that γ and p satisfy (B0) and (B3) when $n \geq 3$ or (B0) and (B2) when $n = 2$ or (B0)-(B1) when $n = 1$, and the initial condition in* (4.2) *satisfies*

$$u_0 = w_0 + \mathsf{p} \quad \textit{with} \quad w_0 \in \mathcal{H}^{2-\frac{2}{p}+\delta, \gamma+2-\frac{2}{p}+\delta}(\mathsf{M}, \mathsf{N} - \mathsf{p}) \quad \textit{and} \quad \mathsf{p} \in \mathsf{N}$$

for an arbitrary positive constant δ. Then (4.2) *with initial condition u_0 has a unique solution*

$$u \in L_p(J_T, \mathcal{H}_p^{2,\gamma+2}(\mathsf{M}, \mathbb{R}^L) \oplus \mathbb{R}^L) \cap H_p^1(J_T, \mathcal{H}_p^{0,\gamma}(\mathsf{M}, \mathbb{R}^L))$$

on $J_T = [0, T)$ for some $T > 0$. Moreover, for any $\varepsilon > 0$

$$u \in BC(J_T, \mathcal{H}^{2-\frac{2}{p}-\varepsilon, \gamma+2-\frac{2}{p}-\varepsilon}(\mathsf{M}, \mathbb{R}^L) \oplus \mathbb{R}^L).$$

PROOF. We only need to show the last inclusion. It follows from [**1**, Chapter III: Theorem 4.10.2] that

$$u \in BC(J_T, (\mathcal{H}_p^{0,\gamma}(\mathsf{M}, \mathbb{R}^L), \mathcal{H}_p^{2,\gamma+2}(\mathsf{M}, \mathbb{R}^L) \oplus \mathbb{R}^L)_{1-1/p,p}).$$

Now the inclusion is a direct result of Lemma 2.3. □

REMARK 4.4. Because the argument leading to Theorem 4.3 is independent of the choice $s \geq 0$, once (B0) and (Bj) are assumed with $j = \min\{3, n\}$, for every $s \geq 0$ and

$$u_0 = w_0 + \mathsf{p} \quad \text{with} \quad w_0 \in \mathcal{H}^{s+2-\frac{2}{p}+\delta, \gamma+2-\frac{2}{p}+\delta}(\mathsf{M}, \mathsf{N} - \mathsf{p}) \quad \text{and} \quad \mathsf{p} \in \mathsf{N},$$

(4.2) with initial condition u_0 has a unique solution

$$u \in L_p(J_T, \mathcal{H}_p^{s+2,\gamma+2}(\mathsf{M}, \mathbb{R}^L) \oplus \mathbb{R}^L) \cap H_p^1(J_T, \mathcal{H}_p^{s,\gamma}(\mathsf{M}, \mathbb{R}^L))$$

on $J_T = [0, T)$ for some $T = T(s) > 0$. In addition,

$$u \in BC(J_T, \mathcal{H}^{s+2-\frac{2}{p}-\varepsilon, \gamma+2-\frac{2}{p}-\varepsilon}(\mathsf{M}, \mathbb{R}^L) \oplus \mathbb{R}^L).$$

In the rest of this section, our objective is to show that the solution u obtained in Theorem 4.3 actually solves (4.1). To this end, we put

$$\rho(t) = |\pi_{\mathsf{N}}(u(t)) - u(t)|^2.$$

Direct computations show that

$$\partial_t \rho = 2(\pi_{\mathsf{N}}(u) - u, \partial \pi_{\mathsf{N}}(u) \partial_t u - \partial_t u),$$

where $(\cdot, \cdot)$ is the standard inner product in $\mathbb{R}^L$, and

$$\begin{aligned} \Delta_g \rho =& 2|\nabla(\pi_{\mathsf{N}}(u) - u)|_g^2 + 2(\pi_{\mathsf{N}}(u) - u, \partial \pi_{\mathsf{N}}(u) \Delta_g u - \Delta_g u) \\ &+ 2(\pi_{\mathsf{N}}(u) - u, \partial^2 \pi_{\mathsf{N}}(u)(\nabla u, \nabla u)_{g^*}). \end{aligned}$$

Suppose that $z \in \mathsf{T}_a$ with $\pi_{\mathsf{N}}(z) = z_0 \in \mathsf{N}$. Then there exist some $\nu \in (T_{z_0}\mathsf{N})^\perp$ and $s_z \in (-a, a)$ such that $z = z_0 + s_z \nu$. Since $\pi_{\mathsf{N}}(z_0 + s\nu) = z_0$ for all $s \in (-a, a)$, one can derive that $\partial \pi_{\mathsf{N}}(z)\nu = 0$.

In view of the above consideration, we can infer from the expressions of $\partial_t \rho$ and $\Delta_g \rho$ that

$$\partial_t \rho - \Delta_g \rho = -2|\nabla(\pi_{\mathsf{N}}(u) - u)|_g^2.$$

Since M only has isolated point singularities, multiplying the above equality by ρ and integrating over M yields

$$\partial_t \int_{\mathsf{M}} \rho^2 \, dv_g + 2 \int_{\mathsf{M}} |\nabla \rho|_g^2 \, dv_g \leq 0.$$

Since $\rho(0) \equiv 0$, the above integral inequality implies

$$\rho(t) \equiv 0, \quad t \in J_T.$$

We have thus proved that

$$\pi_{\mathsf{N}}(u(t)) = u(t), \quad t \in J_T.$$

Now it follows from [**19**, Lemma 3.2] that

$$-\partial^2 \pi_{\mathsf{N}}(u)(\nabla u, \nabla u)_{g^*} = A_g(u)(\nabla u, \nabla u).$$

Therefore u is indeed a solution to (4.1).

From now on, we consider the case that Condition (A) holds. In this situation, we build a conic manifold $(\widehat{\mathsf{M}}, \widehat{g})$ up on the manifold $\tilde{\mathsf{M}} \times S^l$ with $l = 3 - n$.

More precisely, let $\widehat{\mathbb{B}} := \mathbb{B} \times S^l$ equipped with the product metric $g_{\widehat{\mathbb{B}}} = g_{\mathbb{B}} + g_{S^l}$, where g_{S^l} is the standard metric on S^l. Near the conic singularities, we equip $(0,1] \times (\mathbb{B} \times S^l)$ with the metric

$$\widehat{g} = dx^2 + x^2 g_{\widehat{\mathbb{B}}},$$

and, outside $(0,1] \times (\mathbb{B} \times S^l)$, $\widehat{\mathsf{M}}$ is endowed with the product metric $\tilde{g} + g_{S^l}$.

For every $u_0 = w_0 + \mathsf{p}$ with $w_0 \in \mathcal{H}^{2-\frac{2}{p}+\delta,\gamma+2-\frac{2}{p}+\delta}(\mathsf{M}, \mathsf{N} - \mathsf{p})$, we can define a new initial datum $\widehat{u}_0(z,\theta) = u_0(z)$, where θ is the coordinates for S^l and $z \in \mathsf{M}$. It is not hard to see that $\widehat{u}_0 = \widehat{w}_0 + \mathsf{p}$ with some $\widehat{w}_0 \in \mathcal{H}^{2-\frac{2}{p}+\delta,\gamma+2-\frac{2}{p}+\delta}(\widehat{\mathsf{M}}, \mathsf{N} - \mathsf{p})$.

Note that the base $\widehat{\mathbb{B}}$ of the conic manifold $(\widehat{\mathsf{M}}, \widehat{g})$ is of dimension 3, and Condition (A) has guaranteed that the new manifold $(\widehat{\mathsf{M}}, \widehat{g})$ satisfies (B0) and (B3). Therefore Theorem 4.3 and the above discussion imply that (4.1) from $\widehat{\mathsf{M}} \to \mathsf{N}$ with initial condition $\widehat{u}_0$ has a unique solution

$$\widehat{u} \in L_p(J_T, \mathcal{H}_p^{2,\gamma+2}(\widehat{\mathsf{M}}, \mathbb{R}^L) \oplus \mathbb{R}^L) \cap H_p^1(J_T, \mathcal{H}_p^{0,\gamma}(\widehat{\mathsf{M}}, \mathsf{N})).$$

Suppose that R is an arbitrary rotation matrix in $\mathbb{R}^{l+1}$, i.e. $R \in SO(l+1)$. Define

$$v_R(z,\theta) := v(z, R\theta); \text{ and } u_R(z,\theta,t) := u(z, R\theta, t) \text{ for } (z,\theta) \in M \times S^l, \ t \geq 0.$$

Since the Dirichlet energy (1.2) is invariant under $R \in SO(l+1)$, i.e. $E(v) = E(v_R)$ and (4.1) is the negative L_2-gradient flow of (1.2), $\widehat{u}_R$ too solves (1.1) for any $R \in SO(l+1)$.

In view of $\widehat{u}_0 = (\widehat{u}_0)_R$ for any $R \in SO(l+1)$, it follows from the uniqueness of solution to (1.1) that $\widehat{u}_R(z,\theta,t) = \widehat{u}(z,\theta,t)$ on $\widehat{M} \times [0,T)$ for any $R \in SO(l+1)$. Hence $\widehat{u}$ is indeed independent of the spherical variable $\theta \in S^l$.

Now we can just look at a hypersurface $\{\theta = \theta_0\}$ of $\widehat{\mathsf{M}}$, the restriction of $\widehat{u}$ on this hypersurface generates a solution

$$u \in L_p(J_T, \mathcal{H}_p^{2,\gamma+2}(\mathsf{M}, \mathbb{R}^L) \oplus \mathbb{R}^L) \cap H_p^1(J_T, \mathcal{H}_p^{0,\gamma}(\mathsf{M}, \mathsf{N}))$$

to (4.1) from $\mathsf{M} \to \mathsf{N}$ with initial value u_0. To prove the uniqueness of solutions, we assume that v is another solution to (4.1) from $\mathsf{M} \to \mathsf{N}$ with initial value u_0 that is different from u. Now $\widehat{v}(\cdot,\theta) = v(\cdot)$ also solves (4.1) from $\widehat{\mathsf{M}} \to \mathsf{N}$ with initial value $\widehat{u}_0$, which is different from $\widehat{u}$. This violates the uniqueness of solutions to (1.1) in the case $n = 3$.

To sum up, we are ready to state the first main theorem of this article.

THEOREM 4.5. *Assume that*

- *γ and p satisfy (B0) and (B3) when $n \geq 3$; or*
- *when $n = 1, 2$, γ and p satisfy (A) or satisfy (B0)-(B1) for $n = 1$ or satisfy (B0) and (B2) for $n = 2$.*

Let

$$u_0 = w_0 + \mathsf{p} \quad \textit{with} \quad w_0 \in \mathcal{H}_p^{2-\frac{2}{p}+\delta,\gamma+2-\frac{2}{p}+\delta}(\mathsf{M}, \mathsf{N} - \mathsf{p}) \quad \textit{and} \quad \mathsf{p} \in \mathsf{N}$$

for an arbitrary positive constant δ. Then (4.1) with initial condition u_0 has a unique solution

$$\begin{aligned} u \in & L_p(J_T, \mathcal{H}_p^{2,\gamma+2}(\mathsf{M}, \mathbb{R}^L) \oplus \mathbb{R}^L) \cap H_p^1(J_T, \mathcal{H}_p^{0,\gamma}(\mathsf{M}, \mathsf{N})) \\ & \cap BC(J_T, \mathcal{H}_p^{2-\frac{2}{p}-\varepsilon,\gamma+2-\frac{2}{p}-\varepsilon}(\mathsf{M}, \mathbb{R}^L) \oplus \mathbb{R}^L) \end{aligned}$$

on $J_T = [0,T)$ *for some* $T > 0$ *and any* $\varepsilon > 0$. *Moreover, for any* $\alpha > 0$ *and* $k \in \mathbb{N}_0$

$$u \in H_p^k((\alpha,T), \mathcal{H}_p^{k+2,\gamma+2}(\mathsf{M},\mathbb{R}^L) \oplus \mathbb{R}^L) \cap H_p^{k+1}((\alpha,T), \mathcal{H}_p^{k,\gamma}(\mathsf{M},\mathsf{N})) \tag{4.8}$$

and

$$u \in BC^\infty((\alpha,T), \mathcal{H}_p^{\infty,\gamma+2-\frac{2}{p}-\varepsilon}(\mathsf{M},\mathbb{R}^L) \oplus \mathbb{R}^L) \cap C^\infty((0,T), \mathcal{H}_p^{\infty,\gamma+2}(\mathsf{M},\mathbb{R}^L) \oplus \mathbb{R}^L). \tag{4.9}$$

PROOF. We have already shown the existence of a unique solution in

$$L_p(J_T, \mathcal{H}_p^{2,\gamma+2}(\mathsf{M},\mathbb{R}^L) \oplus \mathbb{R}^L) \cap H_p^1(J_T, \mathcal{H}_p^{0,\gamma}(\mathsf{M},\mathsf{N})).$$

Based on the argument on [**20**, page 201] and (4.7), we can conclude that for any $\alpha \in (0,T)$ and $k \in \mathbb{N}_0$

$$u \in H_p^k((\alpha,T), \mathcal{H}_p^{2,\gamma+2}(\mathsf{M},\mathbb{R}^L) \oplus \mathbb{R}^L) \cap H_p^{k+1}((\alpha,T), \mathcal{H}_p^{0,\gamma}(\mathsf{M},\mathsf{N})).$$

In view of Remark 4.4, we can apply a bootstrapping argument to improve the spatial regularity to

$$u \in H_p^k((\alpha,T), \mathcal{H}_p^{l+2,\gamma+2}(\mathsf{M},\mathbb{R}^L) \oplus \mathbb{R}^L) \cap H_p^{k+1}((\alpha,T), \mathcal{H}_p^{l,\gamma}(\mathsf{M},\mathsf{N}))$$

for arbitrary $l \geq 0$. Utilizing [**1**, Chapter III: Theorem 4.10.2] and Lemma 2.3, we thus have

$$u \in BC^k((\alpha,T), \mathcal{H}^{l_{\varepsilon,p},\gamma_{\varepsilon,p}}(\mathsf{M},\mathbb{R}^L) \oplus \mathbb{R}^L)$$

for all $\alpha \in (0,T)$, $l \geq 0$, $k \in \mathbb{N}_0$ and $\varepsilon > 0$ small. Therefore, we have obtained the extra temporal and spatial regularity in (4.8) and (4.9). □

5. Global existence of the harmonic map heat flow into manifolds with nonpositive sectional curvature

In this section, we assume that the sectional curvature K_N of N is nonpositive, and the following condition.

(A') When $n = 1, 2$, we assume that

$$4/p < \gamma < \frac{n+1}{2}, \quad p > 6,$$

and $\gamma < \min\{\sqrt{1-\lambda_1} - 1, \sqrt{4-n} - 1\}$.

or (B0) and (Bj) in Section 4 hold for $j = \min\{3, n\}$.

REMARK 5.1. Assumption (A') is slightly stronger than (A), as the upper bound $\frac{n+1}{2}$ of γ in (A') is smaller than 2. We need this stronger upper bound $\frac{n+1}{2}$ to derive an estimate for the nonlinear term $A_g(u)(\nabla u, \nabla u)$ in the space $L_p(J_T, \mathcal{H}_p^{0,\gamma}(\mathsf{M},\mathbb{R}^L))$ (cf. (5.3)) by means of a gradient estimate of the solution u.

Because of the compactness of $(\tilde{\mathsf{M}}, \tilde{g})$, we can find a constant $C > 0$ such that the Ricci curvature $\mathrm{Ric}_{\tilde{g}}$ of $(\tilde{\mathsf{M}}, \tilde{g})$ satisfies

$$\mathrm{Ric}_{\tilde{g}} \geq -C. \tag{5.1}$$

We will assume that the initial condition u_0 satisfies

(IC) $u_0 = w_0 + \mathsf{p}$, for some $\mathsf{p} \in \mathsf{N}$, with $w_0 \in \mathcal{H}_p^{2-\frac{2}{p}+\delta,\gamma+2-\frac{2}{p}+\delta}(\mathsf{M}, \mathsf{N}-\mathsf{p})$ and $\|\nabla w_0\|_\infty < \infty$.

Assume that $\mho$ is a geometric object on (M, g) defined in terms of the metric g. We use $\mho(\mathbb{B})$ to denote the corresponding object on $(\mathbb{B}, g_{\mathbb{B}})$ defined with respect to $g_{\mathbb{B}}$.

Suppose that the local patches in $I \times \mathbb{B}$ are of the form $I \times O_j$, where O_j are the local patches of $\mathbb{B}$.

In local coordinates, the Ricci curvature of (M, g) is of the form $\mathrm{Ric}_g = R_{ik} dx^i \otimes dx^k$, where

$$R_{ik} = \partial_j \Gamma^j_{ki} - \partial_k \Gamma^j_{ji} + \Gamma^j_{jl}\Gamma^l_{ki} - \Gamma^j_{kl}\Gamma^l_{ji}. \tag{5.2}$$

Here Γ^j_{ki} are the Christoffel symbols of g.

Let us first compute the Christoffel symbols of g in $I \times \mathbb{B}$. In the following calculations, we will use i, j, k, l to denote the subscripts of the local coordinates in $(\mathbb{B}, g_{\mathbb{B}})$ and x to denoted the coordinates in I as usual. Careful computations yield

$$\Gamma^j_{ki} = \Gamma^j_{ki}(\mathbb{B}), \quad \Gamma^x_{ij} = -x g_{ij}(\mathbb{B}), \quad \Gamma^x_{xj} = \Gamma^i_{xx} = 0, \quad \Gamma^i_{xj} = \frac{1}{x}\delta^i_j.$$

Plugging the above expressions into (5.2), we can obtain

$$R_{ik} = R_{ik}(\mathbb{B}) - (n-1) g_{ik}(\mathbb{B}), \quad R_{xx} = 0, \quad R_{xk} = 0.$$

Therefore, the Ricci curvature of (M, g) is of the form

$$\mathrm{Ric}_g = \mathrm{Ric}_{g_{\mathbb{B}}} - (n-1) g_{\mathbb{B}}$$

in $I \times \mathbb{B}$, and thus is globally bounded from below on M.

We will prove that the solution u to (4.1) with initial value u_0 satisfying (IC) is global. For, otherwise, we can assume that u has its maximal existence time $0 < T_{\max} < \infty$.

Let $e(u) := |\nabla u|_g^2$ be the energy density. Now we can follow the argument in [**18**, Lemma 5.3.3] and prove that

$$\partial_t e(u) - \Delta_g e(u) \leq M e(u) \quad \text{on } \mathsf{M}_{T_{\max}}$$

for some $M > 0$. It yields

$$(\partial_t - \Delta_g)(e^{-Mt} e(u)) \leq 0 \quad \text{on } \mathsf{M}_{T_{\max}}.$$

Define

$$f_+ = \begin{cases} f & \text{if } f \geq 0; \\ 0 & \text{if } f < 0. \end{cases}$$

Let $v(x,t) = (e^{-Mt} e(u)(x,t) - \|e(u_0)\|_\infty)_+$ for $(x,t) \in \mathsf{M}_{T_{\max}}$. Then

$$(\partial_t - \Delta_g) v \leq 0 \text{ on } \mathsf{M}_{T_{\max}}.$$

Since $v(0) \equiv 0$, multiplying the above inequality by $v(x,t)$ and integrating over M gives us

$$(e^{-Mt} e(u)(x,t) - \|e(u_0)\|_\infty)_+ \equiv 0 \quad \text{for all } (x,t) \in \mathsf{M}_{T_{\max}}.$$

This implies that

$$e^{-Mt} e(u)(x,t)) \leq \|e(u_0)\|_\infty \quad \text{for all } (x,t) \in \mathsf{M}_{T_{\max}},$$

and thus

$$|\nabla u|_g^2(x,t) \le e^{MT_{\max}} \|e(u_0)\|_\infty \quad \text{for all } (x,t) \in \mathsf{M}_{T_{\max}}.$$

We can infer from the above gradient estimate, Conditions (A') and (B0) that

$$A_g(u)(\nabla u, \nabla u) \in L_p(J_{\max}, \mathcal{H}_p^{0,\gamma}(\mathsf{M}, \mathsf{N})). \tag{5.3}$$

Here $J_{\max} = [0, T_{\max}]$.

Now it follows from (3.4), [**1**, Chapter III: Theorem 4.10.2] and Lemma 2.3 that for any $\varepsilon > 0$

$$\begin{aligned}
&\|u\|_{BC(J_{\max}, \mathcal{H}_p^{2-\frac{2}{p}-\varepsilon, \gamma+2-\frac{2}{p}-\varepsilon}(\mathsf{M}, \mathbb{R}^L)\oplus\mathbb{R}^L)} \\
\le &\|u\|_{L_p(J_{\max}, \mathcal{H}_p^{2,\gamma+2}(\mathsf{M},\mathbb{R}^L)\oplus\mathbb{R}^L)\cap H_p^1(J_{\max}, \mathcal{H}_p^{0,\gamma}(\mathsf{M},\mathbb{R}^L))} \\
\le &\|u_0\|_{\mathcal{H}_p^{2-\frac{2}{p}+\delta, \gamma+2-\frac{2}{p}+\delta}(\mathsf{M},\mathbb{R}^L)\oplus\mathbb{R}^L} + \|A_g(u)(\nabla u, \nabla u)\|_{L_p(J_{\max}, \mathcal{H}_p^{0,\gamma}(\mathsf{M},\mathbb{R}^L))} \\
\le &C
\end{aligned} \tag{5.4}$$

for some $C > 0$. From the above inequality, Theorem 2.8(i), Conditions (A') and (B0), we can further infer that

$$A_g(u)(\nabla u, \nabla u) \in L_p(J_{\max}, \mathcal{H}_p^{1-\frac{2}{p}-\varepsilon, \gamma+\tilde{\delta}}(\mathsf{M}, \mathsf{N})).$$

for some $\tilde{\delta} > 0$. We can pick $\tilde{\delta}$ so small that (A') or (B0) and (Bj) still hold for $j = \min\{3, n\}$ with γ replaced by $\gamma + \tilde{\delta}$; and note that

$$u(\alpha) \in \mathcal{H}_p^{2,\gamma+2}(\mathsf{M}, \mathbb{R}^L) \oplus \mathbb{R}^L \hookrightarrow \mathcal{H}_p^{2-\frac{2}{p}+\delta', \gamma+2-\frac{2}{p}+\delta'}(\mathsf{M}, \mathbb{R}^L) \oplus \mathbb{R}^L$$

for all $\alpha \in (0, T_{\max})$ and δ' sufficiently small.

Choose $\tilde{\delta} < \delta'$. Now we can utilize the argument in (5.4) once more to obtain a higher order estimate in both spatial regularity and the weight.

$$\begin{aligned}
&\|u\|_{BC((\alpha,T_{\max}), \mathcal{H}_p^{2-\frac{2}{p}+\tilde{\delta}-\varepsilon, \gamma+2-\frac{2}{p}+\tilde{\delta}-\varepsilon}(\mathsf{M},\mathbb{R}^L)\oplus\mathbb{R}^L)} \\
\le &\|u\|_{L_p((\alpha,T_{\max}), \mathcal{H}_p^{2+\tilde{\delta},\gamma+2+\tilde{\delta}}(\mathsf{M},\mathbb{R}^L)\oplus\mathbb{R}^L)\cap H_p^1((\alpha,T_{\max}), \mathcal{H}_p^{\tilde{\delta},\gamma+\tilde{\delta}}(\mathsf{M},\mathbb{R}^L))} \\
\le &\|u(\alpha)\|_{\mathcal{H}_p^{2-\frac{2}{p}+\delta', \gamma+2-\frac{2}{p}+\delta'}(\mathsf{M},\mathbb{R}^L)\oplus\mathbb{R}^L} + \|A_g(u)(\nabla u, \nabla u)\|_{L_p((\alpha,T_{\max}), \mathcal{H}_p^{\tilde{\delta},\gamma+\tilde{\delta}}(\mathsf{M},\mathbb{R}^L))} \\
\le &C
\end{aligned}$$

for some $C > 0$ and arbitrary $\varepsilon > 0$.

By means of Lemma 2.1(ii), we can thus extract a bounded sequence

$$\{u(t_k)\}_k \subset \mathcal{H}_p^{2-\frac{2}{p}+\tilde{\delta}-\varepsilon, \gamma+2-\frac{2}{p}+\tilde{\delta}-\varepsilon}(\mathsf{M}, \mathbb{R}^L) \oplus \mathbb{R}^L \quad \text{with} \quad t_k \to T_{\max}^-$$

such that

$$u(t_k) \to u^* \quad \text{in } \mathcal{H}_p^{2-\frac{2}{p}+\delta, \gamma+2-\frac{2}{p}+\delta}(\mathsf{M}, \mathbb{R}^L) \oplus \mathbb{R}^L$$

for some $\delta > 0$ and some $u^* \in \mathcal{H}_p^{2-\frac{2}{p}+\delta, \gamma+2-\frac{2}{p}+\delta}(\mathsf{M}, \mathbb{R}^L) \oplus \mathbb{R}^L$. More precisely, it follows from Proposition 2.6 that $u^* = w^* + z$ for some $z \in \mathsf{N}$ and $w^* \in \mathcal{H}_p^{2-\frac{2}{p}+\delta, \gamma+2-\frac{2}{p}+\delta}(\mathsf{M}, \mathsf{N} - z)$. Because of the continuity of u in the space

$$BC((\alpha, T_{\max}), \mathcal{H}_p^{2-\frac{2}{p}+\tilde{\delta}-\varepsilon, \gamma+2-\frac{2}{p}+\tilde{\delta}-\varepsilon}(\mathsf{M}, \mathbb{R}^L) \oplus \mathbb{R}^L),$$

the limit u^* is independent of the choice of the sequence $\{u(t_k)\}_k$.

Therefore, we can apply Theorem 4.5 to extend u beyond $T_{\max}$ smoothly. This leads to a contradiction.

Let us summarize the above discussion in the following theorem.

THEOREM 5.2. *Suppose that the sectional curvature of* N *is nonpositive. Assume that*

- γ *and* p *satisfy (B0) and (B3) when* $n \geq 3$; *or*
- *when* $n = 1, 2$, γ *and* p *satisfy (A') or satisfy (B0)-(B1) for* $n = 1$ *or satisfy (B0) and (B2) for* $n = 2$.

Let

$$u_0 = w_0 + \mathsf{p} \quad \textit{with} \quad \mathsf{p} \in \mathsf{N} \quad \textit{and} \quad w_0 \in \mathcal{H}_p^{2-\frac{2}{p}+\delta, \gamma+2-\frac{2}{p}+\delta}(\mathsf{M}, \mathsf{N} - \mathsf{p})$$

for an arbitrary positive constant δ; *in addition* $\|\nabla w_0\|_\infty < \infty$. *Then* (4.1) *with initial condition* u_0 *has a unique global solution*

$$\begin{aligned} u \in & L_p([0,T), \mathcal{H}_p^{2,\gamma+2}(\mathsf{M}, \mathbb{R}^L) \oplus \mathbb{R}^L) \cap H_p^1([0,T), \mathcal{H}_p^{0,\gamma}(\mathsf{M}, \mathsf{N})) \\ & \cap BC([0,T), \mathcal{H}_p^{2-\frac{2}{p}-\varepsilon, \gamma+2-\frac{2}{p}-\varepsilon}(\mathsf{M}, \mathbb{R}^L) \oplus \mathbb{R}^L) \end{aligned}$$

for any $\varepsilon > 0$ *and* $T > 0$. *Moreover, for any* $\alpha > 0$ *and* $k \in \mathbb{N}_0$

$$u \in H_p^k((\alpha, T), \mathcal{H}_p^{k+2,\gamma+2}(\mathsf{M}, \mathbb{R}^L) \oplus \mathbb{R}^L) \cap H_p^{k+1}((\alpha, T), \mathcal{H}_p^{k,\gamma}(\mathsf{M}, \mathsf{N}))$$

and

$$u \in BC^\infty((\alpha, T), \mathcal{H}_p^{\infty, \gamma+2-\frac{2}{p}-\varepsilon}(\mathsf{M}, \mathbb{R}^L) \oplus \mathbb{R}^L) \cap C^\infty((0,\infty), \mathcal{H}_p^{\infty,\gamma+2}(\mathsf{M}, \mathbb{R}^L) \oplus \mathbb{R}^L).$$

6. Acknowledgements

Parts of the paper were completed while the first author was a Golomb Assistant Professor at Purdue University. He would like to thank the staff and faculty at Purdue University for providing a friendly environment. The first author would also like to express his gratitude to Prof. Elmar Schrohe for helpful discussions on the cone differential operators during his stay in Leibniz University Hanover in 2015. The second author is partially supported by NSF.

References

[1] H. Amann, *Linear and quasilinear parabolic problems. Vol. I*, Monographs in Mathematics, vol. 89, Birkhäuser Boston, Inc., Boston, MA, 1995. Abstract linear theory. MR1345385

[2] H. Amann, *Function spaces on singular manifolds*, Math. Nachr. **286** (2013), no. 5-6, 436–475, DOI 10.1002/mana.201100157. MR3048125

[3] J. Bergh and J. Löfström, *Interpolation spaces. An introduction*, Springer-Verlag, Berlin-New York, 1976. Grundlehren der Mathematischen Wissenschaften, No. 223. MR0482275

[4] J. Brüning and R. Seeley, *An index theorem for first order regular singular operators*, Amer. J. Math. **110** (1988), no. 4, 659–714, DOI 10.2307/2374646. MR955293

[5] P. Clément and S. Li, *Abstract parabolic quasilinear equations and application to a groundwater flow problem*, Adv. Math. Sci. Appl. **3** (1993/94), no. Special Issue, 17–32. MR1287921

[6] S. Coriasco, E. Schrohe, and J. Seiler, *Differential operators on conic manifolds: maximal regularity and parabolic equations*, Bull. Soc. Roy. Sci. Liège **70** (2001), no. 4-6, 207–229 (2002). Hommage à Pascal Laubin. MR1904055

[7] S. Coriasco, E. Schrohe, and J. Seiler, *Bounded imaginary powers of differential operators on manifolds with conical singularities*, Math. Z. **244** (2003), no. 2, 235–269, DOI 10.1007/s00209-003-0495-1. MR1992538

[8] R. Denk, M. Hieber, and J. Prüss, *$\mathcal{R}$-boundedness, Fourier multipliers and problems of elliptic and parabolic type*, Mem. Amer. Math. Soc. **166** (2003), no. 788, viii+114, DOI 10.1090/memo/0788. MR2006641

[9] J. Eells Jr. and J. H. Sampson, *Harmonic mappings of Riemannian manifolds*, Amer. J. Math. **86** (1964), 109–160, DOI 10.2307/2373037. MR0164306

[10] A. Fardoun and R. Regbaoui, *Équation de la chaleur pour les applications harmoniques entre variétés riemanniennes complètes* (French, with English and French summaries), C. R. Acad. Sci. Paris Sér. I Math. **331** (2000), no. 4, 299–304, DOI 10.1016/S0764-4442(00)01628-1. MR1787188

[11] J. B. Gil, T. Krainer, and G. A. Mendoza, *Resolvents of elliptic cone operators*, J. Funct. Anal. **241** (2006), no. 1, 1–55, DOI 10.1016/j.jfa.2006.07.010. MR2264246

[12] R. S. Hamilton, *Harmonic maps of manifolds with boundary*, Lecture Notes in Mathematics, Vol. 471, Springer-Verlag, Berlin-New York, 1975. MR0482822

[13] M. Lesch, *Operators of Fuchs type, conical singularities, and asymptotic methods*, Teubner-Texte zur Mathematik [Teubner Texts in Mathematics], vol. 136, B. G. Teubner Verlagsgesellschaft mbH, Stuttgart, 1997. MR1449639

[14] J. Li, *The heat flows and harmonic maps of complete noncompact Riemannian manifolds*, Math. Z. **212** (1993), no. 2, 161–173, DOI 10.1007/BF02571650. MR1202804

[15] P. Li and L.-F. Tam, *The heat equation and harmonic maps of complete manifolds*, Invent. Math. **105** (1991), no. 1, 1–46, DOI 10.1007/BF01232256. MR1109619

[16] G. G. Liao and L.-F. Tam, *On the heat equation for harmonic maps from noncompact manifolds*, Pacific J. Math. **153** (1992), no. 1, 129–145. MR1145918

[17] F. Lin and C. Wang, *Harmonic and quasi-harmonic spheres*, Comm. Anal. Geom. **7** (1999), no. 2, 397–429, DOI 10.4310/CAG.1999.v7.n2.a9. MR1685578

[18] F. Lin and C. Wang, *The analysis of harmonic maps and their heat flows*, World Scientific Publishing Co. Pte. Ltd., Hackensack, NJ, 2008. MR2431658

[19] R. Moser, *Partial regularity for harmonic maps and related problems*, World Scientific Publishing Co. Pte. Ltd., Hackensack, NJ, 2005. MR2155901

[20] J. Prüss and G. Simonett, *Moving interfaces and quasilinear parabolic evolution equations*, Monographs in Mathematics, vol. 105, Birkhäuser/Springer, [Cham], 2016. MR3524106

[21] N. Roidos and E. Schrohe, *The Cahn-Hilliard equation and the Allen-Cahn equation on manifolds with conical singularities*, Comm. Partial Differential Equations **38** (2013), no. 5, 925–943, DOI 10.1080/03605302.2012.736913. MR3046298

[22] N. Roidos and E. Schrohe, *Bounded imaginary powers of cone differential operators on higher order Mellin-Sobolev spaces and applications to the Cahn-Hilliard equation*, J. Differential Equations **257** (2014), no. 3, 611–637, DOI 10.1016/j.jde.2014.04.004. MR3208087

[23] N. Roidos and E. Schrohe, *Existence and maximal L^p-regularity of solutions for the porous medium equation on manifolds with conical singularities*, Comm. Partial Differential Equations **41** (2016), no. 9, 1441–1471, DOI 10.1080/03605302.2016.1219745. MR3551464

[24] E. Schrohe and J. Seiler, *The resolvent of closed extensions of cone differential operators*, Canad. J. Math. **57** (2005), no. 4, 771–811, DOI 10.4153/CJM-2005-031-1. MR2152939

[25] R. Schoen and S. T. Yau, *Harmonic maps and the topology of stable hypersurfaces and manifolds with non-negative Ricci curvature*, Comment. Math. Helv. **51** (1976), no. 3, 333–341, DOI 10.1007/BF02568161. MR0438388

[26] J. Wang, *The heat flow and harmonic maps between complete manifolds*, J. Geom. Anal. **8** (1998), no. 3, 485–514, DOI 10.1007/BF02921799. MR1707741

[27] M. Wang, *The heat flow of harmonic maps from noncompact manifolds*, Nonlinear Anal. **71** (2009), no. 3-4, 1042–1048, DOI 10.1016/j.na.2008.11.029. MR2527523

Department of Mathematical Sciences, Georgia Southern University, 65 Georgia Avenue, P.O. Box 8093, Statesboro, Georgia 30460

Email address: yshao@georgiasouthern.edu

Department of Mathematics, Purdue University, 150 N. University Street, West Lafayette, Indiana 47907-2067

Email address: wang2482@purdue.edu

Contemporary Mathematics
Volume **725**, 2019
https://doi.org/10.1090/conm/725/14552

On the global regularity issue of the two-dimensional magnetohydrodynamics system with magnetic diffusion weaker than a Laplacian

Kazuo Yamazaki

ABSTRACT. In this manuscript, we discuss the recent developments in the research direction concerning whether the solution to the two-dimensional magnetohydrodynamics system with certain velocity dissipation and magnetic diffusion strengths preserves the high regularity of a given initial data for all time or exhibits a blowup in finite time. In particular, we address an open problem in case the magnetic diffusion is weaker than a full Laplacian. In short, we consider this system with both velocity dissipation and magnetic diffusion strength measured in terms of fractional Laplacians with certain powers, and point out the following gap in the results among the current literature. In case the power of the fractional Laplacian representing the magnetic diffusion is one, the global well-posedness follows as long as the velocity dissipation is present regardless of how weak its strength may be, and hence the sum of the two powers need to only be more than one ([Monatsh. Math. 175 (2014), no. 1, 127–131], [*Second proof of the global regularity of the two-dimensional MHD system with full diffusion and arbitrary weak dissipation*, Methods Appl. Anal., to appear]). On the other hand, once the power of the fractional Laplacian representing the magnetic diffusion drops below one, in order to ensure the system's global well-posedness, the sum of the powers from the dissipation and diffusion must be equal to or more than two, improved only logarithmically ([Nonlinear Anal. 85 (2013), 43–51], [J. Math. Fluid Mech. 13 (2011), no. 2, 295–305], [Appl. Math. Lett. 29 (2014), 46–51]). We discuss this issue in more detail, explain its reasoning, and provide some basic regularity criteria to gain better insight to this difficult direction of research.

1. Introduction

One of the most well-known outstanding open problems in mathematical analysis questions whether the solution to the following N-dimensional Navier-Stokes equations (NSE), in case $N = 3$, admits a unique solution with finite kinetic energy for all time or if there exists an initial data with finite kinetic energy such that its

2010 *Mathematics Subject Classification.* Primary 35B65, 35Q35, 35Q61.
Key words and phrases. BMO space, Fourier transform, fractional Laplacian, magnetohydrodynamics system, regularity.

corresponding solution experiences a blowup in finite time:

(1a) $$\frac{\partial u}{\partial t} + (u\cdot\nabla)u + \nabla\pi - \nu\Delta u = 0,$$

(1b) $$\nabla\cdot u = 0, \quad u(x,0) \triangleq u_0(x),$$

where we denoted by $u : \mathbb{R}^N \times \mathbb{R}^+ \mapsto \mathbb{R}^N, \pi : \mathbb{R}^N \times \mathbb{R}^+ \mapsto \mathbb{R}$, the velocity, the pressure fields respectively, and $\nu > 0$ represents viscosity. This system at $\nu = 0$ recovers the Euler equations.

Hereafter let us write $\frac{\partial}{\partial t} = \partial_t, \frac{\partial}{\partial x_i} = \partial_i$ for $i \in \mathbb{N}$, $x = (x_1, \dots, x_N)$, as well as $\int f = \int_{\mathbb{R}^N} f(x)dx$. Furthermore, for brevity we use the notations $A \lesssim_{a,b} B, A \approx_{a,b} B$ to imply that there exists a constant $c(a,b)$ that depends on a, b such that $A \leq cB, A = cB$ respectively. Finally, we denote the fractional Laplacian, $\Lambda^r \triangleq (-\Delta)^{\frac{r}{2}}, r \in \mathbb{R}$, defined through a Fourier symbol of $|\xi|^r$ so that

$$\widehat{\Lambda^r f}(\xi) = |\xi|^r \hat{f}(\xi).$$

Let us now couple (1a) with Maxwell's equation from electromagnetism to write down the generalized magnetohydrodynamics (MHD) system as

(2a) $$\partial_t u + (u\cdot\nabla)u + \nabla\pi + \nu\Lambda^{2\alpha}u = (b\cdot\nabla)b,$$

(2b) $$\partial_t b + (u\cdot\nabla)b + \eta\Lambda^{2\beta}b = (b\cdot\nabla)u,$$

(2c) $$\nabla\cdot u = \nabla\cdot b = 0, \quad (u,b)(x,0) \triangleq (u_0,b_0)(x),$$

where we denoted by $b : \mathbb{R}^N \times \mathbb{R}^+ \mapsto \mathbb{R}^N$, the magnetic field and $\eta \geq 0$ the magnetic diffusivity. Let us call the generalized MHD system at $\nu, \beta > 0, \alpha = \beta = 1$, the classical MHD system. As we will see, it is important to observe here that (2a) at $\alpha = 1$ is precisely (1a) with the forcing term of $(b\cdot\nabla)b$. It is also worth noting that if one achieves in proving that given an arbitrary initial data (u_0, b_0) sufficiently smooth, its corresponding solution (u, b) preserves its regularity for all time, then one may take $b_0 \equiv 0$, and deduce by uniqueness the smooth solution u for the NSE (1a)-(1b). In this way, it may be argued that proving the global well-posedness of the MHD system (2a)-(2c) is more difficult than that of the NSE (1a)- (1b).

Now similarly to many other systems of equations in physics, the solutions to both the NSE (1a)- (1b) and the MHD system (2a)- (2c) admit a rescaling property, specifically that e.g. in the latter case, if $(u,b)(x,t)$ (and $\pi(x,t)$) solves the generalized MHD system (2a)-(2c) for $\gamma = \alpha = \beta$, then so does

(3) $$(u_\lambda, b_\lambda)(x,t) \triangleq \lambda^{2\gamma-1}(u,b)(\lambda x, \lambda^{2\gamma}t), \quad \lambda \in \mathbb{R}^+,$$

(and $\pi_\lambda(x,t) \triangleq \lambda^{4\gamma-2}\pi(\lambda x, \lambda^{2\gamma}t)$). Moreover, taking $L^2(\mathbb{R}^N)$-inner products of (2a)-(2b) with (u,b), making use of the divergence-free property of both u and b from (2c), the following identity that describes that conservation of energy and cumulative energy dissipation and diffusion may be shown:

(4) $$\left(\|u\|_{L^2}^2 + \|b\|_{L^2}^2\right)(t) + \int_0^t \nu\|\Lambda^\alpha u\|_{L^2}^2 + \eta\|\Lambda^\beta b\|_{L^2}^2 d\tau = \|u_0\|_{L^2}^2 + \|b_0\|_{L^2}^2$$

for all t in the time interval $[0,T]$ over which the solution exists. It is well-known that while other bounded quantities of the solutions exist (see [**15**]), those in (4) seem to be the most useful upon *a priori* estimates in the effort to prove higher regularity of the solution. Studying the rescaled solution, e.g. $u_\lambda(x,t)$, in this *most useful* bounded quantity deduces

$$\|u_\lambda(t)\|_{L^2}^2 = \lambda^{4\gamma-2-N}\|u(\lambda^{2\gamma}t)\|_{L^2}^2 \tag{5}$$

by (3), where $4\gamma - 2 - N = 0$ if and only if $\gamma = \frac{1}{2} + \frac{N}{4}$, indicating that $\frac{1}{2} + \frac{N}{4}$ is a kind of a threshold for the powers α, β such that relying only on these bounds from (4) in the *a priori* estimates, one can hope to prove higher regularity only if both α and β are equal to or more than $\frac{1}{2} + \frac{N}{4}$. Indeed, this is precisely what was proven in [**23**]; the result therein at $N = 2$ in particular recovers the pioneering work of [**14, 16**] that the NSE (1a)-(1b), and the classical MHD system respectively are both globally well-posed. Additionally, let us mention that the works of [**19, 22, 24, 27**] proved logarithmic improvement and in particular the results from [**22, 24, 27**] showed that in the case of the generalized MHD system (2a)-(2c), while $\alpha \geq \frac{1}{2} + \frac{N}{4}$ remains necessary, the condition on $\beta \geq \frac{1}{2} + \frac{N}{4}$ can be relaxed to only $\beta \geq 0$ and $\alpha + \beta \geq 1 + \frac{N}{2}$. That is, the lower bound on β may be shifted toward α although not vice versa, rooting from the fact that upon the $H^1(\mathbb{R}^N)$-estimate on the system (2a)-(2c), every non-linear term that must be estimated involves u, but not necessarily b and hence the support from dissipation plays a more important role than the diffusion in most cases, although not all as we will see. Let us formally state the result from [**24**] for completeness:

THEOREM 1.1. *([**24**, Theorem 1.1], see also [**22, 27**] for the case $\eta = 0$) Consider the following N-dimensional generalized MHD system:*

$$\begin{aligned}
&\partial_t u + (u\cdot\nabla)u + \nabla\pi + \nu\mathcal{L}_1^2 u = (b\cdot\nabla)b,\\
&\partial_t b + (u\cdot\nabla)b + \eta\mathcal{L}_2^2 b = (b\cdot\nabla)u,\\
&\nabla\cdot u = \nabla\cdot b = 0, \quad (u,b)(x,0) \triangleq (u_0,b_0)(x),
\end{aligned}$$

where the Fourier operators $\mathcal{L}_i, i \in \{1,2\}$, are defined to have the Fourier symbols of $\widehat{\mathcal{L}_i f}(\xi) = m_i(\xi)\hat{f}(\xi)$ with $m_i(\xi)$ conditioned to obey the lower bounds of

$$m_1(\xi) \geq \frac{|\xi|^\alpha}{g_1(\xi)}, \quad m_2(\xi) \geq \frac{|\xi|^\beta}{g_2(\xi)}$$

for all sufficiently large $|\xi|$ and $g_i : \mathbb{R}^+ \mapsto [1,\infty)$ are radially symmetric, non-decreasing functions. If $\nu > 0$ and $\eta > 0$,

$$\alpha \geq \frac{1}{2} + \frac{N}{4}, \quad \beta > 0, \quad \alpha + \beta \geq 1 + \frac{N}{2},$$

and

$$\int_1^\infty \frac{d\tau}{\tau(g_1^2(\tau) + g_2^2(\tau))^2} = +\infty,$$

then given any initial data $(u_0, b_0) \in H^s(\mathbb{R}^N), s > 1 + \frac{N}{2}$, this system has a unique global classical solution (u,b).

From our previous discussion, we understand that breaking the threshold of $\alpha \geq \frac{1}{2} + \frac{N}{4}, \beta \geq \frac{1}{2} + \frac{N}{4}$ more than logarithmically is extremely difficult and may require some bounded quantity *better* than those in (4). For this reason, it is understood among experts that N-dimensional generalized MHD system (2a)-(2c) is energy-critical if $\alpha + \beta = 1 + \frac{N}{2}$, energy-subcritical if $\alpha + \beta > 1 + \frac{N}{2}$ and energy-supercritical if $\alpha + \beta < 1 + \frac{N}{2}$, which in particular implies that the two-dimensional (2d) NSE as well as the 2d classical MHD system are energy-critical.

A remarkable feature of the NSE (1a)-(1b) in the 2d case is that such a bounded quantity better than those in (4) in fact exists. Indeed, even in the case of the Euler equations, denoting by $w \triangleq \nabla \times u$ and applying a curl operator on (1a) with $\nu = 0$ shows that the evolution of w is governed by a transport equation of

$$\partial_t w + (u \cdot \nabla) w = 0. \tag{6}$$

Thus, multiplying (6) by $|w|^{p-2}w, p \geq 2$, integrating over $\mathbb{R}^2$ and using that

$$\int (u \cdot \nabla) w |w|^{p-2} w = \frac{1}{p} \int (u \cdot \nabla) |w|^p = 0 \tag{7}$$

due to (1b) lead to

$$\frac{1}{p} \partial_t \|w\|_{L^p}^p = 0. \tag{8}$$

Using that $\frac{1}{p}\partial_t \|w\|_{L^p}^p = \|w\|_{L^p}^{p-1} \partial_t \|w\|_{L^p}$, dividing (8) by $\|w\|_{L^p}^{p-1}$, taking $p \nearrow +\infty$ shows that the $\sup_{t\in[0,T]} \|w(t)\|_{L^\infty}$ is bounded by $\|w(0)\|_{L^\infty}$. With this bound attained, applying Λ^s to (1a) with $\nu = 0$, $s > 2$, taking $L^2(\mathbb{R}^2)$-inner products with $\Lambda^s u$, employing (36) from Lemma 3.1 and Lemma 3.2 lead immediately to

$$\begin{aligned} \frac{1}{2}\partial_t \|\Lambda^s u\|_{L^2}^2 =& - \int [\Lambda^s((u\cdot\nabla)u) - (u \cdot \nabla)\Lambda^s u] \cdot \Lambda^s u \\ \lesssim& \|\nabla u\|_{L^\infty} \|\Lambda^s u\|_{L^2}^2 \\ \lesssim& \left(\|u\|_{L^2} + \|w\|_{L^\infty} \log_2(1 + \|\Lambda^s u\|_{L^2}^2) + 1 \right) \|\Lambda^s u\|_{L^2}^2; \end{aligned} \tag{9}$$

hence, the higher regularity is attained via Gronwall's inequality applied to (9). This recovers the classical results from [**35**] (see also [**10**]).

A natural question arises whether the favorable formulation of the vorticity equation may exploited in order to improve the results on the MHD system from [**16, 23**] that required $\alpha \geq 1, \beta \geq 1$ in the 2d case. In this endeavor, applying a curl operator on (2a)-(2b), one arrives at

$$\partial_t w + (u\cdot\nabla)w + \nu\Lambda^{2\alpha} w = (b \cdot \nabla) j, \tag{10a}$$

$$\begin{aligned} &\partial_t j + (u \cdot \nabla) j + \eta \Lambda^{2\beta} j \\ &\quad = (b\cdot\nabla)w + 2[\partial_1 b_1(\partial_1 u_2 + \partial_2 u_1) - \partial_1 u_1(\partial_1 b_2 + \partial_2 b_1)] \end{aligned} \tag{10b}$$

where $j \triangleq \nabla \times b$ represents current density. A quick comparison of (10a) with (6) shows that (10a) is still a transport equation, but now forced by $(b \cdot \nabla)j$ while dissipated fractionally by $\nu\Lambda^{2\alpha}w$. Thus, upon an $L^p(\mathbb{R}^2)$-estimate of w, in contrast to (8), we are faced with

$$\frac{1}{p}\partial_t \|w\|_{L^p}^p + \nu \int \Lambda^{2\alpha} w |w|^{p-2} w = \int (b \cdot \nabla) j |w|^{p-2} w. \tag{11}$$

Therefore, we see heuristically that if $\eta > 0$ and β is sufficiently large so that $(b \cdot \nabla)j$ is sufficiently smooth due to (4) and thus bounded, then the higher regularity should be attained via an analogous computations in (7)-(9). Moreover, even if β is not sufficiently large, if $\nu > 0$ and $\alpha > 0$ is adequately large, then $|w|^{p-2}w$ within $\int (b \cdot \nabla) j |w|^{p-2} w$ may be appropriately handled using (4) and thus higher regularity may follow. Following this intuitive argument, an explosive amount of work appeared very recently improving results one after another.

On one hand, for the case $\nu = 0$ and hence $\alpha = 0$, Tran, Yu and Zhai in [**21**] showed that $\beta > 2$ suffices in order to prove the global well-posedness. Jiu and

Zhao in [**8**] and the author in [**26**] independently improved this result to $\beta > \frac{3}{2}$. Thereafter, Jiu and Zhao in [**9**] and Cao, Wu and Yuan in [**2**] independently showed that $\beta > 1$ in fact suffices; the approach of the former was taking advantage of the property of a heat kernel, while the latter the Besov space techniques (see also [**5**]). On the other hand, for the case $\eta > 0, \beta = 1$, Tran, Yu and Zhai in [**21**] showed that $\alpha \geq \frac{1}{2}$ suffices. Subsequently, Yuan and Bai in [**34**], as well as the author in [**28**], independently improved this result to $\alpha > \frac{1}{3}$. Thereafter, Ye and Xu in [**33**] improved from $\alpha > \frac{1}{3}$ to $\alpha \geq \frac{1}{4}$. Finally, the authors in [**6**] proved that $\alpha > 0$ suffices (also [**31**]). At the time of writing this current manuscript, the most difficult remaining open problem is the case of $\nu = 0$ and hence $\alpha = 0$ while $\eta > 0, \beta = 1$, which would represent a complete extension of the classical results from [**10, 35**] to the MHD system. We also refer to [**20**] for numerical analysis, and [**7, 32**] for regularity criteria results.

A common ingredient in the hypothesis of every one of these recent new results on the 2d generalized MHD system is that $\eta > 0, \beta \geq 1$. This is because applying a curl operator on (2a) not only gives a favorable vorticity formulation in the form of a forced dissipative transport equation in (10a), but furthermore if one combines an $L^2(\mathbb{R}^2)$-estimate of w with that of j, even the forcing term $(b \cdot \nabla) j$ vanishes due to a remarkable cancellation. Indeed, upon taking $L^2(\mathbb{R}^2)$-estimates of (10a)-(10b) with (w, j) respectively, summing them up leads to

$$\begin{aligned}(12)\qquad &\frac{1}{2}\partial_t(\|w\|_{L^2}^2 + \|j\|_{L^2}^2) + \nu\|\Lambda^\alpha w\|_{L^2}^2 + \eta\|\Lambda^\beta j\|_{L^2}^2 \\ &= \int 2[\partial_1 b_1(\partial_1 u_2 + \partial_2 u_1) - \partial_1 u_1(\partial_1 b_2 + \partial_2 b_1)]j\end{aligned}$$

where we made use of the facts that not only

$$(13)\qquad \int (u\cdot\nabla)ww = \frac{1}{2}\int (u\cdot\nabla)|w|^2 = 0, \quad \int (u\cdot\nabla)jj = \frac{1}{2}\int (u\cdot\nabla)|j|^2 = 0,$$

but also

$$(14)\qquad \int (b\cdot\nabla)jw + \int (b\cdot\nabla)wj = \int (b\cdot\nabla)(wj) = 0$$

due to (2c). The right-hand side of (12) may be estimated in the following simple and in fact an optimal fashion:

$$\begin{aligned}(15)\qquad &\frac{1}{2}\partial_t(\|w\|_{L^2}^2 + \|j\|_{L^2}^2) + \nu\|\Lambda^\alpha w\|_{L^2}^2 + \eta\|\Lambda^\beta j\|_{L^2}^2 \\ &= \int 2[\partial_1 b_1(\partial_1 u_2 + \partial_2 u_1) - \partial_1 u_1(\partial_1 b_2 + \partial_2 b_1)]j \\ &\lesssim \|\nabla b\|_{L^4}\|\nabla u\|_{L^2}\|j\|_{L^4} \\ &\lesssim \|j\|_{L^2}\|\nabla j\|_{L^2}\|w\|_{L^2}\end{aligned}$$

by Hölder's inequality, Gagliardo-Nirenberg's inequality and Lemma 3.3. This implies that if $\eta > 0, \beta = 1$ so that from (4) we would have $\int_0^T \|\nabla b\|_{L^2}^2 d\tau \lesssim 1$, then Young's inequality applied in (15) separating $\frac{\eta}{2}\|\nabla j\|_{L^2}^2$ would allow us to close this $(\|w\|_{L^2}^2 + \|j\|_{L^2}^2)(t)$-estimate over $[0, T]$. Let us point out, very importantly for our subsequent discussion, that from the first inequality of (15) we chose $\|\nabla u\|_{L^2}$ to play the role of "being estimated" because $\|\nabla u\|_{L^2} \approx \|w\|_{L^2}$, and we chose $\|\nabla b\|_{L^4}$ and $\|j\|_{L^4}$ to share the role of "to be absorbed to the diffusive term $\eta\|\Lambda^\beta j\|_{L^2}^2$" and "to be square-integrable in time." This immediate $H^1(\mathbb{R}^2)$-bound of (u, b) in the

2d case is precisely the reason why the door was opened for all the recent results in [**2, 6–9, 21, 26, 28, 31–34**].

Now let us consider the same estimate in (15) with $\beta < 1$ and $\alpha + \beta < 2$ so that a global well-posedness in such a case will be an improvement of the work in [**10, 35**]. One may follow the analogous idea in (15) and use Hölder's inequality to bound the right-hand side of (12) by

$$\int 2[\partial_1 b_1(\partial_1 u_2 + \partial_2 u_1) - \partial_1 u_1(\partial_1 b_2 + \partial_2 b_1)]j \lesssim \|\nabla b\|_{L^4}\|\nabla u\|_{L^4}\|j\|_{L^2}. \tag{16}$$

However, we immediately realize that $\|\nabla b\|_{L^4}$ from (16) cannot possibly be interpolated between $\|\nabla b\|_{L^2}$ "being estimated," and $\|\Lambda^\beta b\|_{L^2}$ which is "to be square-integrable in time," because $\beta < 1$. Therefore, this term must be interpolated as

$$\|\nabla b\|_{L^4} \lesssim \|\nabla b\|_{L^2}^{\frac{2\beta-1}{2\beta}} \|\Lambda^\beta \nabla b\|_{L^2}^{\frac{1}{2\beta}}, \tag{17}$$

i.e. partially "being estimated" and partially "to be absorbed to the diffusive term $\eta\|\Lambda^\beta j\|_{L^2}^2$." However, this application of Gagliardo-Nirenberg's inequality in (17) requires $\beta \geq \frac{1}{2}$. On the other hand, since the interpolation of (17) did not make use of terms that are square-integrable in time from (4), it seems most efficient to interpolate $\|\nabla u\|_{L^4}$ from (16) between the "being estimated" and "to be square-integrable in time" as follows:

$$\|\nabla u\|_{L^4} \lesssim \|\nabla u\|_{L^2}^{\frac{2\alpha-3}{2\alpha-2}} \|\Lambda^\alpha u\|_{L^2}^{\frac{1}{2\alpha-2}}. \tag{18}$$

However, this application of Gagliardo-Nirenberg's inequality requires $\alpha \geq \frac{3}{2}$ and thus together with the requirement from (17), we already have the restriction of $\alpha+\beta \geq 2$. Various other options of Hölder's inequality in (16) are possible; however, upon many attempts, every one of them seems to require $\alpha + \beta \geq 2$; in particular, the extreme case of

$$\int 2[\partial_1 b_1(\partial_1 u_2 + \partial_2 u_1) - \partial_1 u_1(\partial_1 b_2 + \partial_2 b_1)]j \lesssim \|\nabla b\|_{L^2}\|\nabla u\|_{L^\infty}\|j\|_{L^2}$$

immediately requires $\alpha > 2$ so that $\|\nabla u\|_{L^\infty} \lesssim \|u\|_{L^2} + \|\Lambda^\alpha u\|_{L^2}$ becomes integrable over $[0, T]$.

Heuristically the problem is that the right-hand side of (12) is more or less $|\nabla b|^2|\nabla u|$; if this were $|\nabla u|^2|\nabla b|$, then $H^1(\mathbb{R}^2)$-estimate of (u, b) may be immediately attained with just $\nu > 0, \alpha = 1$ and $\eta = 0$. Unfortunately upon many attempts to make some cancellations to reverse such roles of u and b, we found this task very difficult. Indeed, although in the literature there have been various identities due to unique cancellations via appropriate additions and subtractions, taking advantage of divergence-free conditions (e.g. [**13**, Lemma 2.3], [**30**, Proposition 1.1]), vorticity formulation is really remarkable in that upon applying a curl operator on the non-linear term,

$$\begin{aligned}
\nabla \times ((u \cdot \nabla)u) =& \partial_1((u \cdot \nabla)u_2) - \partial_2((u \cdot \nabla)u_1) \\
=& \partial_1 u_1 \partial_1 u_2 + \partial_1 u_2 \partial_2 u_2 + (u \cdot \nabla)\partial_1 u_2 \\
& - \partial_2 u_1 \partial_1 u_1 - \partial_2 u_2 \partial_2 u_1 - (u \cdot \nabla)\partial_2 u_1 = (u \cdot \nabla)w,
\end{aligned}$$

cancellations occur within itself. One does not even need to add or subtract with another equation to make this happen!

One noteworthy idea, that has proven to be useful in the literature, is to consider a symmetric form of $S \triangleq w + j, \overline{S} \triangleq w - j$ so that the roles of u and b are indeed somehow reversible. That is, we assume $\gamma = \alpha = \beta, \nu = \eta = \lambda$, estimate

$$\begin{aligned}&\partial_t S + ((u-b)\cdot\nabla)S + \lambda\Lambda^{2\gamma}S\\ &\qquad = -\partial_1 u\cdot\nabla b_2 + \partial_2 u\cdot\nabla b_1 + \partial_1 b\cdot\nabla u_2 - \partial_2 b\cdot\nabla u_1,\end{aligned} \tag{19a}$$

$$\begin{aligned}&\partial_t \overline{S} + ((u+b)\cdot\nabla)\overline{S} + \lambda\Lambda^{2\gamma}\overline{S}\\ &\qquad = \partial_1 u\cdot\nabla b_2 - \partial_2 u\cdot\nabla b_1 - \partial_1 b\cdot\nabla u_2 + \partial_2 b\cdot\nabla u_1,\end{aligned} \tag{19b}$$

where we used that we may rewrite

$$\begin{aligned}&2[\partial_1 b_1(\partial_1 u_2 + \partial_2 u_1) - \partial_1 u_1(\partial_1 b_2 + \partial_2 b_1)]\\ &= -\partial_1 u\cdot\nabla b_2 + \partial_2 u\cdot\nabla b_1 + \partial_1 b\cdot\nabla u_2 - \partial_2 b\cdot\nabla u_1,\end{aligned} \tag{20}$$

and hope that we can complete the estimate of $\|S\|_{L^2}^2 + \|\overline{S}\|_{L^2}^2$ with $\gamma < 1$ so that by triangle inequality we may conclude the bound of $\|w\|_{L^2}^2 + \|j\|_{L^2}^2$. Unfortunately this strategy also seems to require $\gamma \geq 1$; e.g. assuming $\gamma < 1$, taking $L^2(\mathbb{R}^2)$-inner products with S in (19a), we may estimate

$$\begin{aligned}\frac{1}{2}\partial_t\|S\|_{L^2}^2 + \lambda\|\Lambda^\gamma S\|_{L^2}^2 \lesssim&(\|\partial_1 u\cdot\nabla b_2\|_{L^{\frac{2}{1+\gamma}}} + \|\partial_2 u\cdot\nabla b_1\|_{L^{\frac{2}{1+\gamma}}}\\ &+ \|\partial_1 b\cdot\nabla u_2\|_{L^{\frac{2}{1+\gamma}}} + \|\partial_2 b\cdot\nabla u_1)\|_{L^{\frac{2}{1+\gamma}}}\|\Lambda^\gamma S\|_{L^2}\\ \leq&\frac{\lambda}{4}\|\Lambda^\gamma S\|_{L^2}^2 + c\|\nabla u\|_{L^{\frac{2}{\gamma}}}^2\|\nabla b\|_{L^2}^2\\ \leq&\frac{\lambda}{4}\|\Lambda^\gamma S\|_{L^2}^2 + c\|\Lambda^{2-\gamma}u\|_{L^2}^2\|\nabla b\|_{L^2}^2\end{aligned}$$

by Hölder's inequality, Hardy-Littlewood-Sobolev theorem (e.g. [**18**, pg. 119]), Young's inequality and Sobolev embedding of $\dot{H}^{1-\gamma}(\mathbb{R}^2) \hookrightarrow L^{\frac{2}{\gamma}}(\mathbb{R}^2)$. For this estimate to be closed, in comparison with (4), we will need $2-\gamma \leq \alpha$ which is if and only if $1 \leq \alpha$ and because we assumed $\gamma = \alpha = \beta$, this requires $\alpha + \beta \geq 2$.

Therefore, we conclude that there has been this "large gap," for a technical reason, that if $\eta > 0, \beta \geq 1$, the best results in the literature (e.g. [**2,5,6,9,31**]) requires only $\alpha + \beta > 1$. But once we reduce our hypothesis to $\eta > 0, \beta < 1$, then the requirement suddenly jumps to $\alpha + \beta \geq 2$ from the classical results ([**16,23**]), improvable only logarithmically in [**22,24,27**]. In the rest of this paper we provide a basic Serrin-type regularity criterion for the solution to be smooth ([**17**]) to gain better insight to this difficult case when $\eta > 0, \beta < 1$ and $\alpha + \beta < 2$:

THEOREM 1.2. *Let $N = 2, \nu > 0, \eta > 0$, $\beta \in [\frac{1}{2}, 1)$, $\alpha \in [\frac{1}{2}, 2-\beta)$ and*

$$\begin{aligned}(u,b) \in \in\, &C([0,T]; H^s(\mathbb{R}^2)) \cap L^2([0,T]; H^{s+\alpha}(\mathbb{R}^2))\\ &\times C([0,T]; H^s(\mathbb{R}^2)) \cap L^2([0,T]; H^{s+\beta}(\mathbb{R}^2))\end{aligned} \tag{21}$$

be the solution pair to the generalized MHD system (2a)*-*(2c) *for a given $(u_0, b_0) \in H^s(\mathbb{R}^2), s > 2$. Suppose the solution (u,b) over the time interval of $[0,T]$ satisfies*

$$\begin{aligned}&\int_0^T \|\nabla u\|_{L^p}^r d\tau < \infty\\ &\textit{where } \frac{2}{p} + \frac{2}{r} \leq 2 + \frac{2}{p}\left(1 - \frac{1}{\beta}\right), \quad \frac{2}{\beta} \leq p \leq \infty,\end{aligned} \tag{22}$$

where the case $p = +\infty$ amounts to the condition of $r \geq 1$. Then (u, b) remains in the same regularity class as (21) on $[0, T']$ for some $T' > T$.

THEOREM 1.3. *Let $N = 2, \nu > 0, \eta > 0$, $\beta \in [\frac{1}{2}, 1)$, $\alpha \in [\frac{1}{2}, 2 - \beta)$ and (u, b) in the regularity class of (21) be the solution pair to the generalized MHD system (2a)-(2c) for a given $(u_0, b_0) \in H^s(\mathbb{R}^2), s > 2$. Suppose the solution (u, b) over the time interval of $[0, T]$ satisfies*

$$\begin{aligned} &\int_0^T \|\nabla b\|_{L^p}^r d\tau < \infty \\ &\text{where } \frac{2}{p} + \frac{2}{r} \leq \begin{cases} 2 + \frac{2}{p}\left(1 - \frac{1}{\beta}\right), & \text{for } \frac{2}{\beta} \leq p < \infty, \\ 2 + \frac{2}{p}\left(1 - \frac{1}{\alpha}\right), & \text{for } \max\{2, \frac{2}{\alpha}\} < p < \infty, \end{cases} \\ &\text{or } \int_0^T \|j\|_{BMO} d\tau < \infty. \end{aligned} \tag{23}$$

Then (u, b) remains in the same regularity class as (21) on $[0, T']$ for some $T' > T$.

REMARK 1.1. Improvements on the space of initial data from $H^s(\mathbb{R}^2)$ for $s > 2$ is certainly possible. We choose not to pursue this direction of research here. We also emphasize that $\lim_{\beta \to 1^-} 2 + \frac{2}{p}(1 - \frac{1}{\beta}) = 2$, indicating again that heuristically the well-posedness of the generalized MHD system (2a)-(2c) under the hypotheses of Theorem 1.2 and Theorem 1.3 may not be too far out of reach.

REMARK 1.2. A natural question is whether we may obtain a regularity criteria in terms of u or b instead of their gradients as in (22) and (23) respectively. We choose to leave this in the future direction of research, only commenting that naive attempts actually seem to fail immediately. This is because a standard procedure of obtaining such a regularity criteria e.g. in terms of u instead of ∇u for the MHD system starts with first integrating by parts and taking away the derivative on u. However, if we work on the $H^1(\mathbb{R}^2)$-estimate of (u, b), then although by taking $L^2(\mathbb{R}^2)$-inner products on (2a)-(2b) with $(-\Delta u, -\Delta b)$ respectively, one may compute

$$\begin{aligned} &\frac{1}{2}\partial_t(\|\nabla u\|_{L^2}^2 + \|\nabla b\|_{L^2}^2) + \nu\|\Lambda^\alpha \nabla u\|_{L^2}^2 + \eta\|\Lambda^\beta \nabla b\|_{L^2}^2 \\ =& \int (u \cdot \nabla)u \cdot \Delta u + (u \cdot \nabla)b \cdot \Delta b - \int (b \cdot \nabla)b \cdot \Delta u + (b \cdot \nabla)u \cdot \Delta b \\ =& \sum_{i,j,k=1}^2 \int u_i \partial_k(\partial_i b_j \partial_k b_j) - \int u_j \partial_k(\partial_k b_i \partial_i b_j) + u_j \partial_i(\partial_k b_i \partial_k b_j), \end{aligned}$$

it will be difficult to close this estimate due to the second-order derivatives on b and the hypothesis that $\beta < 1$. Similarly in an attempt to estimate $L^2(\mathbb{R}^2)$-norm of w and j, one immediately realizes that integration by parts in the right side of (12) leads to second-order derivatives on b which we will not be able to estimate because $\beta < 1$:

$$\begin{aligned} &\int [2\partial_1 b_1(\partial_1 u_2 + \partial_2 u_1) - \partial_1 u_1(\partial_1 b_2 + \partial_2 b_1)]j \\ =& -2\int u_2 \partial_1((\partial_1 b_1)j) + u_1 \partial_2((\partial_1 b_1)j) + 2\int u_1 \partial_1((\partial_1 b_2 + \partial_2 b_1)j). \end{aligned}$$

2. Proofs of Theorem 1.2 and Theorem 1.3

Local theory may be proven using mollifiers in a standard way (e.g. [**15**]); thus, in this manuscript, we choose to focus on *a priori* estimates.

2.1. $H^1(\mathbb{R}^2)$-bound.

PROPOSITION 2.1. *Under the hypothesis of Theorem 1.2, the solution* (u, b) *over the time interval* $[0, T]$ *satisfies*

$$\sup_{t\in[0,T]} (\|w\|_{L^2}^2 + \|j\|_{L^2}^2)(t) + \int_0^T \|\Lambda^\alpha w\|_{L^2}^2 + \|\Lambda^\beta j\|_{L^2}^2 d\tau < \infty. \tag{24}$$

PROOF. Firstly, let us assume $p \in (\frac{2}{\beta}, \infty)$ in (22) and continue from (15), and estimate

$$\begin{aligned} &\int [2\partial_1 b_1(\partial_1 u_2 + \partial_2 u_1) - \partial_1 u_1(\partial_1 b_2 + \partial_2 b_1)]j \\ \lesssim &\|\nabla u\|_{L^p} \|\nabla b\|_{L^2} \|\nabla b\|_{L^{\frac{2p}{p-2}}} \\ \lesssim &\|\nabla u\|_{L^p} \|\nabla b\|_{L^2} \|\nabla b\|_{L^2}^{\frac{\beta p-2}{\beta p}} \|\Lambda^\beta \nabla b\|_{L^2}^{\frac{2}{\beta p}} \\ \le &\frac{\eta}{2}\|\Lambda^\beta j\|_{L^2}^2 + c\|\nabla u\|_{L^p}^{\frac{\beta p}{\beta p-1}} \|\nabla b\|_{L^2}^2 \end{aligned} \tag{25}$$

by Hölder's inequality, Gagliardo-Nirenberg's inequality and Young's inequality. Applying (25) in (15), subtracting $\frac{\eta}{2}\|\Lambda^\beta j\|_{L^2}^2$ from both sides leads to

$$\partial_t(\|w\|_{L^2}^2 + \|j\|_{L^2}^2) + 2\nu\|\Lambda^\alpha w\|_{L^2}^2 + \eta\|\Lambda^\beta j\|_{L^2}^2 \lesssim \|\nabla u\|_{L^p}^{\frac{\beta p}{\beta p-1}} \|\nabla b\|_{L^2}^2 \tag{26}$$

from which Gronwall's inequality completes the proof of Proposition 2.1 in case $p \in (\frac{2}{\beta}, \infty)$. Next, the case $p = \frac{2}{\beta}$ is immediate as well because

$$\begin{aligned} &\int [2\partial_1 b_1(\partial_1 u_2 + \partial_2 u_1) - \partial_1 u_1(\partial_1 b_2 + \partial_2 b_1)]j \\ \lesssim &\|\nabla u\|_{L^{\frac{2}{\beta}}} \|\nabla b\|_{L^2} \|\Lambda^\beta \nabla b\|_{L^2} \le \frac{\eta}{2}\|\Lambda^\beta \nabla b\|_{L^2}^2 + c\|\nabla u\|_{L^{\frac{2}{\beta}}}^2 \|\nabla b\|_{L^2}^2 \end{aligned}$$

by Hölder's inequality, Sobolev embedding $\dot{H}^\beta(\mathbb{R}^2) \hookrightarrow L^{\frac{2}{1-\beta}}(\mathbb{R}^2)$ and Young's inequality. Finally, in the case $p = +\infty$ we immediately obtain

$$\int [2\partial_1 b_1(\partial_1 u_2 + \partial_2 u_1) - \partial_1 u_1(\partial_1 b_2 + \partial_2 b_1)]j \lesssim \|\nabla u\|_{L^\infty} \|\nabla b\|_{L^2}^2$$

by Hölder's inequality. After applying these estimates in (15), Gronwall's inequality deduces (24) and completes the proof of Proposition 2.1. □

PROPOSITION 2.2. *Under the hypothesis of Theorem 1.3, the solution* (u, b) *over the time interval* $[0, T]$ *satisfies*

$$\sup_{t\in[0,T]} (\|w\|_{L^2}^2 + \|j\|_{L^2}^2)(t) + \int_0^T \|\Lambda^\alpha w\|_{L^2}^2 + \|\Lambda^\beta j\|_{L^2}^2 d\tau < \infty. \tag{27}$$

PROOF. Firstly, let us consider the case $\frac{2}{p}+\frac{2}{r}\leq 2+\frac{2}{p}\left(1-\frac{1}{\beta}\right), \frac{2}{\beta}<p<\infty$. We continue again from (15) and estimate

$$\begin{aligned}
&\int [2\partial_1 b_1(\partial_1 u_2+\partial_2 u_1)-\partial_1 u_1(\partial_1 b_2+\partial_2 b_1)]j\\
&\lesssim \|\nabla u\|_{L^2}\|\nabla b\|_{L^p}\|\nabla b\|_{L^{\frac{2p}{p-2}}}\\
&\lesssim \|\nabla u\|_{L^2}\|\nabla b\|_{L^p}\|\nabla b\|_{L^2}^{\frac{\beta p-2}{\beta p}}\|\Lambda^\beta \nabla b\|_{L^2}^{\frac{2}{\beta p}}\\
&\leq \frac{\eta}{2}\|\Lambda^\beta j\|_{L^2}^2+c\|\nabla b\|_{L^p}^{\frac{\beta p}{\beta p-1}}(\|\nabla u\|_{L^2}^2+\|\nabla b\|_{L^2}^2)
\end{aligned}\tag{28}$$

by Hölder's inequality, Gagliardo-Nirenberg's inequality and Young's inequality. After applying (28) in (15), Gronwall's inequality leads to (27) in this case. The case in which $p=\frac{2}{\beta}$ is immediately done similarly:

$$\begin{aligned}
&\int [2\partial_1 b_1(\partial_1 u_2+\partial_2 u_1)-\partial_1 u_1(\partial_1 b_2+\partial_2 b_1)]j\\
&\lesssim \|\nabla u\|_{L^2}\|\nabla b\|_{L^{\frac{2}{\beta}}}\|\nabla b\|_{L^{\frac{2}{1-\beta}}}\leq \frac{\eta}{2}\|\Lambda^\beta \nabla b\|_{L^2}^2+c\|\nabla u\|_{L^2}^2\|\nabla b\|_{L^{\frac{2}{\beta}}}^2
\end{aligned}$$

by Hölder's inequality, Sobolev embedding $\dot{H}^\beta(\mathbb{R}^2)\hookrightarrow L^{\frac{2}{1-\beta}}(\mathbb{R}^2)$ and Young's inequality.

Secondly, let us consider the case $\frac{2}{p}+\frac{2}{r}\leq 2+\frac{2}{p}\left(1-\frac{1}{\alpha}\right), \max\{2,\frac{2}{\alpha}\}<p<\infty$. We estimate from (15):

$$\begin{aligned}
&\int [2\partial_1 b_1(\partial_1 u_2+\partial_2 u_1)-\partial_1 u_1(\partial_1 b_2+\partial_2 b_1)]j\\
&\lesssim \|\nabla u\|_{L^{\frac{2p}{p-2}}}\|\nabla b\|_{L^2}\|\nabla b\|_{L^p}\\
&\lesssim \|\nabla u\|_{L^2}^{\frac{\alpha p-2}{\alpha p}}\|\Lambda^\alpha \nabla u\|_{L^2}^{\frac{2}{\alpha p}}\|\nabla b\|_{L^2}\|\nabla b\|_{L^p}\\
&\leq \frac{\nu}{2}\|\Lambda^\alpha \nabla u\|_{L^2}^2+c(\|\nabla u\|_{L^2}^2+\|\nabla b\|_{L^2}^2)\|\nabla b\|_{L^p}^{\frac{\alpha p}{\alpha p-1}}
\end{aligned}\tag{29}$$

by Hölder's inequality, Gagliardo-Nirenberg's inequality and Young's inequality. After applying (29) in (15), Gronwall's inequality deduces (27).

Lastly, for the case $\int_0^T\|j\|_{BMO}d\tau<\infty$, we estimate

$$\begin{aligned}
&\int [2\partial_1 b_1(\partial_1 u_2+\partial_2 u_1)-\partial_1 u_1(\partial_1 b_2+\partial_2 b_1)]j\\
&=\int [-\partial_1 u\cdot\nabla b_2+\partial_2 u\cdot\nabla b_1+\partial_1 b\cdot\nabla u_2-\partial_2 b\cdot\nabla u_1]j\\
&\lesssim \|\nabla u\|_{L^2}\|\nabla b\|_{L^2}\|j\|_{BMO}\lesssim \|j\|_{BMO}(\|\nabla u\|_{L^2}^2+\|\nabla b\|_{L^2}^2)
\end{aligned}\tag{30}$$

by (20), duality of BMO space and the Hardy space $\mathcal{H}^1$, [**4**, Theorem II.1] and Young's inequality. After applying (30) to (15), Gronwall's inequality deduces (27) and completes the proof of Proposition 2.2. □

2.2. Higher regularity from $H^1(\mathbb{R}^2)$-bound. In this subsection, we raise the regularity from $H^1(\mathbb{R}^2)$-bound to $H^s(\mathbb{R}^2)$. As we emphasized in the Section 1, this is the relatively *easier* part of the proof; the difficulty in case $\beta<1$ is the $H^1(\mathbb{R}^2)$-estimate which we already accomplished in Proposition 2.1 and Proposition 2.2.

PROPOSITION 2.3. *Under the hypothesis of Theorem 1.2 or Theorem 1.3, the solution (u,b) over the time interval $[0,T]$ satisfies*

$$\sup_{t\in[0,T]}(\|\nabla w\|_{L^2}^2+\|\nabla j\|_{L^2}^2)(t)+\int_0^T\|\Lambda^\alpha\nabla w\|_{L^2}^2+\|\Lambda^\beta\nabla j\|_{L^2}^2 d\tau<\infty. \tag{31}$$

PROOF. Let us first assume the more difficult case $\alpha<1$. We take $L^2(\mathbb{R}^2)$-inner products on (10a)-(10b) with $(-\Delta w,-\Delta j)$ to deduce

$$\begin{aligned}
&\frac{1}{2}\partial_t(\|\nabla w\|_{L^2}^2+\|\nabla j\|_{L^2}^2)+\nu\|\Lambda^\alpha\nabla w\|_{L^2}^2+\eta\|\Lambda^\beta\nabla j\|_{L^2}^2\\
&=-\int\nabla u\cdot\nabla w\cdot\nabla w-\int\nabla u\cdot\nabla j\cdot\nabla j+2\int\nabla b\cdot\nabla j\cdot\nabla w\\
&\quad+2\int\Lambda^{1-\beta}[\partial_1 b_1(\partial_1 u_2+\partial_2 u_1)-\partial_1 u_1(\partial_1 b_2+\partial_2 b_1)]\Lambda^{1+\beta}j
\end{aligned}\tag{32}$$

after integration by parts. For clarify, let us estimate the first three integrals and the last integral separately. Firstly,

$$\begin{aligned}
&-\int\nabla u\cdot\nabla w\cdot\nabla w-\int\nabla u\cdot\nabla j\cdot\nabla j+2\int\nabla b\cdot\nabla j\cdot\nabla w\\
\lesssim&\|\nabla u\|_{L^{\frac{2}{\alpha}}}\|\nabla w\|_{L^2}\|\nabla w\|_{L^{\frac{2}{1-\alpha}}}+\|\nabla u\|_{L^{\frac{2}{\beta}}}\|\nabla j\|_{L^2}\|\nabla j\|_{L^{\frac{2}{1-\beta}}}\\
&+\|\nabla b\|_{L^{\frac{2}{\alpha}}}\|\nabla j\|_{L^2}\|\nabla w\|_{L^{\frac{2}{1-\alpha}}}\\
\lesssim&\|\Lambda^{1-\alpha}\nabla u\|_{L^2}\|\nabla w\|_{L^2}\|\Lambda^\alpha\nabla w\|_{L^2}+\|\Lambda^{1-\beta}\nabla u\|_{L^2}\|\nabla j\|_{L^2}\|\Lambda^\beta\nabla j\|_{L^2}\\
&+\|\Lambda^{1-\alpha}\nabla b\|_{L^2}\|\nabla j\|_{L^2}\|\Lambda^\alpha\nabla w\|_{L^2}\\
\leq&\frac{\nu}{8}\|\Lambda^\alpha\nabla w\|_{L^2}^2+\frac{\eta}{8}\|\Lambda^\beta\nabla j\|_{L^2}^2\\
&+c(\|\nabla w\|_{L^2}^2+\|\nabla j\|_{L^2}^2)(\|\Lambda^{1-\alpha}w\|_{L^2}^2+\|\Lambda^{1-\beta}w\|_{L^2}^2+\|\Lambda^{1-\alpha}j\|_{L^2}^2)
\end{aligned}\tag{33}$$

by Hölder's inequalities, Sobolev embedding of $\dot{H}^\gamma(\mathbb{R}^2)\hookrightarrow L^{\frac{2}{1-\gamma}}(\mathbb{R}^2)$, $\dot{H}^{1-\gamma}(\mathbb{R}^2)\hookrightarrow L^{\frac{2}{\gamma}}(\mathbb{R}^2)$, Lemma 3.3 and Young's inequality. Secondly,

$$\begin{aligned}
&2\int\Lambda^{1-\beta}[\partial_1 b_1(\partial_1 u_2+\partial_2 u_1)-\partial_1 u_1(\partial_1 b_2+\partial_2 b_1)]\Lambda^{1+\beta}j\\
\lesssim&\||\nabla b||\nabla u|\|_{\dot{H}^{1-\beta}}\|\Lambda^\beta\nabla j\|_{L^2}\\
\lesssim&(\|\Lambda^{1-\beta}\nabla b\|_{L^{\frac{2}{1-\beta}}}\|\nabla u\|_{L^{\frac{2}{\beta}}}+\|\nabla b\|_{L^{\frac{2}{\beta}}}\|\Lambda^{1-\beta}\nabla u\|_{L^{\frac{2}{1-\beta}}})\|\Lambda^\beta\nabla j\|_{L^2}\\
\leq&\frac{\eta}{8}\|\Lambda^\beta\nabla j\|_{L^2}^2+c(\|\nabla w\|_{L^2}^2+\|\nabla j\|_{L^2}^2)(\|\Lambda^{1-\beta}w\|_{L^2}^2+\|\Lambda^{1-\beta}j\|_{L^2}^2)
\end{aligned}\tag{34}$$

by Hölder's inequality, Lemma 3.4, Lemma 3.3, Sobolev embeddings of $\dot{H}^\beta(\mathbb{R}^2)\hookrightarrow L^{\frac{2}{1-\beta}}(\mathbb{R}^2)$ and $\dot{H}^{1-\beta}(\mathbb{R}^2)\hookrightarrow L^{\frac{2}{\beta}}(\mathbb{R}^2)$) and Young's inequality. Applying (33) and (34) in (32), using that $\alpha,\beta\geq\frac{1}{2}$ by hypothesis, Gronwall's inequality type argument using Proposition 2.1 and Proposition 2.2 deduces (31).

The case in which $\alpha\in[1,2-\beta)$ is easier and it suffices for us to modify within the estimate of (33) as follows:

$$\begin{aligned}
&-\int\nabla u\cdot\nabla w\cdot\nabla w+2\int\nabla b\cdot\nabla j\cdot\nabla w\\
\lesssim&\|\nabla u\|_{L^4}\|\nabla w\|_{L^2}\|\nabla w\|_{L^4}+\|\nabla b\|_{L^{\frac{2}{1-\beta}}}\|\nabla j\|_{L^2}\|\nabla w\|_{L^{\frac{2}{\beta}}}\\
\leq&\frac{\nu}{8}\|\Lambda^\alpha\nabla w\|_{L^2}^2+\frac{\eta}{8}\|\Lambda^\beta\nabla j\|_{L^2}^2+c(\|\nabla w\|_{L^2}^2+\|\nabla j\|_{L^2}^2)(1+\|\Lambda^\alpha w\|_{L^2}^2+\|\Lambda^\beta j\|_{L^2}^2)
\end{aligned}$$

by Hölder's inequalities, the Sobolev embeddings of $H^{\alpha}(\mathbb{R}^2) \hookrightarrow L^4(\mathbb{R}^2), \dot{H}^{\beta}(\mathbb{R}^2) \hookrightarrow L^{\frac{2}{1-\beta}}(\mathbb{R}^2), H^{\alpha}(\mathbb{R}^2) \hookrightarrow L^{\frac{2}{\beta}}(\mathbb{R}^2)$, and Young's inequality. Applying this estimate, along with the estimate of

$$-\int \nabla u \cdot \nabla j \cdot \nabla j \leq \frac{\eta}{8}\|\Lambda^{\beta}\nabla j\|_{L^2}^2 + c\|\Lambda^{1-\beta}w\|_{L^2}^2\|\nabla j\|_{L^2}^2$$

which may be seen from (33), and (34) to (32) deduces (31) via Gronwall's inequality. The proof of Proposition 2.3 is now complete. □

PROOFS OF THEOREM 1.2 AND THEOREM 1.3. The proofs of Theorem 1.2 and Theorem 1.3 follow immediately from Proposition 2.3, Lemma 3.2 and (37) from Lemma 3.1. Indeed, taking $L^2(\mathbb{R}^2)$-inner products of (2a)-(2b) with $(\Lambda^{2s}u, \Lambda^{2s}b)$, results in

$$\begin{aligned}
&\frac{1}{2}\partial_t(\|\Lambda^s u\|_{L^2}^2 + \|\Lambda^s b\|_{L^2}^2) + \nu\|\Lambda^{s+\alpha}u\|_{L^2}^2 + \eta\|\Lambda^{s+\beta}b\|_{L^2}^2\\
=&-\int \Lambda^s((u\cdot\nabla)u)\cdot\Lambda^s u + \int \Lambda^s((b\cdot\nabla)b)\cdot\Lambda^s u\\
&-\int \Lambda^s((u\cdot\nabla)b)\cdot\Lambda^s b + \int \Lambda^s((b\cdot\nabla)u)\cdot\Lambda^s b\\
\leq&\|\Lambda^s((u\cdot\nabla)u) - (u\cdot\nabla)\Lambda^s u\|_{L^2}\|\Lambda^s u\|_{L^2}\\
&+\|\Lambda^s((b\cdot\nabla)b) - (b\cdot\nabla)\Lambda^s b\|_{L^2}\|\Lambda^s u\|_{L^2}\\
&+\|\Lambda^s((u\cdot\nabla)b) - (u\cdot\nabla)\Lambda^s b\|_{L^2}\|\Lambda^s b\|_{L^2}\\
&+\|\Lambda^s((b\cdot\nabla)u) - (b\cdot\nabla)\Lambda^s u\|_{L^2}\|\Lambda^s b\|_{L^2}\\
\lesssim&\|\nabla u\|_{L^\infty}\|\Lambda^s u\|_{L^2}^2 + \|\nabla b\|_{L^\infty}\|\Lambda^s b\|_{L^2}\|\Lambda^s u\|_{L^2} + \|\nabla u\|_{L^\infty}\|\Lambda^s b\|_{L^2}^2\\
\lesssim&(\|\Delta u\|_{L^2} + \|\Delta b\|_{L^2})\ln(e + \|\Lambda^s u\|_{L^2}^2 + \|\Lambda^s b\|_{L^2}^2)(\|\Lambda^s u\|_{L^2}^2 + \|\Lambda^s b\|_{L^2}^2),
\end{aligned} \tag{35}$$

where we used that

$$\int (u\cdot\nabla)\Lambda^s u\cdot\Lambda^s u = 0, \quad \int (u\cdot\nabla)\Lambda^s b\cdot\Lambda^s b = 0,$$
$$\int (b\cdot\nabla)\Lambda^s b\cdot\Lambda^s u + \int (b\cdot\nabla)\Lambda^s u\cdot\Lambda^s b = 0,$$

due to (2c). Gronwall's inequality applied to (35), along with Proposition 2.3 completes the proofs of Theorem 1.2 and Theorem 1.3. □

3. Appendix

For completeness, we state lemmas that have been used:

LEMMA 3.1. *([**1**]; see also [**26**, Lemma 2.3 and Appendix], and [**29**, Lemma 2.7 and Appendix]) Let* $f \in H^s(\mathbb{R}^2), s > 2$, *satisfy* $\nabla \cdot f = 0, \nabla \times f \in L^\infty(\mathbb{R}^2)$. *Then*

$$\|\nabla f\|_{L^\infty} \lesssim (\|f\|_{L^2} + \|\nabla \times f\|_{L^\infty}\log_2(2 + \|f\|_{H^s}) + 1). \tag{36}$$

Similarly let $f \in L^2(\mathbb{R}^2) \cap W^{s,p}(\mathbb{R}^2)$ *where* $s \in \mathbb{R}$ *such that* $p \in [2, \infty), \frac{2}{p} < s$. *Then*

$$\|f\|_{L^\infty} \lesssim (\|f\|_{L^2} + \|f\|_{H^1}\log_2(2 + \|f\|_{W^{s,p}}) + 1). \tag{37}$$

LEMMA 3.2. *([**12**]) Let* f, g *be smooth such that* $\nabla f \in L^{p_1}(\mathbb{R}^2), \Lambda^{s-1}g \in L^{p_2}(\mathbb{R}^2), \Lambda^s f \in L^{p_3}(\mathbb{R}^2), g \in L^{p_4}(\mathbb{R}^2)$, *where* $p \in (1, \infty), \frac{1}{p} = \frac{1}{p_1} + \frac{1}{p_2} = \frac{1}{p_3} + \frac{1}{p_4}$, $p_2, p_3 \in (1, \infty)$, *and* $s > 0$. *Then*

$$\|\Lambda^s(fg) - f\Lambda^s g\|_{L^p} \lesssim (\|\nabla f\|_{L^{p_1}}\|\Lambda^{s-1}g\|_{L^{p_2}} + \|\Lambda^s f\|_{L^{p_3}}\|g\|_{L^{p_4}}).$$

LEMMA 3.3. *(e.g. Theorem 3.1.1* [**3**]*) Suppose f satisfies $\nabla \cdot f = 0$, $\nabla f \in L^p(\mathbb{R}^2), p \in (1, \infty)$. Then*

$$\|\nabla f\|_{L^p} \lesssim \frac{p^2}{p-1} \|\nabla \times f\|_{L^p}.$$

LEMMA 3.4. *(e.g. Lemma A.2* [**11**]*) Let $f \in W^{\delta,p_1}(\mathbb{R}^2) \cap L^{q_2}(\mathbb{R}^2)$ and $g \in W^{\delta,p_2}(\mathbb{R}^2) \cap L^{q_1}(\mathbb{R}^2)$ where $\delta \geq 0, 1 < p_k < \infty, 1 < q_k \leq \infty, \frac{1}{p_k} + \frac{1}{q_k} = \frac{1}{p}$ and $k = 1, 2$. Then*

$$\|fg\|_{\dot{W}^{\delta,p}} \lesssim (\|f\|_{\dot{W}^{\delta,p_1}} \|g\|_{L^{q_1}} + \|f\|_{L^{q_2}} \|g\|_{\dot{W}^{\delta,p_2}}).$$

Acknowledgment

The author expresses deep gratitude to Prof. Jiahong Wu and Prof. Hongjie Dong for stimulating discussion in this research topic, and the anonymous referee and the Editor for their valuable comments and suggestions which significantly improved this manuscript.

References

[1] H. Brézis and S. Wainger, *A note on limiting cases of Sobolev embeddings and convolution inequalities*, Comm. Partial Differential Equations **5** (1980), no. 7, 773–789, DOI 10.1080/03605308008820154. MR579997

[2] C. Cao, J. Wu, and B. Yuan, *The 2D incompressible magnetohydrodynamics equations with only magnetic diffusion*, SIAM J. Math. Anal. **46** (2014), no. 1, 588–602, DOI 10.1137/130937718. MR3163239

[3] J.-Y. Chemin, *Perfect incompressible fluids*, Oxford Lecture Series in Mathematics and its Applications, vol. 14, The Clarendon Press, Oxford University Press, New York, 1998. Translated from the 1995 French original by Isabelle Gallagher and Dragos Iftimie. MR1688875

[4] R. Coifman, P.-L. Lions, Y. Meyer, and S. Semmes, *Compensated compactness and Hardy spaces* (English, with English and French summaries), J. Math. Pures Appl. (9) **72** (1993), no. 3, 247–286. MR1225511

[5] B.-Q. Dong, J. Li and J. Wu, *Global regularity for the 2D MHD equations with partial hyperresistivity*, Int. Math. Res. Not. IMRN, to appear.

[6] J. Fan, H. Malaikah, S. Monaquel, G. Nakamura, and Y. Zhou, *Global Cauchy problem of 2D generalized MHD equations*, Monatsh. Math. **175** (2014), no. 1, 127–131, DOI 10.1007/s00605-014-0652-0. MR3249890

[7] Z. Jiang, Y. Wang, and Y. Zhou, *On regularity criteria for the 2D generalized MHD system*, J. Math. Fluid Mech. **18** (2016), no. 2, 331–341, DOI 10.1007/s00021-015-0235-4. MR3503188

[8] Q. Jiu and J. Zhao, *A remark on global regularity of 2D generalized magnetohydrodynamic equations*, J. Math. Anal. Appl. **412** (2014), no. 1, 478–484, DOI 10.1016/j.jmaa.2013.10.074. MR3145815

[9] Q. Jiu and J. Zhao, *Global regularity of 2D generalized MHD equations with magnetic diffusion*, Z. Angew. Math. Phys. **66** (2015), no. 3, 677–687, DOI 10.1007/s00033-014-0415-8. MR3347406

[10] T. Kato, *On classical solutions of the two-dimensional nonstationary Euler equation*, Arch. Rational Mech. Anal. **25** (1967), 188–200, DOI 10.1007/BF00251588. MR0211057

[11] T. Kato, *Liapunov functions and monotonicity in the Navier-Stokes equation*, Functional-analytic methods for partial differential equations (Tokyo, 1989), Lecture Notes in Math., vol. 1450, Springer, Berlin, 1990, pp. 53–63, DOI 10.1007/BFb0084898. MR1084601

[12] T. Kato and G. Ponce, *Commutator estimates and the Euler and Navier-Stokes equations*, Comm. Pure Appl. Math. **41** (1988), no. 7, 891–907, DOI 10.1002/cpa.3160410704. MR951744

[13] I. Kukavica and M. Ziane, *One component regularity for the Navier-Stokes equations*, Nonlinearity **19** (2006), no. 2, 453–469, DOI 10.1088/0951-7715/19/2/012. MR2199398

[14] J. Leray, *Sur le mouvement d'un liquide visqueux emplissant l'espace* (French), Acta Math. **63** (1934), no. 1, 193–248, DOI 10.1007/BF02547354. MR1555394

[15] A. J. Majda and A. L. Bertozzi, *Vorticity and incompressible flow*, Cambridge Texts in Applied Mathematics, vol. 27, Cambridge University Press, Cambridge, 2002. MR1867882

[16] M. Sermange and R. Temam, *Some mathematical questions related to the MHD equations*, Comm. Pure Appl. Math. **36** (1983), no. 5, 635–664, DOI 10.1002/cpa.3160360506. MR716200

[17] J. Serrin, *On the interior regularity of weak solutions of the Navier-Stokes equations*, Arch. Rational Mech. Anal. **9** (1962), 187–195, DOI 10.1007/BF00253344. MR0136885

[18] E. M. Stein, *Singular integrals and differentiability properties of functions*, Princeton Mathematical Series, No. 30, Princeton University Press, Princeton, N.J., 1970. MR0290095

[19] T. Tao, *Global regularity for a logarithmically supercritical hyperdissipative Navier-Stokes equation*, Anal. PDE **2** (2009), no. 3, 361–366, DOI 10.2140/apde.2009.2.361. MR2603802

[20] C. V. Tran, X. Yu, and L. A. K. Blackbourn, *Two-dimensional magnetohydrodynamic turbulence in the limits of infinite and vanishing magnetic Prandtl number*, J. Fluid Mech. **725** (2013), 195–215, DOI 10.1017/jfm.2013.193. MR3062190

[21] C. V. Tran, X. Yu, and Z. Zhai, *On global regularity of 2D generalized magnetohydrodynamic equations*, J. Differential Equations **254** (2013), no. 10, 4194–4216, DOI 10.1016/j.jde.2013.02.016. MR3032302

[22] C. V. Tran, X. Yu, and Z. Zhai, *Note on solution regularity of the generalized magnetohydrodynamic equations with partial dissipation*, Nonlinear Anal. **85** (2013), 43–51, DOI 10.1016/j.na.2013.02.019. MR3040347

[23] J. Wu, *Generalized MHD equations*, J. Differential Equations **195** (2003), no. 2, 284–312, DOI 10.1016/j.jde.2003.07.007. MR2016814

[24] J. Wu, *Global regularity for a class of generalized magnetohydrodynamic equations*, J. Math. Fluid Mech. **13** (2011), no. 2, 295–305, DOI 10.1007/s00021-009-0017-y. MR2805867

[25] J. Wu, *The 2D magnetohydrodynamic equations with partial or fractional dissipation*, Lectures on the analysis of nonlinear partial differential equations. Part 5, Morningside Lect. Math., vol. 5, Int. Press, Somerville, MA, 2018, pp. 283–332. MR3793019

[26] K. Yamazaki, *Remarks on the global regularity of the two-dimensional magnetohydrodynamics system with zero dissipation*, Nonlinear Anal. **94** (2014), 194–205, DOI 10.1016/j.na.2013.08.020. MR3120685

[27] K. Yamazaki, *Global regularity of the logarithmically supercritical MHD system with zero diffusivity*, Appl. Math. Lett. **29** (2014), 46–51, DOI 10.1016/j.aml.2013.10.014. MR3133570

[28] K. Yamazaki, *On the global regularity of two-dimensional generalized magnetohydrodynamics system*, J. Math. Anal. Appl. **416** (2014), no. 1, 99–111, DOI 10.1016/j.jmaa.2014.02.027. MR3182750

[29] K. Yamazaki, *On the global regularity of N-dimensional generalized Boussinesq system*, Appl. Math. **60** (2015), no. 2, 109–133, DOI 10.1007/s10492-015-0087-5. MR3320341

[30] K. Yamazaki, *Horizontal Biot-Savart law in general dimension and an application to the 4D magneto-hydrodynamics*, Differential Integral Equations **31** (2018), no. 3-4, 301–328. MR3738200

[31] K. Yamazaki, *Second proof of the global regularity of the two-dimensional MHD system with full diffusion and arbitrary weak dissipation*, Methods Appl. Anal., to appear.

[32] Z. Ye, *Remarks on regularity criteria for 2D generalized MHD equations*, Rocky Mountain J. Math. **47** (2017), no. 4, 1333–1353, DOI 10.1216/RMJ-2017-47-4-1333. MR3689957

[33] Z. Ye and X. Xu, *Global regularity of two-dimensional incompressible generalized magnetohydrodynamics system*, Nonlinear Anal., **100** (2014), 86–96.

[34] B. Yuan and L. Bai, *Remarks on global regularity of 2D generalized MHD equations*, J. Math. Anal. Appl. **413** (2014), no. 2, 633–640, DOI 10.1016/j.jmaa.2013.12.024. MR3159793

[35] V. I. Judovič, *Non-stationary flows of an ideal incompressible fluid* (Russian), Ž. Vyčisl. Mat. i Mat. Fiz. **3** (1963), 1032–1066. MR0158189

Department of Mathematics, University of Rochester, Rochester, New York 14627
Email address: kyamazak@ur.rochester.edu

Contemporary Mathematics
Volume **725**, 2019
https://doi.org/10.1090/conm/725/14561

Orbital stability of standing waves for fractional Hartree equation with unbounded potentials

Jian Zhang, Shijun Zheng, and Shihui Zhu

ABSTRACT. We prove the existence of the set of ground states in a suitable energy space $\Sigma^s = \{u : \int_{\mathbb{R}^N} \bar{u}(-\Delta + m^2)^s u + V|u|^2 < \infty\}$, $s \in (0, \frac{N}{2})$ for the mass-subcritical nonlinear fractional Hartree equation with unbounded potentials. As a consequence we obtain, as a priori result, the orbital stability of the set of standing waves. The main ingredient is the observation that Σ^s is compactly embedded in L^2. This enables us to apply the concentration compactness argument in the works of Cazenave-Lions and Zhang, namely, relative compactness for any minimizing sequence in the energy space.

1. Introduction

Consider the nonlinear fractional Hartree equation with an unbounded potential in the following form: For $0 < s < N/2$ and $(t, x) \in \mathbb{R}^{1+N}$

$$iu_t = (-\Delta + m^2)^s u + V(x)u - (\frac{1}{|x|^\gamma} * |u|^2)u \tag{1.1}$$

$$u(0, x) = u_0 \in \Sigma^s, \tag{1.2}$$

where $u = u(t, x)$: $\mathbb{R} \times \mathbb{R}^N \to \mathbb{C}$ is a complex valued function and the convolution $*$ is defined by $W * |u|^2 := \left(\frac{1}{|\cdot|^\gamma} * |u|^2\right)(x) = \int \frac{|u(y)|^2}{|x-y|^\gamma} dy$ with $0 < \gamma < \min\{4s, N\}$. The operator $(-\Delta + m^2)^s$ is defined by

$$(-\Delta + m^2)^s u = \mathcal{F}^{-1}[(|\xi|^2 + m^2)^s \mathcal{F}[u](\xi)],$$

where $m \geq 0$, $\mathcal{F}$ and $\mathcal{F}^{-1}$ are the Fourier transform and the inverse in $\mathbb{R}^N$, respectively. Here the potential function $V \in C^\infty(\mathbb{R}^N)$ is bounded from below $V(x) \geq -c_0$ for some $c_0 > 0$ and satisfies

$$V(x) \to \infty \quad \text{as} \quad |x| \to \infty. \tag{1.3}$$

Such class of V includes unbounded potentials arising in physics, for instance, the harmonic potential $V(x) = |x|^2$, and more generally, polynomial functions that are bounded from below. Then $H_{s,V} := (-\Delta + m^2)^s + V$ is essentially self-adjoint in the Hilbert space $\Sigma^s := \{v \in L^2 : (-\Delta + m^2)^{s/2} v \in L^2 \textit{ and } |V|^{1/2} v \in L^2\}$, as is given in Section 2.

2010 *Mathematics Subject Classification.* Primary 35Q55, 35B35; Secondary 35C08.
Key words and phrases. Fractional Hartree equation, standing wave, orbital stability.

The Hartree equation is non-integrable analog of the NLS system, which arises naturally in large quantum systems that describe the wave motion for the bosonic or fermionic particles. When $s = 1$, it can be derived from the mean-field limit of N-body GPE hierarchy, where $W(x) = |x|^{-\gamma}$ represents the long-range two-body interaction potential. A special feature of the Hartree equation lies in the convolution kernel W that preserves the fine structure of micro two-body interactions of particles. This is in contrast to the NLS, which can be viewed as the limiting case $W \to \delta$ and where the two-body interactions are modeled from the scattering length. When $s = \frac{1}{2}$, (1.1) arises as an effective description of pseudo-relativistic boson stars in the mean field limit, where $u(t, x)$ is a complex-valued wave field. The study of such a perturbed system in the presence of external potential V is both physically and mathematically very important.

When V is zero or bounded, the dynamical properties of solutions for the fractional Hartree equation (1.1) have been considered in e.g., [**1, 7, 8, 12, 13, 16, 24, 25**]. In [**7, 24**], the authors studied the orbital stability for standing wave solutions, that is, ground states (the set of minimizers) for (1.1) with zero potential by profile decomposition method. Furthermore, [**24**] showed the strong instability in the mass-critical case. The analogous results on orbital stability for nonlinear fractional Schrödinger equation (FNLS) were obtained in [**5, 11, 26**]. The paper [**5**] studied the FNLS with power nonlinearity and a decaying potential, while [**11, 26**] studied FNLS with power nonlinearity in the form $|u|^{p-1}u + |u|^{q-1}u$.

In this paper, we are concerned with the orbital stability of standing wave solutions for (1.1) in the mass-subcritical regime $\gamma < 2s$. Note that when $V = 0$, $\gamma = 2s$ corresponds to the mass-critical case and $\gamma = 4s$ corresponds to the energy-critical case by scaling invariance argument. The presence of an unbounded V brings in technical difficulties and the methods in [**5, 11, 24**] do not directly apply to (1.1). Motivated by the treatment for the classical NLS in [**22, 23**], we prove a new compactness lemma in Section 3, adapted to the energy space for (1.1), which leads to the solution to the variational problem 4.1 in Proposition 4.1. Thus follows the main result Theorem 4.3. The proof in Section 4 can be viewed as an adaption to the potential case, also see [**2, 3**] for the original treatment of NLS type equations via variational method.

2. Preliminaries

We will use the notations $L^q := L^q(\mathbb{R}^N)$, $\|\cdot\|_q := \|\cdot\|_{L^q(\mathbb{R}^N)}$, $H^s := H^s(\mathbb{R}^N)$ and $\dot{H}^s := \dot{H}^s(\mathbb{R}^N)$, the latter two denoting the usual Sobolev space and its homogeneous version. The various positive constants will be simply denoted by C. From now on throughout the paper, without loss of generality we may assume there is some positive constant c_1 such that for all x,

$$V(x) \geq c_1 > 0 \quad \textit{and} \quad V(x) \to \infty \quad \textit{as } |x| \to \infty. \tag{2.1}$$

If otherwise, the lower bound of V is $-c_0 < 0$, then one can always apply the substitution $u \to e^{itC}u$ for any $C \geq c_0 + c_1$ to convert equation (1.1) into one with V bounded from below by a positive constant; see also Remark 4.4.

For V satisfying (2.1) define the energy space Σ^s as

$$\Sigma^s := \{v \in L^2 | \int \overline{v}(-\Delta + m^2)^s v + V|v|^2 < \infty\}. \tag{2.2}$$

Then Σ^s is a Hilbert space equipped with the norm

$$\|v\|_{\Sigma^s} := \left(\int \overline{v}(-\Delta + m^2)^s v + V|v|^2\right)^{\frac{1}{2}}, \tag{2.3}$$

which means that $H_{s,V} = (-\Delta+m^2)^s+V$ is essentially self-adjoint in the quadratic form on $\Sigma^s \times \Sigma^s$ given by

$$q(u,v) := \int \overline{(-\Delta+m^2)^{s/2}v}\,(-\Delta+m^2)^{s/2}u + \int \overline{v}Vu.$$

Note that $\|v\|_{\Sigma^s} < \infty$ implies $\|v\|_2$ is finite. When $s \in (0,1)$ and $m = 0$, one can also express

$$(-\Delta)^s u = c_{s,N}\, p.v. \int_{\mathbb{R}^N} \frac{u(x) - u(y)}{|x-y|^{N+2s}} dy,$$

where $c_{s,N} = \frac{2^{2s}}{\pi^{N/2}} \frac{\Gamma(s+\frac{N}{2})}{|\Gamma(-s)|}$.

Define the energy functional $E : \Sigma^s \to \mathbb{R}$ as

$$E(v) = \frac{1}{2}\int \overline{v}(-\Delta+m^2)^s v + \frac{1}{2}\int V|v|^2 - \frac{1}{4}\int (\frac{1}{|x|^\gamma} * |v|^2)|v|^2. \tag{2.4}$$

From the Hardy inequality (3.1), the functional $E(v)$ is well-defined in Σ^s. Note that if $V = 0$ and $m = 0$, $\gamma = 4s < N$ corresponds to the energy-critical case due to that the scaling $u \mapsto \frac{1}{\lambda^{(N-2s)/2}} u(\frac{t}{\lambda^{2s}}, \frac{x}{\lambda})$ leaves invariant the energy and the solution to (1.1). When $\gamma = 2s < N$, it corresponds to to the mass-critical case due to that the scaling $u \mapsto \frac{1}{\lambda^{N/2}} u(\frac{t}{\lambda^{2s}}, \frac{x}{\lambda})$ leaves invariant the mass $M(u) := \int |u|^2$ and (1.1).

In this paper, we assume that the Cauchy problem (1.1)-(1.2) is well-posed in Σ^s, namely, the following global in time existence, uniqueness and conservation laws hold in the mass-subcritical regime $\gamma < 2s$.

HYPOTHESIS 1. *Let $N \geq 1$ and $\frac{\gamma}{2} < s < N/2$. If the initial data $u_0 \in \Sigma^s$, then there exists a unique solution $u(t,x)$ of the Cauchy problem (1.1)-(1.2) on $\mathbb{R}$ such that $u \in C(\mathbb{R};\Sigma^s)\bigcap C^1(\mathbb{R};H^{-s})$. Moreover, for all $t \in \mathbb{R}$, $u(t,x)$ satisfies the following conservation laws:*

(i) (mass)

$$M(u(t)) = M(u_0).$$

(ii) (energy)

$$E(u(t)) = E(u_0).$$

REMARK 2.1. For local well-posedness in the above conjectured proposition, so far we only know about numerical result on cubic FNLS with a quadratic potential by Zhang and Kirkpatrick [**15**], where is suggested the long-time existence along with its dynamics for some special initial data. The theoretical proof has remained an *open question* concerning FNLS and Hartree equations with harmonic potentials. The main difficulty is the lack of proper dispersive or Strichartz estimates because the fractional Laplacian for $0 < s < 1$ does not hold a control over the harmonic potential, which is shown in the deformed trajectories for the associated hamiltonian, see [**15**] and [**19**]. This is in sharp contrast to the classical NLS ($s = 1$) with a harmonic potential, where the existence and stability problem has been studied quite extensively [**6, 9, 10, 14, 18, 22, 23**].

Heuristically the Laplacian $-\Delta$ and $V = |x|^2$ have balanced strength or effect so the $L^1 \to L^\infty$ time decay $t^{-N/2}$ holds locally for $e^{it(\Delta - V)}$. However, for $0 < s < 1$, in the phase space the bound energy for fractional laplacian $(-\Delta + m^2)^s$ is less than $|x|^2$, or, $H_{s,0}$ relative to V is like the laplacian vs. anharmonic potential of higher order. It falls within the quantum situation "a particle at a higher altitude falls down to the bottom of the potential in a shorter time than one at a lower altitude", which obstructs Fujiwara's theorem, cf. [**17**].

3. Proof of main result

For estimating the Hartree nonlinearity we will need Hardy's inequality, see e.g, [**20**]: If $s \in (0, N/2)$

$$\sup_x \int \frac{|u(y)|^2}{|x-y|^{2s}} dy \leq c(s,N)\|u\|^2_{\dot{H}^s}. \tag{3.1}$$

LEMMA 3.1. *Let $v \in \dot{H}^s$ and $0 < s < N/2$. If $0 < \gamma < 2s$, then there exists a positive constant $C > 0$ such that*

$$\int (\frac{1}{|x|^\gamma} * |v|^2)|v|^2 dx \leq C\|v\|^{\frac{\gamma}{s}}_{\dot{H}^s} \|v\|_2^{\frac{4s-\gamma}{s}}. \tag{3.2}$$

PROOF. Note that

$$\begin{aligned} \int (\tfrac{1}{|x|^\gamma} * |v|^2)|v(x)|^2 dx &= \int\int \tfrac{|v(x)|^2}{|x-y|^\gamma}|v(y)|^2 dxdy \\ &\leq \|\int \tfrac{|v(x)|^2}{|x-y|^\gamma} dx\|_\infty \|v(y)\|_2^2. \end{aligned} \tag{3.3}$$

Using Hölder inequality with $1 = \frac{\gamma}{2s} + \frac{2s-\gamma}{2s}$, we obtain

$$\begin{aligned} \int \frac{|v(x)|^2}{|x-y|^\gamma} dx &= \int \frac{|v(x)|^{\frac{\gamma}{s}}}{|x-y|^\gamma} |v(x)|^{\frac{2s-\gamma}{s}} dx \\ &\leq C\||x-y|^{-s}v(x)\|_2^{\frac{\gamma}{s}} \|v\|_2^{\frac{2s-\gamma}{s}} \\ &\leq C\|v\|^{\frac{\gamma}{s}}_{\dot{H}^s} \|v\|_2^{\frac{2s-\gamma}{s}}. \end{aligned} \tag{3.4}$$

In the last step of the above, we have employed (3.1). Thus, (3.2) follows from (3.3) and (3.4). □

LEMMA 3.2. *Let $0 < \gamma < 2s$, $0 < s < N/2$ and V satisfy (2.1). Suppose $\{v_n\}_{n=1}^\infty$ converges weakly to U in Σ^s. Then there exists a subsequence (still denoted by $\{v_n\}$) such that*

$$\|v_n\|_2^2 \to \|U\|_2^2 \quad \text{as} \quad n \to \infty, \tag{3.5}$$

and

$$\iint \frac{|v_n(x)|^2|v_n(y)|^2}{|x-y|^\gamma} dxdy \to \iint \frac{|U(x)|^2|U(y)|^2}{|x-y|^\gamma} dxdy \quad \text{as} \quad n \to \infty. \tag{3.6}$$

REMARK 3.3. Equation (3.5) indeed implies that for $s > 0$, Σ^s is compactly embedded in L^2.

PROOF. Since $\Sigma^s \subset L^2$, it is easy to see that for $U \in \Sigma^s \cap L^2$, $v_n \rightharpoonup U$ weakly in Σ^s and

$$v_n \rightharpoonup U \quad \text{weakly in} \quad L^2 \tag{3.7}$$

as $n \to \infty$. For an elementary proof of (3.7), see Proposition 5.1 in the Appendix.

Since $\{v_n\}$ is weakly convergent in Σ^s, $\|v_n\|_{\Sigma^s}$ is uniformly bounded. So there is a positive constant K such that

$$\sup_n \int V(x)|v_n(x)|^2 dx \leq \sup_n \|v_n\|_{\Sigma^s}^2 < K. \tag{3.8}$$

Furthermore, we will show that there is a subsequence of $\{v_n\}$ that strongly converges in L^2 and satisfies (3.6).

(1) First, we consider the case where $U = 0$. From (3.8) with V satisfying (2.1), we have for arbitray $\epsilon > 0$, there exists a constant $B = B_\epsilon > 0$ (large enough) such that $\frac{1}{V(x)} \leq \epsilon$ when $|x| > B$. We see that

$$\int_{|x|>B} |v_n|^2 dx = \int_{|x|>B} \frac{1}{V(x)} V(x)|v_n|^2 dx \leq K\epsilon.$$

For the fixed B, compact embedding property for Sobolev space on bounded domain gives that one can extract a subsequence (still denoted by $\{v_n\}$) such that

$$v_n \to 0 \qquad \text{strongly in } L^2(\{|x| \leq B\}). \tag{3.9}$$

Hence, there exists $L = L_\epsilon$ such that for all $n > L$, $\int_{|x|\leq B} |v_n|^2 dx \leq \epsilon$. Thus, we have, if $n > L$,

$$\int |v_n|^2 dx = \int_{|x|\leq B} |v_n|^2 dx + \int_{|x|>B} |v_n|^2 dx \leq (K+1)\epsilon. \tag{3.10}$$

Now, taking $\epsilon = \epsilon_k \to 0$ as $k \to \infty$, a standard diagonal argument shows that there exists a subsequence of $\{v_n\}$, still denoted by $\{v_n\}$, such that (3.5) is true in the case $U = 0$. Then, it follows from (3.2) in Lemma 3.1 that

$$\int\int \frac{|v_n(x)|^2|v_n(y)|^2}{|x-y|^\gamma} dxdy \leq C\|v_n\|_{\dot{H}^s}^{\frac{\gamma}{s}} \|v_n\|_2^{\frac{4s-\gamma}{s}} \to 0 \quad \text{as} \quad n \to \infty.$$

This proves (3.6) when $U = 0$.

(2) Secondly, we consider the case $U \neq 0$. Take $w_n = v_n - U$. We have $w_n \rightharpoonup 0$ weakly in Σ^s and in L^2 as $n \to \infty$. Then, from the above discussion, we see that there is a subsequence (still denoted by $\{w_n\}$) such that $w_n \to 0$ strongly in L^2. In other words, we have

$$\|v_n - U\|_2 \to 0 \quad \text{as} \quad n \to \infty. \tag{3.11}$$

To prove (3.6), we will apply (3.4). Indeed, if $0 < \gamma < 2s$, we deduce that

$$\begin{aligned}
&\left|\iint \frac{|v_n(x)|^2|v_n(y)|^2}{|x-y|^\gamma} dxdy - \iint \frac{|U(x)|^2|U(y)|^2}{|x-y|^\gamma} dxdy\right| \\
\leq &\left|\iint \frac{|v_n(x)|^2|v_n(y)|^2}{|x-y|^\gamma} dxdy - \iint \frac{|v_n(x)|^2|U(y)|^2}{|x-y|^\gamma} dxdy\right| \\
+ &\left|\iint \frac{|v_n(x)|^2|U(y)|^2}{|x-y|^\gamma} dxdy - \iint \frac{|U(x)|^2|U(y)|^2}{|x-y|^\gamma} dxdy\right| \\
\leq &\left\| \int \frac{|v_n(x)|^2}{|x-y|^\gamma} dx \right\|_\infty \left|\|v_n\|_2^2 - \|U\|_2^2\right| + \left\| \int \frac{|U(y)|^2}{|x-y|^\gamma} dy \right\|_\infty \left|\|v_n\|_2^2 - \|U\|_2^2\right| \\
\leq &C\|v_n\|_{\dot{H}^s}^{\frac{\gamma}{s}} \|v_n\|_2^{\frac{2s-\gamma}{s}} \left|\|v_n\|_2^2 - \|U\|_2^2\right| + \|U\|_{\dot{H}^s}^{\frac{\gamma}{s}} \|U\|_2^{\frac{2s-\gamma}{s}} \left|\|v_n\|_2^2 - \|U\|_2^2\right| \\
\leq &C\left|\|v_n\|_2^2 - \|U\|_2^2\right| \to 0 \text{ as } n \to \infty.
\end{aligned}$$

Therefore, (3.6) is true for $U \neq 0$, which completes the proof. □

4. Orbital stability of standing waves

Given $0<\gamma<2s$, $s\in(0,N/2)$, $N\geq 1$, let $m\geq 0$ and $M>0$. Consider the following variational problem

$$d_M := \inf_{\{v\in\Sigma^s \mid \|v\|_2^2=M\}} E(v), \tag{4.1}$$

where $E(v)=\frac{1}{2}\|v\|_{\Sigma^s}^2-\frac{1}{4}\iint\frac{|v(x)|^2|v(y)|^2}{|x-y|^\gamma}dxdy$, as defined in (2.4). The following proposition constructs a minimizer, called ground state, to the problem (4.1).

PROPOSITION 4.1. *Let $M>0$. Suppose $0<\gamma<2s$ and V satisfies (2.1), then the infimum in the variational problem can be attained. That is, there exists $U\in\Sigma^s$ such that*

$$E(U)=d_M=\min_{\{v\in\Sigma^s \mid \|v\|_2^2=M\}} E(v). \tag{4.2}$$

Moreover, any minimizing sequence to (4.1) must be relatively compact in Σ^s.

PROOF. Firstly, we prove the variational problem (4.1) is well-defined. That is, $E(v)$ has a lower bound in $\{v\in\Sigma^s \mid \|v\|_2^2=M\}$. From Lemma 3.1, we deduce that

$$\begin{aligned} E(v) &\geq \frac{1}{2}\int(|\xi|^2+m^2)^s|\widehat{v}|^2d\xi+\frac{1}{2}\int V|v|^2dx-C\|v\|_{\dot{H}^s}^{\frac{\gamma}{s}}\|v\|_2^{\frac{4s-\gamma}{s}} \\ &\geq \frac{1}{2}\int|\xi|^{2s}|\widehat{v}|^2d\xi-C_M'\left(\int|\xi|^{2s}|\widehat{v}|^2d\xi\right)^{\frac{\gamma}{2s}} \\ &\geq -C_M. \end{aligned} \tag{4.3}$$

In the last step, we have used the elementary inequality $\frac{1}{2}X-C_M'X^{\gamma/2s}\geq -C_M$ for all $X>0$ and some constants $C_M', C_M>0$. Thus, $E(v)$ is bounded from below and the infimum d_M exists.

Secondly, take any minimizing sequence $\{v_n\}$ of Problem (4.1) satisfying

$$E(v_n)\to d_M \quad \text{and} \quad \|v_n\|_2^2\to M, \quad \text{as } n\to\infty. \tag{4.4}$$

We see that there exists an L such that for all $n\geq L$,

$$E(v_n)<d_M+1. \tag{4.5}$$

This, together with (4.3) implies for all $n\geq L$,

$$\frac{1}{2}\|v_n\|_{\Sigma^s}^2\leq C_M\|v_n\|_{\dot{H}^s}^{\gamma/s}+d_M+1. \tag{4.6}$$

Hence, $\{v_n\}$ must be bounded in Σ^s by virtue of the condition $\frac{\gamma}{s}<2$.

Thirdly, from the boundedness of $\{v_n\}$, we know there exists a subsequence (still denoted by $\{v_n\}$) and $U\in\Sigma^s$ such that

$$v_n(x)\rightharpoonup U(x) \ \text{ weakly in } \ \Sigma^s. \tag{4.7}$$

By the lower semi-continuity of norm Σ^s, we have

$$\|U\|_{\Sigma^s}^2\leq\liminf_{n\to\infty}\|v_n\|_{\Sigma^s}^2. \tag{4.8}$$

Combining (4.8), Lemma 3.2 and (4.4) we obtain that $\|v_n\|_2^2\to\|U\|_2^2=M$ and

$$E(U)\leq\liminf_{n\to\infty}E(v_n)=d_M.$$

But from the definition of d_M, we must have $E(U) = d_M$. That is, U is a minimizer of (4.1). To prove the statement on relative compactness, observe that the last argument shows

$$E(U) = \lim_{n\to\infty} E(v_n).$$

This, along with Lemma 3.2 implies there exists a subsequence $\{v_{n_k}\}$ such that $\lim_{k\to\infty}\|v_{n_k}\|_{\Sigma^s} = \|U\|_{\Sigma^s}$. Therefore, in view of (4.7), the strong convergence $v_{n_k} \to U$ in Σ^s follows. $\square$

REMARK 4.2. The existence of ground states for FNLS with unbounded V was obtained in [**4**] via Nehari's manifold approach. The existence and symmetry of ground state solutions were studied in [**21**] for Hartree equation with zero potential.

Define the set

$$S_M := \{v \in \Sigma^s \big|\ v \text{ is the minimizer of Problem (4.1) }\}. \tag{4.9}$$

From the Euler-Lagrange Theorem (see [**2**]), for any $v \in S_M \subset \Sigma^s$, there exists $\omega \in \mathbb{R}$ such that

$$(-\triangle + m^2)^s v + V(x)v + \omega v - \Big(\frac{1}{|x|^\gamma} * |v|^2\Big)v = 0. \tag{4.10}$$

Moreover, $u(t,x) = e^{i\omega t}v(x)$ is a standing wave solution for (1.1). Thus, each v in S_M is called an orbit. It is easy to check that for any $t > 0$, if $v \in S_M$, then $e^{i\omega t}v(x) \in S_M$. Applying Proposition 4.1, we now prove the following orbital stability for (1.1). More precisely, assuming Hypothesis 1, we show that if the initial data is close to an orbit $v \in S_M$, then the solution of evolution system (1.1)-(1.2) remains close to S_M, the set of ground states, for all time.

THEOREM 4.3 (Orbital stability of standing waves). *Let $M > 0$ and V satisfy* (2.1). *Let $0 < s < N/2$ and $0 < \gamma < 2s$. Then, for arbitrary $\varepsilon > 0$, there exists $\delta > 0$ such that if the initial data u_0 in Σ^s satisfies*

$$\inf_{v\in S_M} \|u_0 - v\|_{\Sigma^s} < \delta, \tag{4.11}$$

it holds for the solution u of the Cauchy problem (1.1)-(1.2),

$$\inf_{v\in S_M} \|u(t,x) - v(x)\|_{\Sigma^s} < \varepsilon \tag{4.12}$$

for all t, where S_M is defined in (4.9).

PROOF. We prove the theorem by contradiction following the standard method for NLS (see [**2, 22, 23**]). Let $u_0 \in \Sigma^s$ and u be the unique solution for (1.1). From the proof of Proposition 4.1 and the conservation laws in Hypothesis 1, we obtain that for all t,

$$\frac{1}{2}\|u(t,\cdot)\|^2_{\Sigma^s} \le E(u_0) + C(\|u_0\|_2)\|u(t,\cdot)\|^{\gamma/s}_{\Sigma^s}.$$

This suggests that the Σ^s-norm of u is uniformly bounded for all t.

Assume that the conclusion in the theorem is false, then there exist $\varepsilon_0 > 0$ and a sequence of initial data $\{u_{0,n}\}_{n=1}^\infty$ such that

$$\inf_{v\in S_M} \|u_{0,n} - v\|_{\Sigma^s} < \frac{1}{n}, \tag{4.13}$$

and there exists a sequence of time $\{t_n\}_{n=1}^{\infty}$ such that for all n,

$$\inf_{v\in S_M} \|u_n(t_n,x)-v\|_{\Sigma^s} \geq \varepsilon_0. \tag{4.14}$$

But from (4.13), the conservation laws and Lemma 3.2, it follows that there exist $w\in\Sigma^s$ and a subsequence of $\{u_n\}$ (still denoted by $\{u_n\}$) such that as $n\to\infty$

$$\|u_n(t_n,x)\|_2^2 = \|u_{0,n}\|_2^2 \to \|w\|_2^2 = M$$

and

$$E(u_n(t_n,x)) = E(u_{0,n}) \to E(w) = d_M.$$

Hence, $\{u_n(t_n,\cdot)\}_{n=1}^{\infty}$ is a minimizing sequence of (4.1). According to Proposition 4.1, $w\in S_M$ is a minimizer such that, when passing to a subsequence if necessary,

$$\|u_n(t_n,x)-w(x)\|_{\Sigma^s} \to 0 \quad \text{as} \quad n\to\infty. \tag{4.15}$$

This contradicts (4.14), which proves the theorem. □

REMARK 4.4. The results in Theorem 4.3 and Proposition 4.1 continue to hold if V satisfies the slightly more general condition (1.3). To see this, it suffices to observe the following properties. Let $E_V(u) := E(u) = \frac{1}{2}\|u\|_{\Sigma^s}^2 - \frac{1}{4}\iint \frac{|u(x)|^2|u(y)|^2}{|x-y|^\gamma}dxdy$ be given as in (2.4) and $\Sigma_V^s := \Sigma^s$ as in (2.2). Let $C\geq c_0+c_1$ be a fixed constant. Then we have:

(1) $M(e^{i\theta}u) = M(u)$
(2) $E_V(e^{i\theta}u) = E_V(u)$
(3) $e^{i\theta}S_M = S_M$
(4) $E_{V+C}(u) = E_V(u) + \frac{C}{2}M(u)$
(5) The set of minimizers S_M is independent of $V+C$
(6) Denote by u_V the solution of (1.1)-(1.2), then we have $u_{V+C}(t,x) = e^{-itC}u_V(t,x)$.

REMARK 4.5. The case $s=1$, $\gamma=2$ (mass-critical) was studied in [**14**], where the threshold for the stability of standing waves for (1.1) are obtained. Earlier result on the l.w.p. and mass and energy conservation laws for (1.1) can be found in [**2**] for $s=1$, and $\gamma\leq 2$. Let Q_0 be the unique radial positive ground state solution of (4.1), where $V=0$. The paper [**14**, Theorem 4.1] followed a quite standard variational approach and proved that if $s=1$, $\gamma=2$ and $M<\|Q_0\|_2^2$, then there exist ground state solutions of the minimization problem (4.1), where $V=|x|^2$. Moreover, these ground state solutions are orbitally stable.

In the absence of V, when $\gamma=2s$ ($0<s<1$ and $N\geq 2$), Zhang and Zhu [**24**] proved the orbital stability via profile decomposition method. They also showed strong instability of (1.1) by constructing blowup solutions with initial data arbitrarily close to Q_0 in H^s.

5. Appendix: Uniqueness of weak convergence

For the proof of Lemma 3.2 we need the following.

PROPOSITION 5.1. *Let $s>0$ and $\Sigma^s\subset L^2$ be defined as in (2.2) with V satisfying (2.1). Let $\{v_n\}$ be a sequence in Σ^s. If $\{v_n\}$ converges weakly to f in Σ^s and $\{v_n\}$ converges weakly to g in L^2. Then f is identical to g in $\Sigma^s\cap L^2$.*

The domain of $H = H_{s,V}$ is given by

$$D(H) = \{\phi \in L^2 : (-\Delta + m^2)^s \phi \in L^2, V\phi \in L^2\}.$$

It is easy to verify that $D(H)$ is a complete Hilbert subspace of L^2 with respect to the norm

$$\|\phi\|_{D(H)} := \left(\|(-\Delta + m^2)^s \phi\|_2^2 + \|V\phi\|_2^2\right)^{1/2}.$$

The form domain of H is defined as: $v \in Q(H) \iff v \in L^2(\mathbb{R}^N)$ and

$$(-\Delta + m^2)^{s/2} v \in L^2 \quad \textit{and} \quad V^{1/2} v \in L^2, \tag{5.1}$$

which is complete with respect to the norm $\sqrt{q(v,v)} = \|v\|_{\Sigma^s}$ given in (2.3). We have $D(H) \subset Q(H)$. The form definition for H is equivalent to the following weak form definition.

DEFINITION 5.2. A function $f \in \Sigma^s = H^s \cap D(\sqrt{V})$ is in $Q(H)$ if and only if for all $\phi \in H^s(\mathbb{R}^N)$, there exists an $h \in L^2 \cap H^{-s}$ s.t.

$$\begin{aligned}\int_{\mathbb{R}^N} f(x) H\phi \, dx &= \int h(x)\phi(x) dx \quad \forall \phi \in H^s \\ &= \int f(x)(-\Delta + m^2)^s \phi \, dx + \int f(x) V\phi \, dx.\end{aligned}$$

We see that $Q(H) = D(H) = H^{2s} \cap D(V)$. Since on $D(H)$ it holds

$$\langle (-\Delta + m^2)^s \phi, \phi \rangle + \langle V\phi, \phi \rangle \geq (m^{2s} + c_1)\langle \phi, \phi \rangle,$$

we obtain that the spectrum $\sigma(H) \subset [c_1, \infty)$, which implies 0 is in the resolvent set for H. Hence $H^{-1} = (H - 0)^{-1}$: $L^2 \to D(H) \subset \Sigma^s$ exists and is a continuous mapping.

PROOF OF PROPOSITION 5.1. For all $\phi \in D(H) \subset \Sigma^s \subset L^2$, $s > 0$, the inner product on Σ^s

$$\begin{aligned}&\langle f, \phi \rangle_{\Sigma^s} = \lim_n \langle v_n, \phi \rangle_{\Sigma^s} \\ =&\lim_n \langle (-\Delta + m^2)^{s/2} v_n, (-\Delta + m^2)^{s/2} \phi \rangle_{L^2} + \langle V^{1/2} v_n, V^{1/2} \phi \rangle_{L^2} \\ =&\lim_n q(v_n, \phi) = \lim_n \langle v_n, H\phi \rangle_{L^2} = \langle g, H\phi \rangle_2,\end{aligned}$$

where $H = H_{s,V}$ is a positive self-adjoint operator in L^2, cf. (2.3) and the quadratic form defined there. On the other hand, observe that

$$\begin{aligned}&\langle f, \phi \rangle_{\Sigma^s} = \langle (-\Delta + m^2)^{s/2} f, (-\Delta + m^2)^{s/2} \phi \rangle_{L^2} + \langle V^{1/2} v_n, V^{1/2} \phi \rangle_{L^2} \\ =&\langle f, H\phi \rangle_2.\end{aligned}$$

Now as functions (or distributions) it follows that $f = g$ in L^2 by taking $\phi = H^{-1}\varphi$ for all $\varphi \in L^2$. This proves Proposition 5.1. □

REMARK 5.3. The proof here relies on the fact that the spectrum of $H_{s,V}$ is bounded below by zero, whence one sees that $H_V = (-\Delta + m^2)^s + V$ has an inverse that continuously maps L^2 onto $D(H)$.

Acknowledgment. This paper was partially done when Shihui Zhu visited School of Mathematics at Georgia Institute of Technology, who would like to thank the hospitality of the School of Mathematics. This work was supported by the National Natural Science Foundation of China grant No. 11501395, No. 11371267 and Excellent Youth Foundation of Sichuan Scientific Committee grant No. 2014JQ0039 in China.

References

[1] W. Bao and X. Dong, *Numerical methods for computing ground states and dynamics of nonlinear relativistic Hartree equation for boson stars*, J. Comput. Phys. **230** (2011), no. 13, 5449–5469, DOI 10.1016/j.jcp.2011.03.051. MR2799519

[2] T. Cazenave, *Semilinear Schrödinger equations*, Courant Lecture Notes in Mathematics, vol. 10, New York University, Courant Institute of Mathematical Sciences, New York; American Mathematical Society, Providence, RI, 2003. MR2002047

[3] T. Cazenave and P.-L. Lions, *Orbital stability of standing waves for some nonlinear Schrödinger equations*, Comm. Math. Phys. **85** (1982), no. 4, 549–561. MR677997

[4] M. Cheng, *Bound state for the fractional Schrödinger equation with unbounded potential*, J. Math. Phys. **53** (2012), no. 4, 043507, 7, DOI 10.1063/1.3701574. MR2953151

[5] Y. Cho, H. Hajaiej, G. Hwang, and T. Ozawa, *On the orbital stability of fractional Schrödinger equations*, Commun. Pure Appl. Anal. **13** (2014), no. 3, 1267–1282, DOI 10.3934/cpaa.2014.13.1267. MR3177699

[6] B. Feng, *Sharp threshold of global existence and instability of standing wave for the Schrödinger-Hartree equation with a harmonic potential*, Nonlinear Anal. Real World Appl. **31** (2016), 132–145, DOI 10.1016/j.nonrwa.2016.01.012. MR3490836

[7] B. Feng and H. Zhang, *Stability of standing waves for the fractional Schrödinger-Hartree equation*, J. Math. Anal. Appl. **460** (2018), no. 1, 352–364, DOI 10.1016/j.jmaa.2017.11.060. MR3739909

[8] J. Fröhlich and E. Lenzmann, *Blowup for nonlinear wave equations describing boson stars*, Comm. Pure Appl. Math. **60** (2007), no. 11, 1691–1705, DOI 10.1002/cpa.20186. MR2349352

[9] R. Fukuizumi, *Stability and instability of standing waves for the nonlinear Schrödinger equation with harmonic potential*, Discrete Contin. Dynam. Systems **7** (2001), no. 3, 525–544, DOI 10.3934/dcds.2001.7.525. MR1815766

[10] R. Fukuizumi and M. Ohta, *Stability of standing waves for nonlinear Schrödinger equations with potentials*, Differential Integral Equations **16** (2003), no. 1, 111–128. MR1948875

[11] B. Guo and D. Huang, *Existence and stability of standing waves for nonlinear fractional Schrödinger equations*, J. Math. Phys. **53** (2012), no. 8, 083702, 15, DOI 10.1063/1.4746806. MR3012653

[12] Q. Guo and S. Zhu, *Sharp threshold of blow-up and scattering for the fractional Hartree equation*, J. Differential Equations **264** (2018), no. 4, 2802–2832, DOI 10.1016/j.jde.2017.11.001. MR3737855

[13] C. Hainzl, E. Lenzmann, M. Lewin, and B. Schlein, *On blowup for time-dependent generalized Hartree-Fock equations*, Ann. Henri Poincaré **11** (2010), no. 6, 1023–1052, DOI 10.1007/s00023-010-0054-3. MR2737490

[14] J. Huang, J. Zhang, and X. Li, *Stability of standing waves for the L^2-critical Hartree equations with harmonic potential*, Appl. Anal. **92** (2013), no. 10, 2076–2083, DOI 10.1080/00036811.2012.716512. MR3169148

[15] K. Kirkpatrick and Y. Zhang, *Fractional Schrödinger dynamics and decoherence*, Phys. D **332** (2016), 41–54, DOI 10.1016/j.physd.2016.05.015. MR3529611

[16] E. H. Lieb and H.-T. Yau, *The Chandrasekhar theory of stellar collapse as the limit of quantum mechanics*, Comm. Math. Phys. **112** (1987), no. 1, 147–174. MR904142

[17] Y.-G. Oh, *Cauchy problem and Ehrenfest's law of nonlinear Schrödinger equations with potentials*, J. Differential Equations **81** (1989), no. 2, 255–274, DOI 10.1016/0022-0396(89)90123-X. MR1016082

[18] H. A. Rose and M. I. Weinstein, *On the bound states of the nonlinear Schrödinger equation with a linear potential*, Phys. D **30** (1988), no. 1-2, 207–218, DOI 10.1016/0167-2789(88)90107-8. MR939275

[19] S. Secchi and M. Squassina, *Soliton dynamics for fractional Schrödinger equations*, Appl. Anal. **93** (2014), no. 8, 1702–1729, DOI 10.1080/00036811.2013.844793. MR3219678

[20] T. Tao, *Nonlinear dispersive equations*, CBMS Regional Conference Series in Mathematics, vol. 106, Published for the Conference Board of the Mathematical Sciences, Washington, DC; by the American Mathematical Society, Providence, RI, 2006. Local and global analysis. MR2233925

[21] D. Wu, *Existence and stability of standing waves for nonlinear fractional Schrödinger equations with Hartree type nonlinearity*, J. Math. Anal. Appl. **411** (2014), no. 2, 530–542, DOI 10.1016/j.jmaa.2013.09.054. MR3128411

[22] J. Zhang, *Stability of standing waves for nonlinear Schrödinger equations with unbounded potentials*, Z. Angew. Math. Phys. **51** (2000), no. 3, 498–503, DOI 10.1007/s000330050011. MR1762704

[23] J. Zhang, *Stability of attractive Bose-Einstein condensates*, J. Statist. Phys. **101** (2000), no. 3-4, 731–746, DOI 10.1023/A:1026437923987. MR1804895

[24] J. Zhang and S. Zhu, *Stability of standing waves for the nonlinear fractional Schrödinger equation*, J. Dynam. Differential Equations **29** (2017), no. 3, 1017–1030, DOI 10.1007/s10884-015-9477-3. MR3694819

[25] S. Zhu, *On the blow-up solutions for the nonlinear fractional Schrödinger equation*, J. Differential Equations **261** (2016), no. 2, 1506–1531, DOI 10.1016/j.jde.2016.04.007. MR3494404

[26] S. Zhu, *Existence of stable standing waves for the fractional Schrödinger equations with combined nonlinearities*, J. Evol. Equ. **17** (2017), no. 3, 1003–1021, DOI 10.1007/s00028-016-0363-1. MR3707306

School of Mathematical Science, University of Electronic Science and Technology of China, Chengdu, Sichuan, 611731, China

Email address: zhangjiancdv@sina.com

Department of Mathematical Sciences, Georgia Southern University, Statesboro, Georgia 30460-8093

Email address: szheng@GeorgiaSouthern.edu

Department of Mathematics, Sichuan Normal University, Chengdu, Sichuan, 610066, China

Email address: shihuizhumath@163.com

Selected Published Titles in This Series

725 **Shijun Zheng, Marius Beceanu, Jerry Bona, Geng Chen, Tuoc Van Phan, and Avy Soffer, Editors,** Nonlinear Dispersive Waves and Fluids, 2019

722 **Yves Aubry, Everett W. Howe, and Christophe Ritzenthaler, Editors,** Arithmetic Geometry: Computation and Applications, 2019

720 **Alexandre Girouard, Editor,** Spectral Theory and Applications, 2018

719 **Florian Sobieczky, Editor,** Unimodularity in Randomly Generated Graphs, 2018

718 **David Ayala, Daniel S. Freed, and Ryan E. Grady, Editors,** Topology and Quantum Theory in Interaction, 2018

717 **Federico Bonetto, David Borthwick, Evans Harrell, and Michael Loss, Editors,** Mathematical Problems in Quantum Physics, 2018

716 **Alex Martsinkovsky, Kiyoshi Igusa, and Gordana Todorov, Editors,** Surveys in Representation Theory of Algebras, 2018

715 **Sergio R. López-Permouth, Jae Keol Park, S. Tariq Rizvi, and Cosmin S. Roman, Editors,** Advances in Rings and Modules, 2018

714 **Jens Gerlach Christensen, Susanna Dann, and Matthew Dawson, Editors,** Representation Theory and Harmonic Analysis on Symmetric Spaces, 2018

713 **Naihuan Jing and Kailash C. Misra, Editors,** Representations of Lie Algebras, Quantum Groups and Related Topics, 2018

712 **Nero Budur, Tommaso de Fernex, Roi Docampo, and Kevin Tucker, Editors,** Local and Global Methods in Algebraic Geometry, 2018

711 **Thomas Creutzig and Andrew R. Linshaw, Editors,** Vertex Algebras and Geometry, 2018

710 **Raphaël Danchin, Reinhard Farwig, Jiří Neustupa, and Patrick Penel, Editors,** Mathematical Analysis in Fluid Mechanics, 2018

709 **Fernando Galaz-García, Juan Carlos Pardo Millán, and Pedro Solórzano, Editors,** Contributions of Mexican Mathematicians Abroad in Pure and Applied Mathematics, 2018

708 **Christian Ausoni, Kathryn Hess, Brenda Johnson, Ieke Moerdijk, and Jérôme Scherer, Editors,** An Alpine Bouquet of Algebraic Topology, 2018

707 **Nitya Kitchloo, Mona Merling, Jack Morava, Emily Riehl, and W. Stephen Wilson, Editors,** New Directions in Homotopy Theory, 2018

706 **Yeonhyang Kim, Sivaram K. Narayan, Gabriel Picioroaga, and Eric S. Weber, Editors,** Frames and Harmonic Analysis, 2018

705 **Graham J. Leuschke, Frauke Bleher, Ralf Schiffler, and Dan Zacharia, Editors,** Representations of Algebras, 2018

704 **Alain Escassut, Cristina Perez-Garcia, and Khodr Shamseddine, Editors,** Advances in Ultrametric Analysis, 2018

703 **Andreas Malmendier and Tony Shaska, Editors,** Higher Genus Curves in Mathematical Physics and Arithmetic Geometry, 2018

702 **Mark Grant, Gregory Lupton, and Lucile Vandembroucq, Editors,** Topological Complexity and Related Topics, 2018

701 **Joan-Carles Lario and V. Kumar Murty, Editors,** Number Theory Related to Modular Curves, 2018

700 **Alexandre Girouard, Dmitry Jakobson, Michael Levitin, Nilima Nigam, Iosif Polterovich, and Frédéric Rochon, Editors,** Geometric and Computational Spectral Theory, 2017

699 **Mark L. Agranovsky, Matania Ben-Artzi, Catherine Bénéteau, Lavi Karp, Dmitry Khavinson, Simeon Reich, David Shoikhet, Gilbert Weinstein, and Lawrence Zalcman, Editors,** Complex Analysis and Dynamical Systems VII, 2017

For a complete list of titles in this series, visit the AMS Bookstore at **www.ams.org/bookstore/conmseries/**.